Engineering Materials and Processes

Series Editor

Professor Brian Derby, Professor of Materials Science
Manchester Materials Science Centre, Grosvenor Street, Manchester M1 7HS, UK

Other titles published in this series:

Fusion Bonding of Polymer Composites
C. Ageorges and L. Ye

Composite Materials
D.D.L. Chung

Titanium
G. Lütjering and J.C. Williams

Corrosion of Metals
H. Kaesche

Corrosion and Protection
E. Bardal

Intelligent Macromolecules for Smart Devices
L. Dai

Microstructure of Steels and Cast Irons
M. Durand-Charre

Phase Diagrams and Heterogeneous Equilibria
B. Predel, M. Hoch and M. Pool

Computational Mechanics of Composite Materials
M. Kamiński

Materials for Information Technology
E. Zschech, C. Whelan and T. Mikolajick
Publication due July 2005

Thermoelectricity
J.P. Heremans, G. Chen and M.S. Dresselhaus
Publication due August 2005

Computer Modelling of Sintering at Different Length Scales
J. Pan
Publication due October 2005

Computational Quantum Mechanics for Materials Engineers
L. Vitos
Publication due January 2006

Fuel Cell Technology
N. Sammes
Publication due January 2006

Stephen J. Pearton, Cammy R. Abernathy
and Fan Ren

Gallium Nitride Processing for Electronics, Sensors and Spintronics

With 241 Figures

Stephen J. Pearton, PhD
Department of Materials Science and Engineering, P.O. Box 116400,
University of Florida, Gainesville, FL 32611, USA

Cammy R. Abernathy, PhD
Department of Materials Science and Engineering, P.O. Box 116400,
University of Florida, Gainesville, FL 32611, USA

Fan Ren, PhD
Department of Chemical Engineering, P.O. Box 116005, University of Florida,
Gainesville, FL 32611, USA

British Library Cataloguing in Publication Data
Pearton, S.J.
  Gallium nitride processing for electronics, sensors and
  spintronics. — (Engineering materials and processes)
  1. Semiconductors   2. Gallium nitride   3. Detectors—
  Technological innovations
  I. Title   II. Abernathy, C.R.   III. Ren, F.
  621.3′8152
  ISBN-10: 1852339357

Library of Congress Control Number: 2005926297

Engineering Materials and Processes ISSN 1619-0181
ISBN-10: 1-85233-935-7
ISBN-13: 978-1-85233-935-7

Printed on acid-free paper

© Springer-Verlag London Limited 2006

Apart from any fair dealing for the purposes of research or private study, or criticism or review, as permitted under the Copyright, Designs and Patents Act 1988, this publication may only be reproduced, stored or transmitted, in any form or by any means, with the prior permission in writing of the publishers, or in the case of reprographic reproduction in accordance with the terms of licences issued by the Copyright Licensing Agency. Enquiries concerning reproduction outside those terms should be sent to the publishers.

The use of registered names, trademarks, etc., in this publication does not imply, even in the absence of a specific statement, that such names are exempt from the relevant laws and regulations and therefore free for general use.

The publisher makes no representation, express or implied, with regard to the accuracy of the information contained in this book and cannot accept any legal responsibility or liability for any errors or omissions that may be made.

Printed in the United States of America   (TB/SBA)

9 8 7 6 5 4 3 2 1

Springer Science+Business Media
springeronline.com

Preface

The GaN-based materials system has provided over a decade of surprises, from the initial breakthroughs with visible light-emitting diodes (LEDs) to laser diodes, solar-blind ultraviolet (UV) detectors to microwave power electronics and then to solid-state UV light sources and white lighting. Even recently, the bandgap of InN was determined to be closer to 0.7 eV rather than the value of 1.9 eV accepted for many years. The areas mentioned above have been extensively covered in other books and in the literature. The purpose of this volume is to cover some of the newer areas of research and development for GaN, such as sensors, megawatt electronics, and gate dielectrics for transistors and spin-transport electronics (or spintronics), along with advances in processing of the material. GaN-based visible LEDs and laser diodes are already commercialized for a variety of lighting and data storage applications. This materials system is also showing promise for microwave and high-power electronics intended for radar, satellite, wireless base stations, and utility grid applications, for biological detection systems, and for a new class of spintronics in which the spin of charge carriers is exploited. The explosive increase in interest in the AlGaInN family of materials in recent years has been fueled by the application of blue–green–UV LEDs in full-color displays, traffic lights, automotive lighting, and general room lighting using so-called white LEDs [1]. In addition, blue–green laser diodes can be used in high storage-capacity digital versatile disks (DVDs) systems. AlGaN-based photodetectors are also useful for solar-blind UV detection and have applications as flame sensors for control of gas turbines or for detection of missiles. There are currently major development programs in the United States for three newer applications for GaN-based materials and devices, namely:

  i. UV optical sources capable of operation down to 280 nm for use in airborne chemical and biological sensing systems, allowing direct multi-wavelength spectroscopic identification and monitoring of UV-induced reactions.
  ii. Power amplifiers and monolithic microwave integrated circuits (MMICs) for use in high-performance radar units and wireless broadband communication links and ultra-high-power (>1 MW) switches for control of distribution on electricity grid networks.

iii. Room-temperature, ferromagnetic semiconductors for use in electrically controlled magnetic sensors and actuators, high-density, ultra-low-power memory and logic, spin-polarized light emitters for optical encoding, advanced optical switches and modulators, and devices with integrated magnetic, electronic, and optical functionality. There is currently a lot of interest in the science and potential technological applications of spintronics, in which the spin of charge carriers (electrons or holes) is exploited to provide new functionality for microelectronic devices. The phenomena of giant magnetoresistance and tunnelling magnetoresistance have been exploited in all-metal or metal–insulator–metal magnetic systems for read/write heads in computer hard drives, magnetic sensors, and magnetic random access memories. The development of magnetic semiconductors with practical ordering temperatures could lead to new classes of device and circuits, including spin transistors, ultra-dense non-volatile semiconductor memory, and optical emitters with polarized output.

In addition, there is increasing interest in use of GaN-based structures for increasing the sensitivity, selectivity, and reliability of sensor devices while keeping their fabrication at low cost. There is still a lack of fundamental understanding of the physical/chemical/biological phenomena at the origin of the sensing mechanism in most cases. The GaN has potential for chemical sensors and field effect transistor (FET) devices, magnetic sensors, radiation sensors, acoustic sensors, mechanical sensors, and biosensors.

It is hoped that this volume will prove useful to researchers entering these new areas.

Acknowledgments

We wish to thank our many collaborators over the past few years, including G. Thaler, B.P. Gila, R. Frazier, A.H. Onstine, J. Kim, S. Kim, B.S. Kang, A.P. Zhang, X.A. Cao, J.W. Johnson, B. Luo, K. Baik, K. Ip, R. Khanna, D.P. Norton, J.M. Zavada, R.G. Wilson, J. Lin, M.E. Law, K.S. Jones, W.M. Chen, I.A. Buyanova, A.Y. Polyakov, Y. Irokawa, J.R. LaRoche, M.E. Overberg, R.J. Shul, A.G. Baca, J.C. Zolper, K.E. Waldrip, M. Stavola, S.N.G. Chu, M.W. Cole, W.S. Hobson, J.W. Lee, and C. Vartuli.

References

1. Nakamura S, Pearton SJ, Fasal G, The Blue Laser Diodes (Springer, Berlin, 2000)

Gainesville, Florida, USA   Stephen J. Pearton
                                                  Cammy R. Abernathy
                                                  Fan Ren

Contents

Preface v

1 Advanced Processing of Gallium Nitride for Electronic Devices 1
  1.1 Abstract 1
  1.2 Introduction 2
  1.3 Results and Discussion 16
      1.3.1 Ultra-High-Temperature Activation of Implant Doping
           in Gallium Nitride 16
           1.3.1.1 High-Temperature Annealing and Aluminum
                 Nitride Encapsulation 17
           1.3.1.2 n-Type Implant Doping 22
           1.3.1.3 p-Type Implant Doping 25
           1.3.1.4 Dopant Redistribution 26
           1.3.1.5 Residual Damage 31
  1.4 Implant Isolation 32
      1.4.1 Oxygen Implantation for Selective Area Isolation 34
      1.4.2 Creation of High-Resistivity Gallium Nitride by
           Ti, Iron, and Chromium Implantation 38
  1.5 Electrical Contacts to Gallium Nitride 41
      1.5.1 Effects of Interfacial Oxides on Schottky Contact 44
      1.5.2 Interfacial Insulator Model 49
      1.5.3 Thermally Stable Tungsten-Based Ohmic Contact 51
      1.5.4 Behavior of Tungsten and Tungsten Silicide
           Contacts on p-Gallium Nitride 54
  1.6 Dry Etch Damage in Gallium Nitride 60
      1.6.1 Plasma Damage in n-Gallium Nitride 61
      1.6.2 Effect of Etching Chemistries on Damage 66
      1.6.3 Thermal Stability of Damage 71
      1.6.4 Plasma Damage in p-Gallium Nitride 75
      1.6.5 Thermal Stability of Damage 80
      1.6.6 Determination of Damage Profile in Gallium Nitride 82
  1.7 Conclusions and Future Trends 86
References 89

2 Dry Etching of Gallium Nitride and Related Materials ... 97
- 2.1 Abstract ... 97
- 2.2 Introduction ... 97
- 2.3 Plasma Reactors ... 97
  - 2.3.1 Reactive Ion Etching ... 98
  - 2.3.2 High-Density Plasmas ... 100
  - 2.3.3 Chemically Assisted Ion Beam Etching ... 101
  - 2.3.4 Reactive Ion Beam Etching ... 102
  - 2.3.5 Low-Energy Electron Enhanced Etching ... 103
- 2.4 Plasma Chemistries ... 104
  - 2.4.1 Chlorine-Based Plasmas ... 104
  - 2.4.2 Iodine- and Bromine-Based Plasmas ... 116
  - 2.4.3 Methane–Hydrogen–Argon Plasmas ... 121
- 2.5 Etch Profile and Etched Surface Morphology ... 122
- 2.6 Plasma-Induced Damage ... 124
  - 2.6.1 n-Gallium Nitride ... 126
  - 2.6.2 p-Gallium Nitride ... 133
  - 2.6.3 Schottky Diodes ... 141
  - 2.6.4 p-n Junctions ... 148
- 2.7 Device Processing ... 152
  - 2.7.1 Microdisk Lasers ... 152
  - 2.7.2 Ridge Waveguide Lasers ... 153
  - 2.7.3 Heterojunction Bipolar Transistors ... 157
  - 2.7.4 Field Effect Transistors ... 161
  - 2.7.5 Ultroviolet Detectors ... 166

References ... 169

3 Design and Fabrication of Gallium Nitride High-Power Rectifiers ... 179
- 3.1 Abstract ... 179
- 3.2 Introduction ... 179
- 3.3 Background ... 180
  - 3.3.1 Temperature Dependence of Bandgap ... 180
    - 3.3.1.1 Gallium Nitride ... 180
    - 3.3.1.2 6H-SiC ... 181
  - 3.3.2 Effective Density of States ... 182
  - 3.3.3 Intrinsic Carrier Concentration ... 182
  - 3.3.4 Incomplete Ionization of Impurity Atoms ... 183
  - 3.3.5 Mobility Models ... 184
    - 3.3.5.1 Analytical Mobility Model ... 184
    - 3.3.5.2 Field-Dependent Mobility Model ... 185
  - 3.3.6 Generation and Recombination ... 186
    - 3.3.6.1 Shockley–Read–Hall Lifetime ... 186

3.3.6.2 Auger Recombination	186
3.3.7 Reverse Breakdown Voltage	186
3.3.8 On-State Resistance	191
3.4 Edge Termination Design	195
3.4.1 Field Plate Termination	195
3.4.2 Junction Termination	198
3.5 Comparison of Schottky and p-n Junction Diodes	201
3.5.1 Reverse Bias	201
3.5.2 Forward Bias	201
3.6 High Breakdown Lateral Diodes	204
3.7 Bulk Diode Arrays	207
3.8 Conclusions	210
References	211
4 Chemical, Gas, Biological, and Pressure Sensing	213
4.1 Abstract	213
4.2 Introduction	214
4.3 Sensors Based on AlGaN–GaN Heterostructures	219
4.3.1 Gateless AlGaN–GaN High Electron Mobility Transistor Response to Block Co-Polymers	219
4.3.2 Hydrogen Gas Sensors Based on AlGaN–GaN-Based Metal-Oxide Semiconductor Diodes	222
4.3.3 Hydrogen-Induced Reversible Changes in Sc_2O_3–AlGaN–GaN High Electron Mobility Transistors	226
4.3.4 Effect of External Strain on Conductivity of AlGaN–GaN High Electron Mobility Transistors	230
4.3.5 Pressure Sensor Fabrication	236
4.3.6 Selective-Area Substrate Removal	239
4.3.7 Biosensors Using AlGaN–GaN Heterostructures	240
4.3.8 Surface Acoustic Wave-Based Biosensors	245
4.4 Surface Acoustic Wave Device Fabrication	247
4.5 Surface Acoustic Wave Device for Gas Sensing	250
4.6 Flexural Plate Wave Device for Liquid Sensing	251
4.7 Surface Acoustic Wave Array	251
4.8 Wireless Sensor Network and Wireless Sensor Array Using Radio Frequency Identification Technology	252
4.9 Summary	255
References	255
5 Nitride-Based Spintronics	261
5.1 Abstract	261
5.2 Introduction	261

- 5.3 Potential Semiconductor Materials for Spintronics 262
- 5.4 Mechanisms of Ferromagnetism 263
- 5.5 (Ga,Mn)N 266
- 5.6 Role of Second Phases 269
- 5.7 Electrical and Optical Properties 273
- 5.8 Transport Properties 281
- 5.9 Contacts to (Ga,Mn)N 285
- 5.10 Aluminum Nitride-Based Ferromagnetic Semiconductors 287
- 5.11 Implanted Aluminum Nitride Films 291
- 5.12 Implanted AlGaN Films 294
- 5.13 Potential Device Applications 299
- 5.14 Issues to Be Resolved 307
- References 307

6 Novel Insulators for Gallium Nitride Metal-Oxide Semiconductor Field Effect Transistors and AlGaN–GaN Metal-Oxide Semiconductor High Electron Mobility Transistors 313

- 6.1 Abstract 313
- 6.2 Introduction 314
- 6.3 Insulators for Gallium Nitride Metal-Oxide Semiconductor and Metal–Insulator–Semiconductor Field Effect Transistors 317
- 6.4 Approach for Gallium Nitride 320
  - 6.4.1 Bixbyite Oxides 321
  - 6.4.2 Scandium Oxide 326
  - 6.4.3 MgO–CaO 327
  - 6.4.4 Amorphous Aluminum Nitride 327
- 6.5 Gate-Controlled Metal-Oxide Semiconductor Diodes 329
  - 6.5.1 Surface Passivation 334
  - 6.5.2 Radiation-Damage Experiments 348
- 6.6 Metal-Oxide Semiconductor Field Effect Transistors 349
  - 6.6.1 n-Channel Depletion Mode Gallium Nitride Metal-Oxide Semiconductor Field Effect Transistors Using Stacked Gate Dielectric of SiO_2–Gd_2O_3 349
  - 6.6.2 MgO–p-GaN Enhancement Mode Metal-Oxide Semiconductor Field Effect Transistors 352
- 6.7 Conclusions 355
- References 355

Index 361

1 Advanced Processing of Gallium Nitride for Electronic Devices

1.1 Abstract

This chapter focuses on understanding and optimization of several key aspects of GaN device processing. For example, to activate ion-implanted dopants it is necessary to preserve the surface during the required high temperature anneal. A novel rapid thermal processing up to 1500°C, in conjunction with AlN encapsulation, is capable of activating high dose implants, although for most applications an anneal temperature of 1100–1150°C is sufficient to achieve reasonable activation. The activation processes of implanted Si or Group VI donors and common acceptors in GaN by using this ultrahigh temperature annealing, along with its effects on surface degradation, dopant redistribution and damage removal are discussed. 1400°C has proven to be the optimum temperature to achieve high activation efficiency and to repair the ion-induced lattice defects. Ion implantation can also be employed to create high resistivity GaN. Damage-related isolation with sheet resistances of 10^{12} $\Omega/\square$ in n-GaN and 10^{10} $\Omega/\square$ in p-GaN have been achieved by implanting O and transition metal elements. Effects of surface cleanliness on characteristics of GaN Schottky contacts have been investigated, and the reduction in barrier height was correlated with removing the native oxide that forms an insulating layer on the conventionally cleaned surface. W alloys have been deposited on Si-implanted samples and Mg-doped epilayers to achieve ohmic contacts with low resistance and better thermal stability than the existing non-refractory contact schemes. Dry etching damage in GaN has been studied systematically using Schottky diode measurements. Wet chemical etching and thermal annealing processes have been developed to restore the ion-degraded material properties.

1.2 Introduction

For the last three decades or so, the III–V semiconductor material system has been viewed as highly promising for semiconductor device applications at blue and UV wavelengths in much the same manner that its highly successful As-based and P-based counterparts have been exploited for infrared, red and yellow wavelengths. As members of the III–V nitrides family, AlN, GaN, InN and their alloys are all wide band-gap materials, and can crystallize in both wurtzite and zincblende polytypes. Wurtzite GaN, AlN and InN have direct room-temperature bandgaps of 3.4 eV, 6.2 eV and 1.9 eV, respectively (Figure 1.1). In cubic form, GaN and InN have direct bandgaps, while AlN is indirect. In view of the available wide range of direct bandgaps, GaN alloyed with AlN and InN may span a continuous range of direct bandgap energies throughout much of the visible spectrum well into the UV wavelengths. This makes the nitride system attractive for optoelectronic device applications, such as LEDs, laser (LDs), and detectors, which are active in the green, blue or UV wavelengths.[1] Although similar applications based on InGaAlP heterostructures have

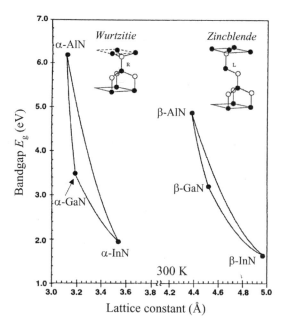

Figure 1.1. Bandgap of hexagonal (α-phase) and cubic (β-phase) InN, GaN, and AlN and their alloys versus lattice constant a_0

been successfully demonstrated, this material system is limited to about 550 nm. The addition of III–V nitrides to the family of device-quality semiconductors is essential for developing full-color displays (Figure 1.2) and coherent sources for high-density optical storage technologies, and very likely devices for signal and illumination application. In particular, the combination of GaN-based blue and green LEDs with GaAs-based red LEDs forms the basis for large-scale full displays and white light illumination. The solid-state white light source generated by mixing the primary colors in a light scrambling configuration would provide not only compactness and high lifetime, but also would reduce power consumption by 80–90% compared with incandescent or fluorescent light sources.

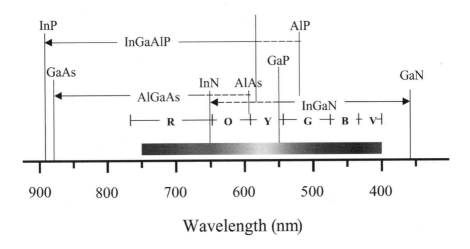

Figure 1.2. The various ternary and quaternary materials used for LEDs with the wavelength ranges indicated

Another area gaining a lot of attention for III–V nitrides is high temperature/high power electronics.[2] The interest stems from two intrinsic properties of this group of semiconductors. The first is their wide bandgap nature. The wide bandgap materials, such as GaN and SiC, are promising for high temperature applications because they go intrinsic at much higher temperatures than materials like Ge, Si and GaAs. This means that GaN power devices can operate with less cooling and fewer high-cost processing steps associated with complicated structures designed to maximize heat extraction. The second attractive property of III–V nitrides is that they have high breakdown fields. The critical electric field of the breakdown scales roughly with the square of the energy band gap, and is estimated to be >4 MV/cm for GaN,[3] compared with 0.2 MV/cm and 0.4 MV/cm for Si and GaAs respectively. Figure 1.3 is a plot of avalanche and punch-through breakdown of GaN Schottky diodes calculated as a function of doping concentration and standoff layer thickness. It can be seen that a 20 kV device may be obtained with an ~100 μm thickness GaN layer with doping concentration <10^{15} cm^{-3}.

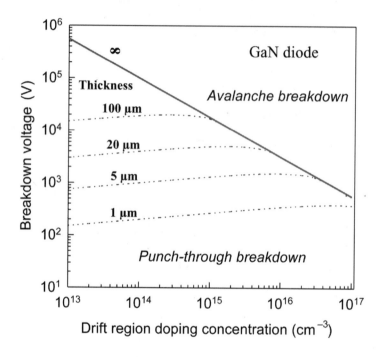

Figure 1.3. Calculated breakdown voltage as a function of doping concentration and thickness of the drift region in GaN M–n$^-$–n$^+$ diodes

GaN also has excellent electron transport properties, including good mobility, and high saturated drift velocity, as shown in Figure 1.4.[4] This makes it adequate for general electronics, and particularly promising for microwave rectifiers, particularly. The material properties associated with high temperature, high power, and high frequency application of GaN and several conventional semiconductors are summarized in Table 1.1. It is anticipated that GaN may eventually prove to be superior to SiC in this area.

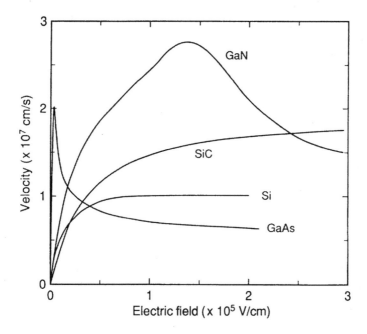

Figure 1.4. Electron drift velocity at 300 K in GaN, SiC, Si and GaAs computed using the Monte Carlo technique

The strongest feature of the III–V nitrides compared with SiC is the heterostructure technology it can support. Quantum well, modulation-doped hetero-interface, and heterojunction structures can all be made in this system, giving access to new spectral regions for optical devices and new operation regimes for electronic devices. From this point of view, III–V nitrides can be considered the wide band-gap equivalent of the AlGaAs–InGaAs system, which has set the modern benchmark for microwave device performance.

Other advantageous properties of III–V nitrides include high mechanical and thermal stability, large piezoelectric constants and the possibility of passivation by forming thin layers of Ga_2O_3 or Al_2O_3 with band gaps of 4.3 eV and 9.2 eV respectively. In addition, AlN has received considerable attention for its insulating property,[5] particularly as a potential isoelectronic insulator for GaAs FETs.

Table 1.1. Comparison of 300 K semiconductor material properties

Property	Si	GaAs	4H-SiC	GaN
Bandgap E_g (eV)	1.12	1.42	3.25	3.25
Breakdown field E_B (MV/cm)	0.25	0.4	3.0	3.0
Electron mobility μ (cm^2/V·s)	1350	6000	800	800
Maximum velocity v_s (10^7 cm/s)	1.0	2.0	2.0	2.0
Thermal conductivity χ (W/cm·K)	1.5	0.5	4.9	4.9
Dielectric constant ε	11.8	12.8	9.7	9.7
CFOM = $\chi\varepsilon\mu v_s E_B^2/(\chi\varepsilon\mu v_s E_B^2)_{Si}$	1	8	458	458

(CFOM=Combined figure of merit for high temperature/high power/high frequency applications)

Substantial research on III–V nitrides growth was initiated in the early 1960s. However, they have trailed way behind the easier-to-grow Si and GaAs semiconductors on the development curve. Nearly 30 years later, Si and GaAs have been pushed to their theoretical limits, while nitrides are just beginning to show their promise. The technological spin-offs came late because ideal substrates could not be found and the consequent growth of GaN thin films contained substantial concentrations of defects and had high n-type backgrounds. Even in films having relatively small background electron concentration, p-type doping could not be achieved until recently.

One particular difficulty in the growth of GaN thin films is the unavailability of sufficiently large (>1 cm) single crystals for use as substrate for homoepitaxial growth. Thus, up to now, heteroepitaxial growth has been a practical necessity and the choice of substrate is critical. Possible substrate materials should have low thermal expansion and lattice mismatch with the crystals grown. Also, they should be unaffected by the growth chemistries (such as NH$_3$ or H$_2$) at high growth temperatures (in excess of 1000°C in some cases). Under these presuppositions, sapphire and SiC are the most popular substrate materials used currently. When hexagonal GaN is grown on the (0001) basal plane of Al$_2$O$_3$, a lattice misfit of ~13% at the growth temperatures can generate a high density of dislocations and defects in the thin film. In the practical case, a large part of the misfit is relaxed through three-dimensional (3D) island growth. The residual strain is comparable to the lattice misfit between 6H-SiC and GaN, and the result is that comparable dislocation densities are observed.[6] Today, SiC substrates, though more costly, are of increasing interest for high temperature, high power devices like transistors due to their good thermal conductivity and possibility of n- and p-type doping. The materials with a close lattice match with GaN, such as LiAlO$_2$ [7] and LiGaO$_2$ [8], were also used for epitaxial substrates. However, the GaN grown lacked the desired electronic properties due to either the rough growth or unintentional contamination from the substrates. The ideal candidate substrate is clearly a GaN wafer. Several research groups are investigating the growth of the bulk GaN crystals and very thick films through various techniques.[9–11] However, commercially available large-area GaN wafers appear to be several years away. The nitride community is, therefore, challenged with growth of heteroepitaxial films.

Many epitaxial thin film growth processes have been developed, including molecular beam epitaxy (MBE),[12,13] hydride vapor phase epitaxy (HVPE),[9–11,14] metalorganic chemical vapor deposition (MOCVD),[15–20] and derivatives of these methods. In the past few years, MOCVD has evolved into a leading technique for production of III–V-nitride optoelectronic and microelectronic devices. One remarkable application worthy mention is the achievement of super-bright blue LEDs.[18] Characteristics of this method include the use of high purity chemical sources, a high degree of composition control and uniformity, high growth rates, large-scale manufacturing potential and the ability to grow abrupt junctions.

Initially, the growth of GaN was performed directly on sapphire and SiC substrate, with large crystalline defects threading vertically from the substrate interface through the newly deposited thin film. The wafer usually had rough surfaces, mainly caused by the 3D growth mode. In 1986, Amano *et al.*[16] succeeded in remarkably improving the GaN surface morphology, as well as the electrical and optical properties, by deposition of a thin low-temperature AlN buffer layer prior to the high temperature growth of GaN. The essential role of this buffer is both to supply nucleation and promote lateral growth of the GaN film due to the decrease in interfacial free energy between the film and the substrate. Although the buffer layer has reduced the effects of the lattice mismatch, the densities of the threading defects in these thin films are still in the range of 10^9–10^{10} cm^{-2}, and on the order of one million times higher than in other semiconductor systems. These defect-laden materials, to date, have had a surprisingly small effect on the performance of both optical and electronic devices, but they may raise major questions as to the long-term stability of these devices. It is unlikely that the full promise of GaN and related alloys can be realized without a major reduction in the defect densities in the as-grown materials.

In 1994, the lateral epitaxial overgrowth (LEO) technique was employed to further improve the quality of the heteroepitaxially grown GaN by a marked reduction in defect density.[19] In this method, a layer of GaN grown by MOCVD is covered with 100–200 nm of amorphous SiO_2 and

Si_3N_4 with *ex situ* techniques. Small circular or rectangular "windows" are then etched through to the underlying GaN. A GaN film is subsequently regrown under conditions such that growth occurs epitaxially only in the windows and not on the mask. If growth continues, then lateral growth over the mask eventually occurs. Since most of the extended dislocations propagate in the growth direction through GaN, very few threading dislocations are visible in the regrown GaN that extends laterally over the mask. Marchand *et al*.[20] observed that the density of dislocation reaching the surface of LEO GaN was in the 10^4–$10^5\,cm^{-2}$ range, while the film over the window regions still contained high levels of the threading defects. Figure 1.5 compares the cross-section transmission electron microscopy (TEM) of a typical MOCVD growth and LEO GaN.

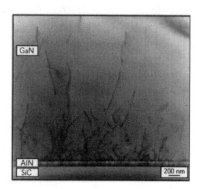 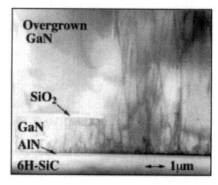

Figure 1.5. Cross-section TEM of typical MOCVD-grown GaN using an AlN buffer layer on SiC (left) and typical LEO GaN (right) (after Ref. [20])

A refined approach to a nearly dislocation-free GaN substrate for devices can be employed by two successive LEO steps with the mask of the second step positioned over the opening defined by the mask of the first step, thus blocking the defects that have grown out of the first windows. This complicated procedure offers the possibility of eliminating the disadvantages of heteroepitaxy, and will be important until GaN substrates become available.

In addition to growing GaN films with low defect densities, another key requirement for fabricating devices is the ability to control the desired electrical properties of the thin film precisely. In general, wide bandgap semiconductors are difficult to dope due to native defects. When the enthalpy for defect formation is lower than the band gap energy, the probability of generating a defect increases with the bandgap, *i.e.* the energy released by a donor-to-acceptor transition. Particularly for GaN, MOCVD-grown material is commonly n-type conductive, and N-vacancy was long believed to be the dominant donor. Many attempts have been made to avoid N-vacancy formation by growing GaN at high pressures and high temperatures.[21,22] Efficient n-type doping of GaN through incorporation of Si during the growth proved relatively easy to achieve. Highly doping can also be created by implantation of Si or Group VI donors. Recently, Burm et al.[23] have shown a shallow Si implant at high dose to produce a doping density of 4×10^{20} cm^{-3}, resulted in an extremely low ohmic contact resistance of 4×10^{-8} $\Omega\cdot$cm^2 using Ti–Au contacts.

Since conductivity is proportional to the product of carrier concentration and Hall mobility, another goal for GaN used in device applications is to obtain the highest Hall mobilities possible. Figure 1.6 summarizes the measured electron mobility in n-type GaN, along with the results obtained from Monte Carlo simulation.[24,25] As can be seen, the experimental data is roughly half of the calculated value, possibly due to significant scattering from impurities and defects in the state-of-the-art materials.

The III–V nitrides are expected to be made p-type by inserting column II elements such as Zn, Mg Be and Ca substitutionally for Ga to form single acceptors. However, all of these divalent elements form deep acceptors, the shallowest being Mg with an ionization level of 0.17 eV, which is still many kT above the valence band edge of GaN.[26] At this acceptor level, one should only expect <10% of the Mg atoms to be ionized at room temperature, which means the Mg concentration needs to be approximately two orders of magnitude larger than the desired hole concentration. When MOCVD is used as the growth method, it has been difficult to obtain p-type conductivity. It was later found that hydrogen plays a crucial role in passivating the Mg acceptors, and creates a neutral complex Mg–H that prevents the formation of holes in GaN.[27] It was first shown by Amano et al.[28] that p-type conductivity can be achieved by activating Mg-doped GaN using low-energy electron irradiation. Nakamura et al. demonstrated subsequently that the activation of Mg can also be realized by thermal annealing at ~700°C.[29] Note that MBE-grown GaN doped with Mg may be p-type without a thermal activation process, because of the absence of hydrogen and H–N radicals during growth. In addition, p-type doping was also achieved by implanting Ca or Mg into GaN, followed by high temperature annealing (~1100°C).[30,31] The highest hole concentration reported so far is ~10^{18} cm^{-3}, and the typical hole mobility is very low, often 10 cm$^2\cdot$V$^{-1}\cdot$s^{-1} or below, but allowing the realization of p–n junctions. Achieving low resistance ohmic contacts to the GaN layers with poor p-type doping concentrations has proved to be troublesome. Recently, Brandt et al.[32] found that by compensating Be with O a neutral dipole is formed that does not scatter the holes. Hence, a record high hole mobility of 150 $cm^2 \cdot V^{-1} \cdot s^{-1}$ was obtained. This may be the ideal contact layer for GaN-based devices.

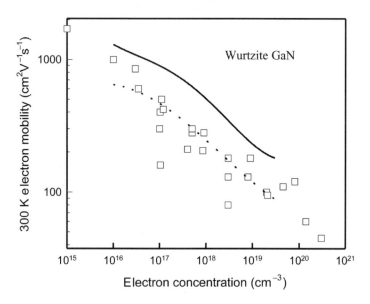

Figure 1.6. A survey of 300 K GaN electron Hall mobility values as reported by various groups. The solid line shows the calculated results for uncompensated n-GaN (after Ref. [24, 25])

The current level of the progress in the development of GaN commercially viable devices, namely GaN-based LEDs, LDs and UV detectors, has been the direct result of the realization of high quality crystals of GaN, AlGaN, InGaN, and the relatively recent achievement of p-type conduction in GaN. The first p–n junction LED was demonstrated by Amano et al.[28] in 1989. Following this, Nichia Chemical Industries announced the commercial availability of blue LEDs with high efficiency and luminous intensities over 1 cd.[18] In the subsequent years, high-brightness single quantum-well-structure blue, green, and yellow InGaN LEDs with luminous intensities above 10 cd have been commercialized.[33,34] In 1996, Nakamura et al.[35] reported the first current-injection GaN-based LDs with separate confinement heterostructure, and subsequently achieved continuous-wave (CW) lasing at room temperature.[36] Figure 1.7 schematically shows the cross-section of the nitride-based laser diode. The active layer is an InGaN multiquantum well with a large number of well layers. GaN and AlGaN were used as the waveguide and cladding layers respectively. The mirror facet was formed by numerous methods, including dry etching, polishing or cleaving.

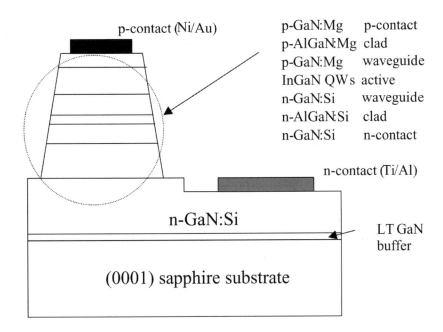

Figure 1.7. Cross-sectional view of a typical structure of GaN-based LD

Surprisingly, the high-density dislocations resulting from the heteroepitaxial growth on sapphire in these optical devices did not appear to be efficient nonradiative centers, as they are in other III–V materials. However, the crystalline defects do affect the device reliability. Nichia employed the LEO growth technique for their blue LDs and achieved an increase in device lifetime from a few hundred hours to an estimated 10,000 h.[37] Another major problem limiting diode performance is high specific contact resistance of the ohmic contact on the p-GaN side of the junction. Present lateral GaN lasers suffer significant *IR* drops due to poor p-type doping and ohmic metallization.

The nitride material growth technology that supports the optical device efforts has also proven to be compatible with the development of electronic devices. In the past several years, the electronic device development has emphasized FET structures, because this important class of devices puts smaller demands on the growth and fabrication technique compared with bipolar transistors. The rapid progress that has been made, especially in modulation-doped FETs (MODFETs), has been sufficient to show that GaN and related alloys will play a significant role in the future development of high temperature, high power, and high frequency electronic devices.[38–41]

Figure 1.8 presents a schematic representation of a GaN–AlGaN heterostructure. Owing to the large conduction band discontinuity, the electrons diffusing from the large bandgap AlGaN into the smaller bandgap GaN form a two-dimensional (2D) electron gas (2DEG) in the triangle quantum well at the interface, which is the hallmark of a MODFET. The sheet carrier density of the 2DEG was found to be further enhanced by the strong pizeoelectronic effect in GaN. Pizeoelectronic coefficients in nitrides were measured to be about an order of magnitude higher than in traditional Group III–V semiconductors.[42] Theoretical simulations have predicted a high peak electron velocity of ~3×10^7 cm/s [25] and an electron mobility of ~2000 $cm^2 \cdot V^{-1} \cdot s^{-1}$ in the GaN channel at room temperature at a carrier concentration of 10^{17} cm^{-3}.[43] Gaska et al.[4] showed that the highest measured Hall mobility at room temperature was 2019 $cm^2 \cdot V^{-1} \cdot s^{-1}$, and this increased approximately fivefold to 10,250 $cm^2 \cdot V^{-1} \cdot s^{-1}$ below 10 K for growth on 6H-SiC substrate.

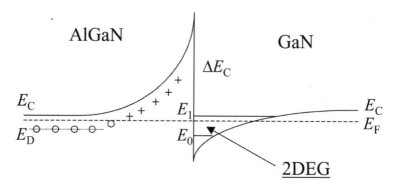

Figure 1.8. Conduction band structure of a modulation-doped structure

In 1993, Khan et al.[38] demonstrated the first AlGaN–GaN MODFET, with a g_m of 23 ms/mm and 2DEG mobility of 563 $cm^2 \cdot V^{-1} \cdot s^{-1}$ at 300 K. They also reported the first microwave results, with f_t of 11 GHz and f_{max} of 14 GHz.[39] In the early stages, the MODFETs exhibited very low transconductances and relatively poor frequency response. This is consistent with the defect-laden nature of the early GaN and AlGaN layers. With improvements in the materials quality, the transconductance, current capacity, and drain breakdown voltage are all increased to the point that GaN-based MODFETs are now strong contenders in the arena of high power devices/amplifiers. To date, the highest power density achieved for a 0.45×125 μm² GaN MODFET is 6.8 W/mm at 10 GHz and associated gain of 10.65 dB.[40] The operation temperature has been pushed to 750°C by employing a thermally stable Pt–Au gate contact.[41]

The published performances of epitaxial GaN-based Metal Semiconductor Field reflect Transistor (MESFETs) demonstrate that all the required components for a MESFET-based technology are in place,[44,45] *i.e.* an appropriate high resistivity buffer–substrate combination has been developed for doped-layer epitaxial growth, FET channels can be grown with thin n^+ contact layers on which ohmic contact with adequate contact resistances have been achieved, gate metalizations which can pitchoff the channel and support a high drain bias have been demonstrated, and it has shown that both mesa etch and implant isolation can be used to define the active device area. Recently, an all implanted GaN junction FET,[46] an Si_3N_4-gated GaN Metal Insulator Semiconductor Field reflect Transistor (MISFET),[47] and a $Ga_2O_3(Gd_2O_3)$ gated GaN MOSFET [48] with reasonable performance were also reported. These types of devices potentially have advantage over MESFET, especially at high temperatures due to low reverse leakage currents.

So far, there are only several reports of development of GaN-based bipolar transistors.[49,50] Basically, the device performance is limited by the difficulty in growth and processing related to the buried p-type layer and the small minority carrier lifetime. Commercialization of these devices, is still far away but their developments will follow the material improvements in the new decade, and much impetus comes from defense applications where ultrawide bandwidth and linearity are desired.

Group III–V nitrides offer a valuable combination of electrical, optical and pizeoelectrical behavior, and enable the fabrication of LEDs, LDs, detectors, and transistors. In the past, the poor quality of the materials, the lack of p-type doping, and the absence of reliable processing procedures thwarted engineers and scientists from fabricating these useful devices. However, the 1990s have brought significant advances in the sophistication of growth techniques, the purity of the chemicals used for film deposition, the controlled introduction and activation of selected impurities, and progress in processing techniques. Most of the aforementioned obstacles have been sufficiently overcome, and the electronic and optical devices have been demonstrated and partially commercialized. Market projections show that GaN-based blue and green LEDs will represent the majority of the estimated $3 billion per year GaN-based device market by 2006. In transistors, GaN can go where no other semiconductors have gone before. The future development in this area will definitely be fueled by the increasing demand for high temperature, high power applications. From materials science to device engineering, from laboratory research to commercial products, III–V nitride technologies have shown a late but exciting blooming.

While further improvements in the III–V nitride materials quality can be expected to enhance device operation, further device advances will also require improved processing technology. Owing to their wide bandgap nature and chemical stability, GaN and related materials present a host of device processing challenges, including poor p-type doping (by implantation), difficulty in achieving reliable low-resistance p-ohmic contacts, high temperatures needed for implant activation, lack of efficient wet etch process, generally low dry etch rates and low selectivity over etching masks, and dry etch damage. These problems constitute a major obstacle to successful demonstration and commercialization of some GaN-based devices, such as bipolar transistors and power switches, whose performance is affected much more by the immature fabrication techniques. To fully exploit these device applications, a number of critical advances are necessary in the areas of implantation doping and isolation, high temperature thermal processing, ohmic contact to p-type material, dry etching process, and device passivation.

Ion implantation is an enabling technology for creating selective-area doping and forming high-resistance regions in device structures. For the development of ion implantation doping for advanced GaN-based electronics, it is important to understand the dopant activation process, and implantation-induced damage generation and removal. Recent studies have showed that quite good activation efficiency can be obtained by annealing at 1100°C, [30,51] but implantation damage cannot be significantly removed at this temperature. Annealing at temperatures >1300°C is suggested to fully remove the damage and further optimize the transport properties of implanted regions in GaN.[51] Since these temperatures are beyond the capability of most rapid thermal annealing systems, new annealing apparatus must be developed. Consequently, there is an urgent need to carry out detailed studies on the dopant activation, impurity redistribution, defect removal, and surface degradation at these elevated temperatures. Efficient surface protection must be developed to prevent material decomposition and N_2 loss from the GaN surfaces.

Ion implantation is also attractive for inter-device isolation and producing current guiding. Efficient compensation has been achieved in the GaN materials by using N or He implantation.[30,52] However, the isolation is not stable at high temperatures, *i.e.* typical implant damage compensation. Implantation in In-containing III–V nitrides has shown that InGaN, as used in LEDs, laser cavities, or transistor channels, is difficult to render highly resistive.[53] The defect level is usually high in the energy gap, not near midgap, as is ideal for implant isolation. There is a strong need for an understanding of the implant isolation process and mechanism in III–V nitride materials because of the emerging applications for high temperature, high power electronics based on this material system. In particular, attempts need to be made to explore thermally stable implant isolation in GaN, and significant compensation must be achieved in the In-containing nitrides.

Considerable progress in the development of contacts to GaN has been made in the past several years.[54] Nevertheless, it is necessary still to improve upon the electrical performance of these contacts, particularly to achieve low contact resistances to p-GaN, and to develop contacts with greater thermal stability, which is critical for high current density devices. It has proven challenging to obtained acceptable low specific contact resistances on p-GaN. Values $\leq 10^{-5}$ $\Omega \cdot cm^2$ would be desirable in general for electronics, but more typical numbers are 10^{-1}–10^{-3} $\Omega \cdot cm^2$. The high contact resistances can be attributed to several factors, including: (1) The absence of a metal with a sufficiently high work function (the bandgap of GaN is 3.4 eV, and the electron affinity is 4.1 eV, but metal work functions are typically ≤ 5 eV). (2) The relatively low hole concentrations in p-GaN due to the deep ionization level of the Mg acceptor (~170 meV). (3) The tendency for the preferential loss of nitrogen from the GaN surface during processing, which may produce surface conversion to n-type conductivity. In order to lower the contact resistances to p-GaN further, it will be necessary to further increase p-type conductivity or to lower the barrier height of the metal contacts, perhaps by growing a more readily contacted compositionally graded semiconductor alloy on the p-GaN.

The thermal stability of the contacts is also noteworthy. Annealing at ~700°C resulted in interfacial reaction, along with serious morphological degradation of the conventional Ti-based or Ni-based contacts.[55,56] In the case of contact to p-GaN, the metalization will heat up as current flows across the interface due to the high series resistance, leading to metal migration down threading dislocations and eventual shorting of the devices. Thermally stable Schottky contacts are also required for power amplifiers and optoelectronics that operate at high temperatures, but the electrical characteristics of the metal–n-GaN diodes have been reported to suffer degradation upon exposure to temperatures as low as 300°C (Pd),[57] 400°C (Pt),[58] 575°C (Au)[59] and 600°C (Ni).[54]

Furthermore, there is a large scatter in the measured results of Schottky barrier height (SBH) and the ohmic contact resistance, suggesting that our understanding of the interface reactions, surface preparation, and non-idealities associated with the metal–GaN contacts is far from complete.

Dry etching has become the dominant patterning technique for the Group III nitrides due to the shortcomings in wet chemical etching. The etch rates in reactive ion etch (RIE) systems are generally low.[60] High density plasma etching, *i.e.* inductively coupled plasma (ICP)[61,62] and electron cyclotron resonance (ECR)[63] etching have been developed in a variety of chlorine- and methane-based chemistries. The etch rates of GaN were reported as high as 9800 Å/min in a Cl_2–H_2–Ar mixture.[62] However, surface roughening, N depletion, and degradation in electrical and optical properties were observed. Some device etching applications, such as gate recessing for FETs and base mesa formation for Hetrojunction Bipolar Transistors (HBTs), can be particularly susceptible to ion-induced damage. In order to minimize the deleterious effects of energetic ions during dry etching, it will be necessary to develop effective low-damage etching processes that utilize ion energies well below 100 eV. At this point, it is also important to determine the nature of the plasma-induced damage in GaN, and to explore the procedures for damage removal.

GaN-based amplifiers and switches are attractive for high power applications in hostile environments. Reliable edge termination and passivation processes are critical to fully exploit these types of devices. There is not much work to date in this area. In addition, as discussed earlier, thermally stable doping, isolation, and metal contacts are all key issues for these special applications.

1.3 Results and Discussion

1.3.1 Ultra-High-Temperature Activation of Implant Doping in Gallium Nitride

Ion implantation is an enabling technology for selected-area doping or isolation of advanced semiconductor devices. Implantation of donors at high doses can be used to decrease source and drain access resistance in FETs, at lower doses to create channel regions for FETs, while sequential implantation of both acceptors and donors may be used to fabricate p–n junctions. In addition, ion implantation is a suitable technological tool to explore doping, compensation effects, and redistribution properties of the potential dopant species.

The first work on implantation in GaN was performed by Pankove and Hutchby in the earlier 1970s.[64] They reported primarily on the photoluminescence properties of a large array of implanted impurities in GaN. The work was successful in identifying Mg as the shallowest acceptor found to date for GaN. In 1995, Pearton *et al.*[30] achieved electrically active n- and p-type dopants in GaN by implantation of Si and Mg respectively. Subsequently, implanted O was also shown to be a donor and implanted Ca an acceptor in GaN.[31] Two different GaN-based device structures have been demonstrated using implant doping, namely a junction FET [46] and a planar homojunction LED.[65]

Tan *et al.*[66] have observed that low dose ($\leq 5 \times 10^{14}$ cm^{-2}) implants in GaN anneal poorly up to 1100°C, leaving a coarse network of extended defects, while high dose ($\geq 2 \times 10^{15}$ cm^{-2}) implants may lead to amorphization. Amorphous layers recrystallize in the range 800–1000°C to form defective polycrystalline material. Quite good activation efficiencies have been obtained for n-type implanted dopants in spite of high residual damage.[30,51] It is also clear that annealing temperatures above 1300°C are desirable for optimal electrical properties in the implanted layers.[51] At these temperatures, the equilibrium N_2 pressure over GaN is very high; thus, surface protection is necessary to prevent dissociation of the GaN. The most effective and convenient approach is use of an AlN encapsulant that can be removed selectively in KOH after annealing. This technique is attractive for processing of GaN devices in a conventional fabrication-line environment, without the need for specialized high-pressure furnaces.

In this section we first introduce a novel rapid thermal processing (RTP) up to 1500°C to the GaN material system, used in conjunction with AlN cap layers. The high temperature activation characteristics of implanted Si and Group VI donors are then examined. The activation of Mg, Be and C for p-type doping has also been investigated, along with the redistribution of all these species in GaN. Finally, the effectiveness of the new RTP for removing lattice damage in implanted GaN was studied by TEM.

1.3.1.1 High-Temperature Annealing and Aluminum Nitride Encapsulation

In the development of advanced electronic devices, the technology of RTP plays a critical role at numerous points, such as implant activation of dopant species, implantation-induced damage removal, alloying of ohmic contacts, maximization of sheet resistance in implant isolation regions, *etc.* The existing commercial RTP equipment typically relies on a series of tungsten–halogen lamps as heat sources to rapidly heat up the semiconductor wafers, and only has modest temperature capacity (<1100°C), primarily because of the point-like nature of the sources and large thermal mass of the systems. Recent interest in developing wide bandgap compound semiconductors such as SiC and GaN has pushed the processing temperature requirements to much higher values (up to 1500°C).

To realize higher temperature capacity, novel molybdenum intermetallic composite heat sources have been employed, which may be used in air at temperatures up to 1900°C. These heaters have high emissivity (up to 0.9) and allow heat-up times of the order of seconds and heat fluxes up to 100 W/cm^2. The Zapper^{TM} unit relies on wafer movement (in/out of the furnace horizontally) to achieve rapid ramp-up or ramp-down rates (~50 °C /s and 25 °C /s respectively). Excellent temperature uniformity ($\leq \pm 4$°C over a 9"×6" area at 1500°C), good reproducibility can be obtained. Figure 1.9 shows the typical temperature–time profiles for annealing cycles at 1400°C and 1500°C. To examine the thermal stability of the nitrides, a variety of undoped GaN and AlN wafers were sealed in quartz ampoules under N_2 gas, and annealed in Zapper ^{TM} unit at different time–temperature combinations. The GaN layers ~ 3 µm thick were grown at ~1050°C by MOCVD using trimethylgallium and ammonia. Growth was preceded by low-temperature deposition of thin (~200 Å) AlN buffers on Al_2O_3 substrates. Some AlN layers were deposited by reactive sputtering of pure AlN targets in 300 mTorr of 20% N_2Ar at 400°C. The others were grown by metal organic MBE (MOMBE) at 750°C using dimethlamine alane and plasma-dissociated nitrogen.[67] The N_2 pressure in the quartz ampoules was ~15 psi. This negative pressure was necessary to prevent blowout of the ampoule at elevated annealing temperatures. The samples were then annealed at 1100–1500°C for a dwell time of ~10 s.

Figure 1.10 (top) shows scanning electron microscopy (SEM) micrographs of the GaN surfaces annealed at 1200–1500°C. The 1200°C annealing does not degrade the surface, and the sample retains the same appearance as the as-grown material. After 1300°C annealing, there is a high density (~10^8 cm^{-2}) of small hexagonal pits due to incongruent evaporation from the surface. The 1400°C treatment produces complete dissociation of the GaN, and only the underlying AlN buffer survives. Annealing at 1500°C also causes loss of this thin buffer layer, and a smooth exposed Al_2O_3 substrate is evident. The corresponding root-mean-square (RMS) surface roughness of these samples measured by atomic force microscopy (AFM) scans is shown in Figure 1.10 (bottom).

In sharp contrast, both the sputtered and MOMBE-grown AlN were found to survive annealing up to 1400–1500 °C. However, for the sput-

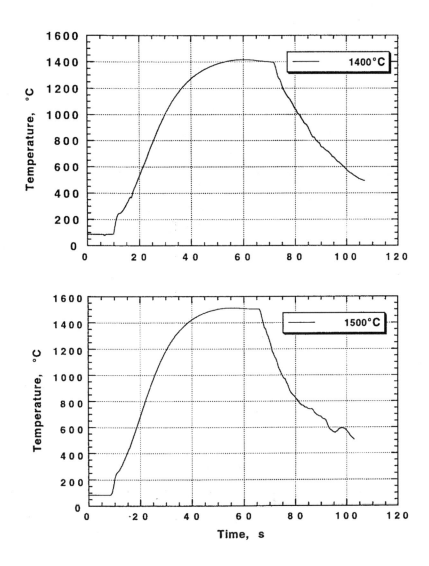

Figure 1.9. Time–temperature profiles for Zapper^{TM} RTP annealing at 1400°C and 1500°C

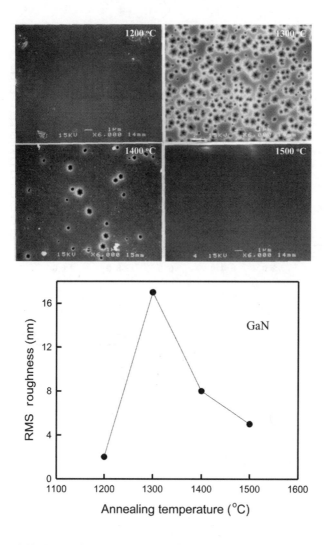

Figure 1.10. SEM micrographs (top) and RMS roughness (bottom) of GaN surfaces annealed at 1200–1500°C

tered materials, we often observed localized failure of the film (Figure 1.11 (top)), possibly due to residual gas (Ar or H_2) agglomeration to form bubbles. The surface roughness tends to go through a maximum at ~1300°C, partially due to some initial localized bubbling, followed by the film densification at temperatures ≥1400°C, as shown in Figure 1.11 (bottom). The film retains its integrity even at 1450°C, and appears to be more thermally stable than the AlN buffer layer under the GaN (Figure 1.10). This can be attributed to the fact that the AlN was deposited at higher temperature (400°C compared with 300°C for the buffer layers) and, therefore, is denser and contains less residual entrapped gas. For the MOMBE grown films, the localized material failure was basically absent.

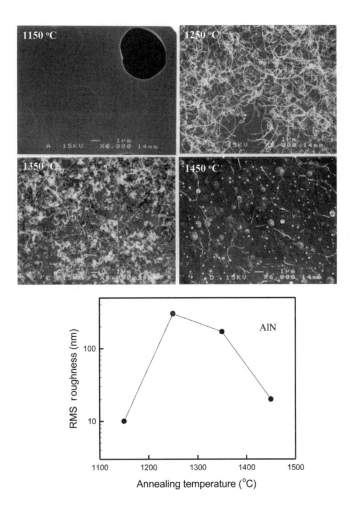

Figure 1.11. SEM micrographs (top) and RMS roughness (bottom) of AlN surfaces annealed at 1150–1450°C

There is clear evidence from previous work that temperatures above 1300°C are required to completely remove implantation damage and to achieve the optimal dopant activation in GaN. However, the above results show that a premium is placed on prevention of loss of nitrogen or even surface dissociation under those conditions. AlN encapsulation can preserve the surface quality to 1500°C, which is quite promising for the high temperature activation processes.

1.3.1.2 n-Type Implant Doping

Nominally undoped ($n{\approx}5{\times}10^{16}$ cm^{-3}) GaN was grown on c-plane Al$_2$O$_3$ substrates by MOCVD. Samples were implanted at room temperature in a non-channeling direction with 150 keV ^{28}Si$^+$ ions at a dose of $5{\times}10^{15}$ cm^{-2}, 200 keV ^{32}S$^+$ and 600 keV ^{128}Te$^+$ ions at doses of $5{\times}10^{14}$ cm^{-2} for n-type doping. The Si implant is a standard condition for producing very high n-type doped regions in GaN when combined with high temperature RTA. All samples were capped with ~1000 Å reactively sputtered AlN, and annealed at 1200–1500 °C under a N$_2$ ambient in the Zapper^{TM} furnace for a dwell time of ~10 s. Following annealing, the AlN was removed in aqueous KOH at ~80 °C, and the samples were characterized. For measurement of the electrical properties, HgIn ohmic contacts were alloyed to the corners of 3×3 mm^2 sections, and Hall effect data was recorded at 25°C in all cases.

Figure 1.12(a) shows an Arrhenius plot of sheet carrier concentration in Si$^+$-implanted material. As a comparison, the electrical results for the unencapsulated samples are also shown. In these uncapped wafers, the sheet electron density increases with annealing temperature up to 1300°C, but this was the highest temperature we would obtain data due to loss of the film. By contrast, the AlN encapsulated samples show a peak in the sheet electron density and the 300 K electron mobility (Figure 1.12(b)) at 1400°C. The activation occurs with an activation energy of ~5.2 eV, which we interpret as the average required to move the interstitial Si atom to a vacant substitutional site by short-range diffusion and to simultaneously remove compensating point defects so that the Si is electrically active. Note that both the electron density and the mobility decreases at 1500°C, indicating that the material is becoming more compensated. This behavior is fairly typical of Si implant activation in III–V materials, and is usually ascribed to self-compensation through Si site-switching, *i.e.* some of the Si$_{Ga}$ donors move to Si$_N$ sites, producing self-compensation. A peak Si activation efficiency of ~90% is obtained at 1400°C (with the ionization level of ~30 meV, Si was fully ionized at room temperature), which corresponds to a peak electron concentration of $\sim 5{\times}10^{20}$ cm^{-3}. This very high doping level can enhance emission over the barrier on metal contacts deposited on the material and reduce contact resistances in GaN electronic devices.

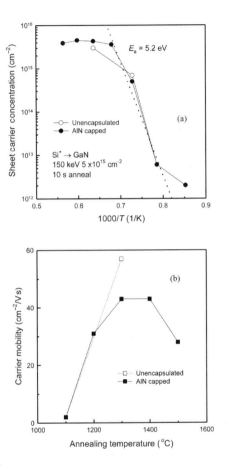

Figure 1.12. Sheet carrier concentration (a) and electron mobility (b) in Si-implanted GaN after uncapped and AlN-capped annealing

There is clear evidence from previous work that temperatures above 1300°C are required to completely remove implantation damage and to achieve the optimal dopant activation in GaN. However, the results in Figure 1.13(a) show an Arrhenius plot of S^+ activation in GaN. The sheet carrier concentration measured at 25°C shows an activation energy of 3.2 eV for the annealing temperature range between 1000 and 1200°C, and the physical origin of this energy is essentially same as that of Si implant activation. The sheet carrier concentration is basically saturated thereafter. The maximum sheet electron density, $\sim 7 \times 10^{13}$ cm^{-2}, corresponds to a peak volume density of ~5×10^{18} cm^{-3}. This is well below that achieved with Si$^+$ implantation and annealing (>10^{20} cm^{-3}). Even though implanted Si$^+$ at the same dose showed evidence of site-switching and self-compensation, it still produces a higher peak doping level than the non-amphoteric donor S, which is only slightly heavier (^{32}S vs ^{28}Si). From temperature-dependent Hall measurements, we find an S$^+$ donor ionization level of 48±10 meV, so that the donors are fully ionized at room temperature.

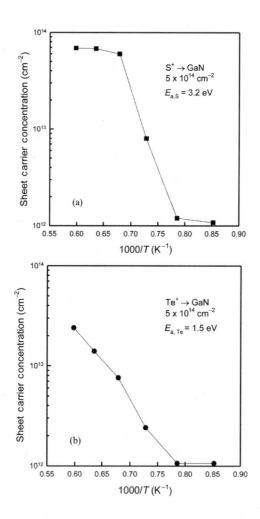

Figure 1.13. Arrhenius plots of sheet electron density in S (a) and Te (b) implanted GaN as a function of annealing temperature

Similar data are shown in Figure 1.13(b) for Te$^+$ implantation. The activation starts around the same temperature as for S, but much lower sheet electron densities are obtained, the activation energy is significantly lower (1.5 eV) and the carrier concentration does not saturate, even at 1400°C. It is likely that because of the much greater atomic weight of ^{128}Te, even higher annealing temperatures would be required to remove all its associated lattice damage, and that the activation characteristics are still being dominated by this defect removal process. Residual lattice damage from the implantation is electrically active in all III–V semiconductors, producing either high resistance behaviors (GaAs) or residual n-type conductivity (InP, GaN). The only data available on Group VI doping in epitaxial material is from Se-doped MOCVD material, where maximum electron concentrations of 2×10^{18}–6×10^{19} cm^{-3} were achieved. These are also below the values reported for Si-doping, and suggests that the Group VI donors do not have any advantage over Si for creation of n-type conductivity in GaN. From limited temperature-dependent Hall data, we estimate the Te ionization level to be 50±20 meV.

1.3.1.3 p-Type Implant Doping

Similar GaN samples were implanted with 80 keV ^{9}Be$^+$, 80 keV ^{12}C$^+$, or 150 keV ^{24}Mg$^+$ at doses of 3–5×10^{14} cm^{-2} for p-type doping. Post-implant annealing was performed at 1000–1400°C under a N$_2$ ambient in the Zapper^{TM} furnace for ~10 s in conjunction with ~1000 Å AlN encapsulation.

The effects of the annealing temperature on the sheet carrier concentrations in Mg$^+$ - and C$^+$ -implanted GaN are shown in Figure 1.14. There are two important features of the data: first, we did not achieve p-type conductivity with carbon; second, only ~1% of the Mg produces a hole at 25°C. Carbon has been predicted previously to have a strong self-compensation effect,[68, 69] and it has been found to produce p-type conductivity only in MOMBE where its incorporation on a N-site is favorable.[70] Based on an ionization level of ~170 meV, the hole density in Mg-doped GaN would be calculated to be ~10% of the Mg acceptor concentration when measured at 25°C. In our case we see an order of magnitude less holes than predicted. This should be related to the existing n-type carrier background in the material and perhaps to residual lattice damage, which is also n-type in GaN. At the highest annealing temperature (1400°C), the hole density falls, which could be due to Mg coming out of solution or to the creation of further compensating defects in the GaN. The results of Be implant are similar with those of C implant, *i.e.* remaining n-type after annealing, probably also due to a strong compensation effect. This data indicates that ion implantation is not so efficient for creating p-type conductivity in state-of-the-art GaN as epitaxial growth doping.

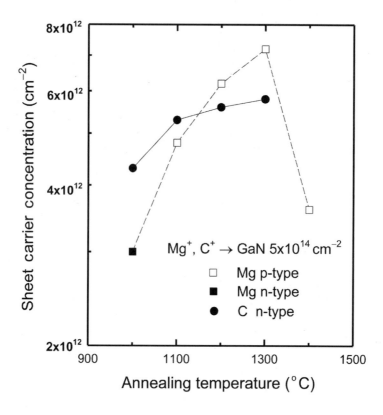

Figure 1.14. Sheet carrier concentrations in Mg^+- and C^+-implanted GaN as a function of annealing temperature

1.3.1.4 Dopant Redistribution

Figure 1.15 shows the calculated, as-implanted Si atomic profile and the secondary ion mass spectroscopy (SIMS) profiles of as-implanted and 1400°C annealed samples. The calculated profile does not produce a good match to the experimental profile, and some work will need to be done to obtain better stopping power data for the ions in GaN. There is little redistribution of the Si at 1400°C, with the diffusivity $\leq 10^{-13}$ cm$^2 \cdot$s^{-1} calculated from the change in width at half-maximum. This result emphasizes the extremely stable nature of Si dopant in GaN even at very high processing temperatures.

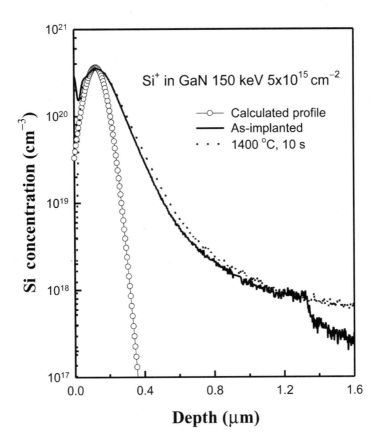

Figure 1.15. Calculated and experimentally measured (by SIMS) profiles of implanted Si (150 keV, 5×10^{15} cm^{-2}) in GaN

Figure 1.16 (top) shows SIMS profiles before and after 1450°C annealing of implanted S in GaN. There is clearly no motion of the sulfur under these conditions and the profiles are essentially coincident. Wilson *et al.* [71] reported some redistribution of implanted S after annealing at 700–1000°C in relatively thin layers of GaN, which might have been influenced by the high crystalline defect density in the material. The samples in the present experiment are much thicker and the extended defect density will be correspondingly lower in the implanted region ($\sim 5\times10^8$ cm^{-2} compared with $\sim 10^{10}$ cm^{-2} in the thin samples). The other Group VI donors, Se and Te, have low diffusion coefficients in all compound semiconductors (e.g. $D_{Se}=5\times10^{-15}$ cm$^2\cdot$s^{-1} at 850°C in GaAs), and we find a similar result for these species implanted into GaN, as shown in Figure 1.16 (bottom).

Given the resolution of SIMS measurement (~200 Å under these conditions), we can obtain the diffusivity at this temperature $\leq 2\times 10^{-13}$ cm$^2 \cdot$s^{-1} using a simple $2\sqrt{Dt}$ estimation.

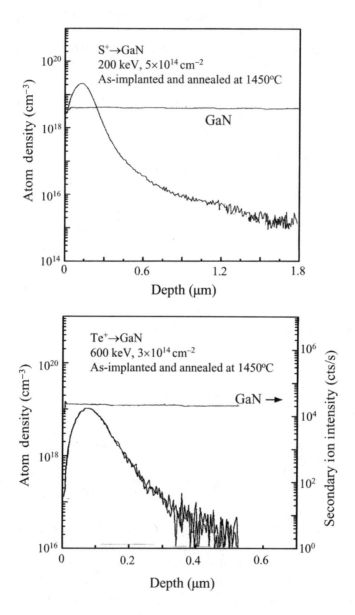

Figure 1.16. SIMS profiles of S$^+$ (top) and Te$^+$ (bottom) implanted GaN before and after annealing at 1450°C (the profiles are essentially coincident)

Figure 1.17 (top) shows the SIMS profiles of implanted Mg in GaN before and after annealing at 1450°C. Again, within the resolution of SIMS, there is no motion of the Mg. This is in sharp contrast to its behavior in GaAs,[72,73] where the rapid diffusion of the Ga-site acceptors during annealing can only be suppressed by co-implanting a Group V element to create a sufficient number of vacant sites for the initially interstitial acceptor ions to occupy upon annealing. This reduces the effective diffusivity of the acceptor and increases its electrical activation. In addition, implanted Mg in GaAs often displays out-diffusion toward the surface (in most cases up, rather than down, the concentration gradient), leading to loss of dopant into the annealing cap.[74] This has been suggested to be due to non-equilibrium levels of Ga interstitials created by the implantation process. This mechanism is clearly absent for implanted Mg in GaN.

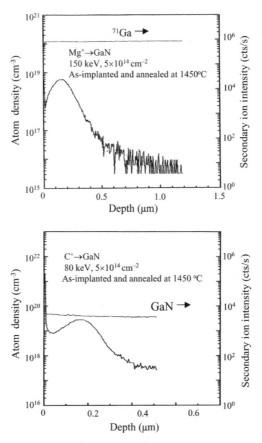

Figure 1.17. SIMS profiles of Mg^+ (top) and C^+ (bottom) implanted GaN before and after annealing at 1450°C (the profiles are essentially coincident)

Carbon is typically a very slow diffuser in all III–V compounds, since it strongly prefers substitutional lattice sites.[74,75] In general, carbon occupies both Ga and N sites in GaN, and the material containing high concentrations of carbon is found to be self-compensated. Figure 1.17 (bottom) shows that it is also an extremely slow diffuser when implanted into GaN, with $D_{eff} \leq 2\times10^{-13}$ cm·s^{-1} at 1400°C.

Figure 1.18 shows a series of profiles for ^{9}Be before and after annealing up to 1200°C. Note that there is an initial broadening of the profile at 900°C, corresponding to an effective diffusivity of ~5×10^{-13} cm^2·s^{-1} at this temperature. However, there is no subsequent redistribution at temperatures up to 1200°C. It appears that, in GaN, the interstitial Be undergoes a type of transient-enhanced diffusion until these excess point defects are removed by annealing, at which stage the Be is basically immobile. Implanted Be shows several types of anomalous diffusion in GaAs, including up-hill diffusion and movement in the tail of the profile, in addition to normal concentration-dependent diffusion,[76] which also result from the non-equilibrium concentrations of point defects created by the nuclear stopping process of the implanted ions.

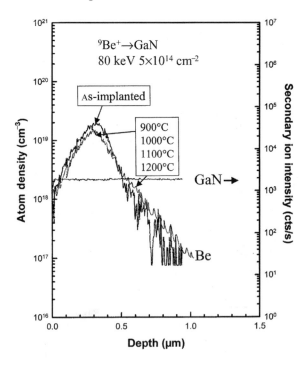

Figure 1.18. SIMS profiles of Be$^+$ implanted GaN before and after annealing at different temperatures

The above results show that most of the common acceptor and donor species implanted into GaN, with the exception of Be, are extremely slow diffusers at high temperatures. This bodes well for the fabrication of GaN-based power devices, such as thyristors and insulated gate bipolar transistors, that will require creation of doped well or source/drain regions by implantation. The low diffusivities of implanted dopants in GaN means that junction placement should be quite precise and there will be less problems with lateral diffusion of the source/drain regions towards the gate. In addition, these results also show the effectiveness of the AlN cap in protecting the GaN surface from dissociation, since if any of the surface was degraded during annealing, then the implant profiles would no longer overlap.

1.3.1.5 Residual Damage

Figure 1.19 shows a plan view TEM micrograph and selected-area electron diffraction pattern from an Si-implanted sample (150 keV, 5×10^{15} cm^{-2}) after annealing at 1100°C for 10 s. This is a high-dose implant of the type used for making n$^+$ ohmic contact regions, and represents a worst-case scenario in terms of damage removal. It also allows comparison with damage expected with other dopants such as S, Ca and Mg due to the representative mass number of Si. The sample is still single crystal as determined by the diffraction pattern, but contains a high density of extended defects ($\sim 10^{10}$ cm^{-2}). This is consistent with past reports of high back-scattering yields in implanted GaN annealed at these conditions.[77] We ascribe these defects to the formation of dislocation loops in the incompletely repaired lattice.

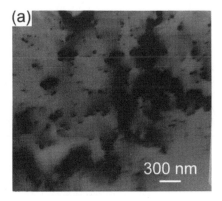
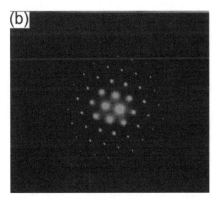

Figure 1.19. TEM plan view (a) and selected-area diffraction pattern from Si$^+$-implanted GaN after 1100°C, 10 s annealing

By sharp contrast, annealing at 1400°C for 10 s brings a substantial reduction in the implant-induced defects, as shown in Figure 1.20. The sample is again single-crystal, but the only contrast in the TEM plan view is due to the lower density (~10^9 cm^{-2}) of threading dislocations arising from lattice mismatch in the heteroepitaxy. This appears to correlate well with the fact that the highest electron mobility and carrier density in these samples was observed for 1400°C annealing. Clearly, the ultra-high temperature annealing is required to completely remove lattice damage in GaN implanted with high doses. However, it may not be needed for the material implanted with lower dose ($\leq 5\times10^{13}$ cm^{-2}), where the amount of damage created is correspondingly less.

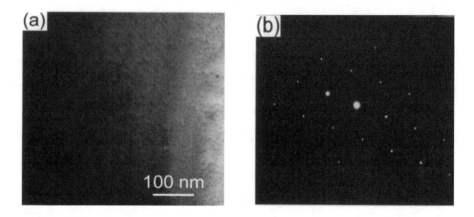

Figure 1.20. (a) TEM plan view and (b) selected-area diffraction pattern from Si$^+$-implanted GaN after 1400°C, 10 s annealing

1.4 Implant Isolation

Implant isolation has been widely used in compound semiconductor devices for inter-device isolation, such as in transistor circuits, or to produce current channeling, such as in lasers. The implantation process can compensate the semiconductor layer either by a damage or chemical mechanism, maintaining a planar device morphology without the need for etched mesa isolation. For the former compensation, high resistivity is created due to introduction of midgap, damage-related levels, which trap the free carriers in the materials. This effect is stable only to the temperature at which the damage is annealed out. For the latter case, the implanted species occupy the substitutional sites and create chemically induced deep levels in the middle of the bandgap. The compensation is thermally stable in the absence of out-diffusion or precipitation of the species. There is a strong need for an understanding of the implant isolation process in GaN because of the emerging applications for high temperature, high power electronics based on this material and its alloys.

To date in the GaN materials system there has only been an examination of damage-induced isolation. Binari *et al.*[52] investigated H^+ and He^+ isolation of n-GaN, with the material remaining compensated to over 850°C with He^+ and 400°C with H^+. Subsequent work focused on N^+ implantation into both n- and p-type GaN, producing defect levels of 0.83 eV and 0.90 eV respectively.[30] The implantation damage was annealed out at 850°C in n-type material and 950°C in p-type material. Very effective isolation of AlGaN heterostructure FET structures has been achieved with a combined P^+/He^+ implantation leading to sheet resistances of $\geq 10^{12}$ $\Omega/\square$ and an activation energy of 0.71 eV for the resistivity.[78] Some work has also been reported for isolation of In-containing nitrides using O^+, F^+ or N^+.[53] Basically, the implantation in InN and InGaN ternaries produced only one or two orders of magnitude increase in sheet resistance after an optimum anneal. The relatively low sheet resistance (~10^4 $\Omega/\square$) achieved is not sufficient for inter-device isolation in electronic circuits.

To create chemically induced isolation, it is necessary to implant impurities with electronic levels in the GaN bandgap, and usually a minimum dose (dependent on the doping level of the sample) is required. In other compound semiconductors, species such as Fe, Cr, Ti and V have been employed, with other examples being O in AlGaAs (where Al–O complexes are thought to form)[79] and N in GaAs (C) (where C–N complexes are thought to form).[80]

In this section, multi-energy O^+, Ti^+, Fe^+ or Cr^+ were implanted into n- and p-type GaN to create high resistivity. The annealing temperature dependence of the sample sheet resistance was measured up to 900°C. The defect levels in the materials of both types were determined by temperature-dependent Transmission Line Method (TLM) measurements. The thermally stable, electrically active concentration of deep states produced by these species was found to be $<7\times10^{17}$ cm^{-3} in both conductivity types of GaN, with the sample resistivity approaching its original, unimplanted value by ~900°C in all cases. The defect levels created in the implanted material are within 0.5 eV of either bandedge.

1.4.1 Oxygen Implantation for Selective Area Isolation

0.3 μm thickness n- (Si-doped) or p-type (Mg-doped) GaN layers were grown on 1 μm thickness undoped GaN on (0001) sapphire substrates by radio frequency (rf) plasma activated MBE. The carrier concentration in the doped layers was 7×10^{17} cm^{-3} in each case. Ohmic contacts were formed in a transmission line pattern (gap spacings of 2, 4, 8, 16 and 32 μm) by e-beam evaporation and lift-off of Ti–Au (n-type) and Ni–Au (p-type) annealed at 800°C and 700°C respectively for 30 s under N_2. The total metal thickness was 4000 Å, so that these regions could act as implanted masks. A schematic of the resultant structure is shown in Figure 1.21.

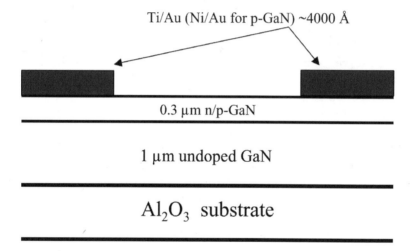

Figure. 1.21. Schematic of GaN structure for measurement of sheet resistance after ion implantation

The samples were then implanted at 25°C using multiple-energy O$^+$ ions at different doses: 50 keV, 1×10^{14} cm^{-2}; 100 keV, 2×10^{14} cm^{-2}; 200 keV, 3×10^{14} cm^{-2}. The total dose was therefore 6×10^{14} cm^{-2}. The sheet resistance was obtained from TLM measurements for both the as-implanted sample and those annealed at increasing temperatures to 900°C. Measurement temperatures in the range of 25–200°C were employed to determine the defect levels in the material.

In the case where the implanted species is chemically active in the GaN it is the ion profiles that are the important feature, since it is the electrically active fraction of these implanted species that determines the isolation behavior. In the case where the isolation simply results from damage-related deep levels, then it is the profile of ion damage that is important. Figure 1.22 shows both the calculated ion profiles (top, from P-CODE™ simulations) and damage profiles (bottom, from transport-of-ions-in-matter (TRIM) simulations) for the multiple-energy O^+ implant scheme. Note that the defect density is generally overstated in these calculations due to recombination of vacancies and interstitials. In any case, the doses are below the amorphization threshold for GaN.

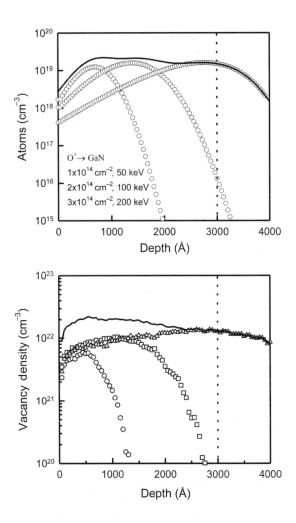

Figure 1.22. Ion (top) and damage (bottom) profiles for multiple-energy O^+ implant sequence into GaN

Figure 1.23 (top) shows the evolution of the sheet resistance for O^+ implanted n- and p-type GaN with annealing temperature. For the n-type sample, the as-implanted sheet resistance was 4×10^9 $\Omega/\square$, and gradually decreased over the entire annealing temperature range. We do not believe this is a result of O-related shallow donor states because these are not activated until annealing temperatures above 1100°C.[30] By contrast, the trend in the sheet resistance in p-type GaN is typical of those observed with damage-related isolation. The as-implanted resistance is five orders of magnitude higher than that of the unimplanted material due to creation of deep traps that remove holes from the valence band. Subsequent annealing tends to further increase the sheet resistance, by reducing the probability for hopping conduction as the average distance between trap sites is increased.[81] Beyond a particular annealing temperature (600°C in this case) the trap density begins to fall below the carrier concentration and holes are returned to the valence band. This produces a decrease in sheet resistance toward the original, unimplanted values. Note that even after a 900°C anneal the sheet resistance is still two orders of magnitude above the unimplanted value. A general rule of thumb for achieving acceptable device isolation is that the implanted region should have a sheet resistance $\geq 10^7$ $\Omega/\square$.[82] In this implant process, sheet resistances above this value were achieved for annealing temperatures up to 600°C for n-type and 800°C for p-type.

Figure 1.23 (bottom) shows Arrhenius plots of the sheet resistance of the implanted n- and p-type GaN annealed at either 300°C (n-type) or 600°C (p-type). The annealing temperature for the p-type sample was chosen to be close to the point where the maximum in the sheet resistance occurs. The activation energies derived from these plots represent the Fermi level positions for the material at the particular annealing temperatures employed. Note that the values are far from midgap (1.7 eV for hexagonal GaN). In particular, the defects created in n-GaN are much shallower and could not produce efficient compensation in this material.

From Figure 1.22 we find that there is a region at the immediate surface (<300 Å) that receives a much lower concentration of cumulative damage, simply because the projected range of the 50 keV O^+ ions is ~400 Å. Inadequate coverage of this near-surface region is a common problem in implant isolation processes, because conventional implant systems typically have minimum operating voltages in the 20–30 keV range. One solution to this problem is to implant through a surface layer, which acts to move the projected range of the lowest energy implant closer to the semiconductor surface. Titanium is a good choice, since it can be readily removed after the implant step with HF solutions. To achieve good compensation in the near-surface region, ~500 Å Ti was deposited on top of the GaN. The implantation performed under same conditions through this Ti overlayer produced a maximum sheet resistance 50 times higher than that without the implant overlayer. The subsequent evolution of the resistance with annealing temperature was similar.

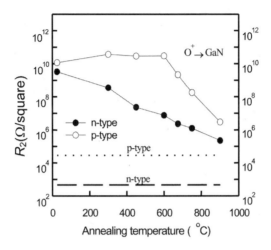

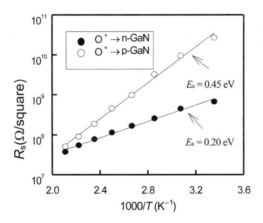

Figure 1.23. Evolution of sheet resistance of n- and p-type GaN with annealing temperature after O^+ implantation (top), and Arrhenius plots of sheet resistance after either 300°C (n-type) or 600°C (p-type) annealing (bottom)

1.4.2 Creation of High-Resistivity Gallium Nitride by Ti, Iron, and Chromium Implantation

For a better understanding of the implant isolation mechanism in GaN, Ti^+, Fe^+ and Cr^+ were implanted into n- and p-type samples at 100 keV (10^{14} cm^{-2}), 300 keV (2×10^{14} cm^{-2}) and 500 keV (3×10^{14} cm^{-2}). The wafer structure is the same as shown in Figure 1.21. The doses and energies were chosen to create an average ion concentration of ~10^{19} cm^{-3} throughout the 0.3 µm thickness doped GaN layers. Both the ion and vacancy profiles for the Fe^+ implants are shown at the top and bottom, respectively, of Figure 1.24. It is very clear that implanting with these heavier ions produced much more cumulative damage than the O^+ implantation.

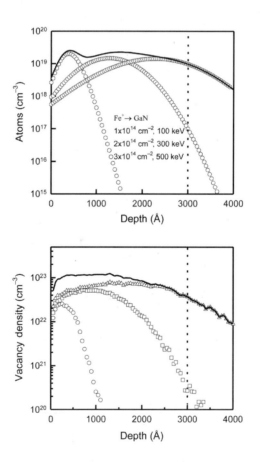

Figure 1.24. Ion (top) and damage (bottom) profiles for multiple-energy Fe^+ implant sequence into GaN

Figure 1.25 shows the annealing temperature dependence of sheet resistance for Cr^+ (top) and Fe^+ (bottom). The as-implanted resistance is six to seven orders of magnitude higher than that of the unimplanted material. The sheet resistances peak at some particular annealing temperatures (500–600°C). There is a significant improvement in the maximum R_s in the n-type sample compared with the O^+ implantation described earlier (~10^{12} vs ~10^{10} Ω/□), because of the higher vacancy and interstitial concentrations created. The trends in the sheet resistance show that the isolations are all typically damage related. If Cr or Fe produced electrically active deep states related to their chemical nature in the bandgap with concentrations greater than the carrier density in the material, then the sheet resistance would remain high for annealing temperatures above 600°C. For these two impurities it is clear that the electrically active concentration of deep states is <7×10^{17} cm^{-3}, otherwise all the carriers would remain trapped beyond an annealing temperature of ~600°C.

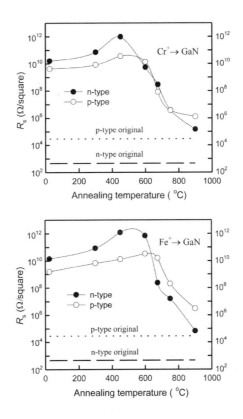

Figure 1.25. Evolution of sheet resistance of GaN with annealing temperature after Cr^+ (top) and Fe^+ (bottom) implantation

Figure 1.26 shows the sheet resistance of Cr^+- (top) or Fe^+ implanted (bottom) n- and p-type GaN annealed at either 450°C (n-type) or 600°C (p-type), as a function of the measurement temperature. The activation energies derived from these plots are 0.49 eV and 0.45 eV in n-type and p-type GaN respectively, which give rough estimates of the positions of the Fermi levels. These implant-induced defect states, although relatively shallow, efficiently compensated the wide bandgap material.

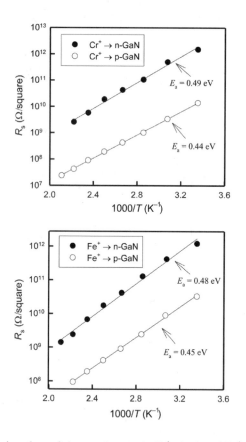

Figure. 1.26. Arrhenius plots of sheet resistance in Cr^+- (top) and Fe^+-implanted (bottom) n- and p-type GaN after annealing at either 450°C (n-type) or 600°C (p-type)

Similar results were obtained from Ti^+ implantation in both n- and p-GaN. Within the experimental error (±0.04 eV), the activation energies are the same for all these implants for both conductivity types. This again suggests that the defect states created are damage related and not chemical in nature. Note that the activation energy obtained with O^+ implant into n-GaN is much smaller. This difference may be related to the lower damage density with O^+ implantation. The defect states in the gap are most likely due to point defect-complexes of vacancies and/or interstitials, and the exact microstructures of these complexes and their resultant energy levels are expected to be very dependent on damage density and creation rate.[83] This might also explain the differences reported in the literature for the activation energies obtained with different implant species.

Figure 1.27 shows a schematic of the energy level positions found in this work for Ti-, Cr-, Fe- and O-implanted p- and n-type GaN annealed to produce the maximum sheet resistance. Although the levels are not at midgap, as is ideal for optimum compensation, they are sufficiently deep to produce high resistivity material ($\sim 10^{12}$ $\Omega/\square$ in n-GaN and $\sim 10^{10}$ $\Omega/\square$ in p-GaN). In GaN contaminated with transition metal impurities, non-phonon-assisted photoluminescence lines attributed to Fe^{3+} at 1.3 eV and Ti^{2+} at 1.19 eV have been reported,[84] but to date there are no electrical measurements.

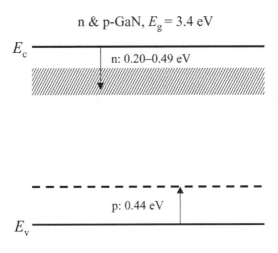

Figure 1.27. Schematic representation of the positions in the energy band gap of defect levels created by O, Fe, Cr and Ti implantation into GaN

1.5 Electrical Contacts to Gallium Nitride

GaN is attractive for optical devices with blue or ultraviolet wavelengths, as well as for high power and high temperature electronics. Critical to the success of these devices, however, are the ohmic contacts, and in some cases Schottky contacts to the semiconductors. In general, making low-resistance ohmic contact is difficult for wide bandgap materials, especially for p-type GaN due to the difficulty in doping. This creates a large voltage drop across the GaN–metal interface for ohmic contacts, which leads to poor device performance and reliability. For most devices, ohmic contacts with resistances lower than 10^{-5} $\Omega \cdot cm^2$ are desired.

Ti-based contacts with contact resistances of 10^{-6}–10^{-5} $\Omega \cdot cm^2$ have been demonstrated on n-type GaN.[23,55,85] These contacts are found to take advantages of the formation of a thin interfacial TiN layer 7(which is refractory), exhibit metallic conductivity, and possess a low work function (<4 eV).[86] However, no satisfying ohmic contact to p-type GaN has been developed so far. In the search for improved contact characteristics, a wide variety of metallizations have been investigated on p-GaN besides the standard Ni–Au,[18,56,87,88] including Ni,[87,88] Au,[87,89] Pd,[87] Pd–Au,[90] Pt–Au,[91] Au–Mg–Au,[92] and Pd–Pt–Au.[91] Typically, metals with a large work function, such as Ni, Pd or Pt, are in direct contact with the GaN, and the structure is annealed at 400–750°C. This produces contact resistances in the 10^{-1}–10^{-3} $\Omega \cdot cm^2$ range. Reports have also appeared on the effect of annealing in O_2 ambient in reducing contact resistance (r_c),[93] oxidation of Ni–Au metallization and its role in improving contact quality,[94] the role of surface treatments in determining contact quality [95,96], and the use of Ta–Ti contacts with low, but unstable contact resistances.[97]

Another approach to achieve low-resistance contact is through semiconductor bandgap engineering. Schubert *et al.*[98] proposed the use of AlGaN–GaN superlattices as a technique to increase the average hole concentration. The periodic oscillation of the valence band edge is superimposed with the oscillation generated by the piezoelectric field; this significantly reduces the Mg acceptor ionization energy and, therefore, increases the hole concentration. Ti–Pt–Au contacts with specific contact resistance of ~4×10^{-4} $\Omega \cdot cm^2$ have been achieved on this structure.[99] However, this scheme may not easily be applied to real devices, since the free carriers are separated into parallel sheets.

The Schottky contacts to n-GaN for a variety of elemental metals have been studied extensively. The reported SBH increases monotonically, but does not scale proportionally with metal work function, with a considerable amount of scatter in the experimental results for a given metal. The *I–V* ideality factor is usually significantly larger than unity, and the measured values of the Richardson constant A^{**} are quite small.[54] This non-ideal behavior of GaN Schottky diodes appears to result from the presence of several transport mechanisms, and to materials and process factors such as defects present in these films, the effectiveness of surface cleans prior to metal deposition, local stoichiometry variations, and variations in surface roughness.

There is not much information on Schottky contacts on p-GaN. Measurement of the barrier height is very difficult, due to the general difficulty in growing high quality p-type material, and the lack of low resistance ohmic contact. The experimental values from conventional *I–V* and *C–V* methods are usually inconsistent, but also appear to be affected by the work function of the metals.[87]

One of the important issues in making high quality Schottky and ohmic contacts to GaN is the surface cleanliness. The ideal metal–semiconductor interface should be oxide- and defect-free, atomically smooth, uniform, and thermally stable, with the metal epitaxial. Analysis by spectroscopic ellipsometry showed that >30 Å of overlayer consisting organic and inorganic, and native oxide is present on air-exposed GaN.[90] *In situ* treatments, such as N-ion sputtering, were reported to remove the oxide layer on the GaN surface without significant modification of the N-to-Ga atomic ratio.[87:101] For practical contacts, the samples are treated with various acidic or basic solutions before deposition of metal layers. It was found that HCl-based solution is more effective in removing oxides and leaves less oxygen residue, but HF is more effective in removing carbon and hydrocarbon contamination.[102] These *ex situ* treatments can remove a significant part of the surface overlayer, but cannot produce atomically clean surfaces. The residual oxide layer is expected to act as a barrier to current flow through the metal–GaN interface and, therefore, has a profound influence on the electrical characteristics of contacts to GaN.

The thermal stability of metal–GaN contacts is also critically important for practical device operation (especially power electronics). The thermal limits of most of the metal–GaN combinations are between 300°C and 600°C.[54] At higher temperatures, severe degradation in contact morphology is observed, usually resulting from the formation of new interfacial phases, such as metal gallides.

In this section, as a first step to explore the improved ohmic contact to p-GaN, the effects of $(NH_4)_2S$ treatment of the GaN surface on the electrical properties were studied. Reductions in SBH by ~0.2 eV were observed. This reduction in SBH was ascribed to removal of the native oxide of thickness 1–2 nm that was present after conventional cleaning, and formed an interfacial insulating layer at the metal–GaN interface. In the second part, the thermal stability of W and WSi contacts on n- and p-type GaN was examined. Specific contact resistances in the range 10^{-5} $\Omega \cdot cm^2$ were obtained for W on high-dose Si-implanted GaN, while true ohmic characteristics could only be achieved at elevated temperatures for contacts on p-GaN.

1.5.1 Effects of Interfacial Oxides on Schottky Contact

Three different types of samples were employed with n-type doping of either 8×10^{16} cm^{-3} or 10^{18} cm^{-3}, or p-type doping (hole concentration) of 10^{17} cm^{-3}. These GaN layers were 1–3 μm thick and were grown on c-plane Al$_2$O$_3$ substrates by rf plasma-assisted MBE. Each of the samples was treated in one of two different ways. The first involved a conventional cleaning process that involved sequential rinsing in acetone, isopropyl alcohol and deionized water prior to lithography for defining the contact areas. After lithography, the samples were rinsed 60 s in 30% HCl (25°C) and 30 s in buffered HF to remove the native oxide and then immediately loaded into the e-beam evaporator. The second cleaning process was the same as the first, but the last step was a 20 min boil in (NH$_4$)$_2$S. This solution does not affect the photoresist mask. The (NH$_4$)$_2$S solution is effective in removing native oxide and prevention of immediate reoxidation, since a Ga–S monolayer is formed on the surface. This should be far less of a hindrance to current flow than the presence of a much thicker native oxide. The samples had Ti–Al or Ni–Au for ohmic contacts, each annealed at 750°C for 30 s to produce low r_c. The Pt (400 Å)–Au (1500 Å) rectifying contacts were e-beam deposited with diameters 50–200 μm. The effective barrier heights were obtained from the forward current–voltage (I–V) characteristics, according to the equation (see Equation (1.1)).

$$J = A^{**}T^2 \exp(-\frac{\phi_b}{kT})[\exp(\frac{eV}{nkT}) - 1] \tag{1.1}$$

where A^{**} is 26.4 A·cm^{-2}·K^{-2} for n-GaN and 96.1 A·cm^{-2}K^{-2} for p-GaN and J is current density, ϕ_4 is barrier height.

Figure 1.28 shows forward (top) and reverse (bottom) I–V characteristics from the n$^+$ GaN diodes, either with or without the (NH$_4$)$_2$S treatment prior to Schottky contact deposition. There are several key points in this data. First, the forward current increases as a result of the (NH$_4$)$_2$S treatment due to a decrease in barrier height from 0.81 eV on the control diode to 0.58 eV on the treated diode. Second, in the linear region of the forward characteristics, the scaling of current with contact diameter is much greater than a power of two. This suggests that there is a non-uniform concentration of defects, which are generation–recombination centers. Finally, the reverse leakage current is also increased by the (NH$_4$)$_2$S treatment.

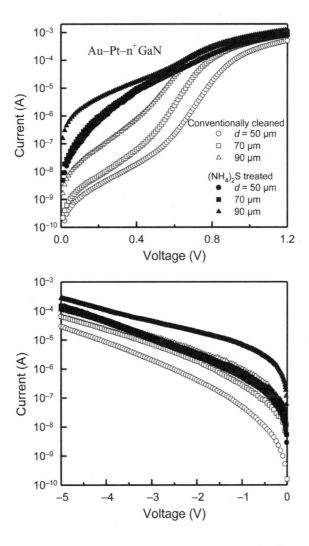

Figure 1.28. Forward (top) and reverse (bottom) I–V characteristics from AuPt–n$^+$-GaN diodes of different diameters, either cleaned in a conventional fashion prior to metal deposition or with an additional (NH$_4$)$_2$S treatment

Similar data are shown in Figure 1.29 for the n-GaN diodes, at two different rectifying contact diameters. The average barrier height decreased from ~0.99 eV on the control diodes to ~0.83 eV on the (NH$_4$)$_2$S-treated samples. Both the forward and reverse currents increased as a result of the (NH$_4$)$_2$S step.

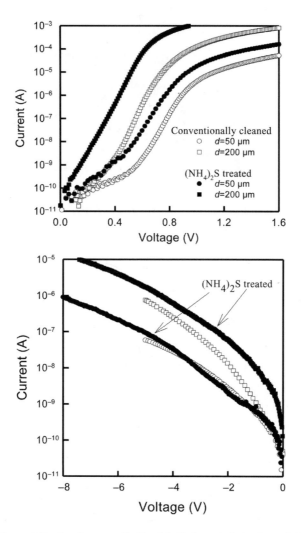

Figure 1.29. Forward (top) and reverse (bottom) I–V characteristics from AuPt–n⁻-GaN diodes of different diameters, either cleaned in a conventional fashion prior to metal deposition or with an additional $(NH_4)_2S$ treatment

Figure 1.30 shows I–V characteristics from the p-GaN diodes. These structures showed high currents due to the difficulty in forming high barriers to p-GaN. The control diodes showed average barrier heights of 0.49 eV, which was slightly reduced to 0.47 eV as a result of the $(NH_4)_2S$ treatment. In these diodes, the low bias current scaled with contact diameter, indicating that surface leakage is important. We did not passivate the perimeters of the devices in this work.

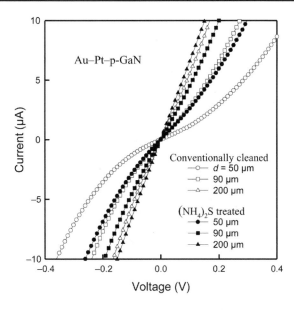

Figure 1.30. I–V characteristics from AuPt–p-GaN diodes of different diameters, either cleaned in a conventional fashion prior to metal deposition or with an additional $(NH_4)_2S$ treatment

Table 1.2 shows a compilation of the barrier heights and ideality factors for the n^+, n and p diodes. The clear effect of the boiling $(NH_4)_2S$ exposure is to decrease ϕ_b, suggesting that it could be promising for fabrication of high quality ohmic contacts on GaN. $(NH_4)_2S$ solution was also employed to treat completed GaN mesa Schottky diodes. Figure 1.31 shows the I–V characteristics from the control and the passivated devices. Both the forward and reverse current densities increase slightly after the treatment. This probably can be ascribed to the same reason as described earlier, *i.e.* $(NH_4)_2S$ passivation reduced the barrier height, and the leakage current on the periphery increased. The forward section of the I–V characteristics shows only a slight improvement in ideality factor, and little change in this parameter is also shown in Table 1.2. Note that the $2kT$ leakage current in a diode of this type has contributions from both the bulk space-charge region and the mesa surface, and that sulfide passivation would only affect the latter. The fact that the diode ideality factors did not improve suggests that the problem for the conventional III–V compound semiconductors, *i.e.* high density of surface states, is absent for GaN, and $(NH_4)_2S$ does not passivate the surface of GaN electronically, as it does for the GaAs or InP surfaces. This is consistent with the fact that no Fermi level pinning has been observed on the GaN surface. However, we generally got ideality factors much larger than unity, which indicates high dislocation densities and high compensation levels in the current state of GaN materials. Particularly for diodes fabricated on MBE-grown wafers, high leakage currents were typically observed, due to the relatively low epi thickness of this growth method.

Table 1.2. Summary of electrical data for test diodes

	n		ϕ_b (eV)	
Sample	Conventionally cleaned	$(NH_4)_2S$ treated	Conventionally cleaned	$(NH_4)_2S$ treated
AuPt–n^+-GaN	1.6–1.8	1.8–1.9	0.81	0.58
AuPt–n-GaN	1.4–1.6	1.3–1.8	0.99	0.83
AuPt–p-GaN	~2	~2	0.49	0.47

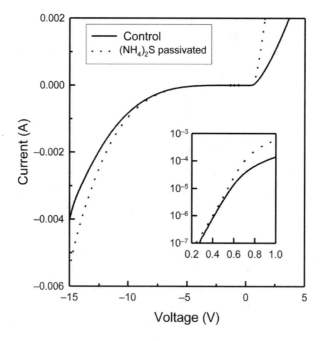

Figure.1.31. I–V characteristics from complete GaN mesa diodes before and after $(NH_4)_2S$ treatment

1.5.2 Interfacial Insulator Model

A contamination layer such as a thin oxide at the metal–semiconductor interface usually behaves as an insulator, and provides energy barrier for carrier injection. It is assumed that the thickness and the mean tunneling barrier height of this thin oxide layer are δ and χ respectively. According to quantum mechanics, the transmission coefficient of an electron with energy E through this barrier is given by

$$T \approx \exp\{-\frac{2}{\hbar}\int_0^\delta (2m)^{1/2}[\chi - E]^{1/2}dx\} = \exp[-\frac{2}{\hbar}(2m)^{1/2}(\chi - E)^{1/2}\delta] \quad (1.2)$$

where m is the tunneling effective mass. Therefore, the current density across the interface can be expressed as (the zero of electron energy is chosen as the conduction band edge at the semiconductor surface)

$$J = \frac{qm_t}{2\pi^2\hbar^3}\exp[-\frac{2}{\hbar}(2m\chi)^{1/2}\delta]\int_0^\infty\int_0^\infty (f_s - f_m)\,dE_t\,dE_x \quad (1.3)$$

where E_x and E_t are the electron energy components normal and parallel to the Schottky barrier; f_s and f_m are the Fermi-Dirac distribution functions for electron states in the semiconductor and metal respectively; and m_t is the effective mass component in the semiconductor transverse to the barrier. In the nondegenerate case, the current density for forward biases $V>3kT/q$ is given by

$$J = A^{**}T^2\exp[-\frac{2}{\hbar}(2m\chi)^{1/2}\delta]\exp(-q\phi_{b0}/kT)\exp(qV/nkT) \quad (1.4)$$

The effective SBH ϕ_b can be obtained from

$$\phi_b = \phi_{b0} + \Delta\phi \quad (1.5)$$

where ϕ_{b0} is the barrier height without the interfacial layer and $\Delta\phi$ is the additional barrier due to the oxide. The parameter $\Delta\phi$ is given by $2kT/\hbar(2m\chi)^{1/2}\delta$. The χ value for the oxide on GaN is calculated to be ~0.2 eV using the free electron mass.[87] In this work we typically observed $\Delta\phi$ to be 0.16–0.23 eV for Pt–Au on n-GaN, which yields an estimate for the oxide layer thickness of 1–2 nm left on the GaN surface after conventional cleaning. Previous work on the effect of different surface treatments on the r_c of ohmic contacts on p-GaN has shown that wet chemical solutions that remove the native oxide improve the contact resistance. For example, solutions of 1.3 HNO_3:HCl were found to produce r_c values of 4.1×10^{-4} $\Omega \cdot cm^2$ for Pd–Au ohmic contacts on p-GaN, whereas untreated samples had values two orders of magnitude larger.[95] Similarly, the use of buffered HF and boiling $(NH_4)_2S$, followed by a final dip in buffered HF prior to metal deposition, was able to lower r_c by three orders of magnitude relative to untreated diodes.[103] The oxide can also be removed by annealing the contact metal at elevated temperatures. Ishikawa *et al.*[87] observed that, for Ni–GaN or Ta–GaN contacts, Ni or Ta diffused into the contamination layer after annealing at 500°C, and grew epitaxially on the GaN surface. However, Au and Pd did not react efficiently with the oxide layer. Our work suggests that the conventional *ex situ* surface treatments using HCl and HF cannot completely remove the native oxide on GaN. This oxide has a strong influence on the contact characteristics on both n- and p-type GaN, and appears to be responsible for some of the wide spread in contact properties reported in the literature.

Figure 1.32 shows the band diagrams for metal–n- or n^+-GaN structures either with an interfacial oxide (top) or after boiling in $(NH_4)_2S$ to remove this oxide (bottom). In the case of n-GaN, the dominant current conduction mechanism probably remains as thermionic field emission (albeit with low barrier height), while for n^+-GaN there may also be a contribution from field emission. Alternative explanations for the increased current in $(NH_4)_2S$-treated samples would be an increase in contact area due to surface roughening. However, AFM measurements showed no significant change in the RMS surface roughness in our samples.

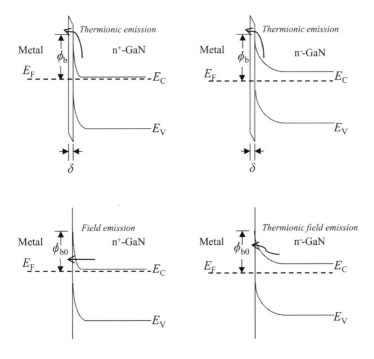

Figure 1.32. Energy band diagrams at the interface of n$^+$-GaN–metal and n$^-$-GaN–metal with or without an interfacial oxide layer

1.5.3 Thermally Stable Tungsten-Based Ohmic Contact

Undoped GaN layers ~3 μm thickness were grown on Al$_2$O$_3$ by MOCV D. These samples were implanted with 150 keV Si$^+$ ions at a dose of 5×10^{15} cm^{-2} and annealed with AlN caps in place to 1400°C for 10 s. W or WSi$_{0.45}$ layers ~1000 Å thickness were deposited using an MRC501 sputtering system. The sample position was biased at 90 V with respect to the Ar discharge. Prior to sputtering, the surface contamination was removed by a rinse in HCl and buffered HF solutions. Transmission line patterns were defined by dry etching the exposed metal with SF$_6$–Ar, and forming mesas around the contact pads using BCl$_3$/N$_2$ dry etching to confine the current flow. The samples were annealed for 60 sn at 500–1100°C under flowing N$_2$.

The Si implantation followed by high-temperature activation yielded a peak n-type doping concentration of ~5×10^{20} cm^{-3}. For as-deposited metal contacts on this highly doped layer, tunneling current dominates the total current. The specific contact resistance is roughly determined by the tunneling mechanism. Figure 1.33 shows the annealing temperature dependence of r_c for W contacts on Si-implanted GaN. The specific contact resistance improves with annealing up to ~950°C, and degrades at high temperatures. Cole et al.[104] observed the formation of the β-W_2N phase at the W–GaN interface at annealing temperatures between 600 and 1000°C, as demonstrated by the x-ray results in Figure 1.34. The as-deposited sample displays a well-oriented (110) W film overlaying a well-oriented (002) GaN epilayer. The samples annealed at 800°C and 1000°C exhibit well-developed face cubic β-W_2N phase, as indicated by diffraction peaks from, the (111) in the x-ray results in Figure 1.34. The as-deposited sample displays a well-oriented (110) W film overlaying a well-oriented (002) GaN epilayer. The onset of the formation of this thin compound was first detected after annealing at 600°C. Annealing at 1000°C caused this phase to disorder, as reflected by the broadening of the W_2N (111) as well as the disappearance of the other W_2N peaks. The β-W_2N phase was no longer stable and subsequently transformed into the defect structure of W–N after annealing at higher temperatures.

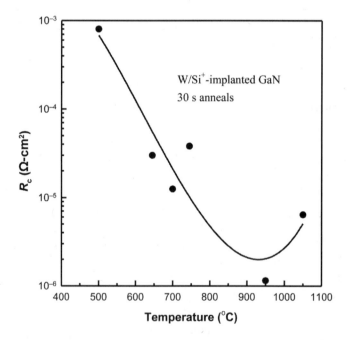

Figure 1.33. Specific contact resistance for W on Si^+-implanted GaN, as a function of post-metallization annealing temperature

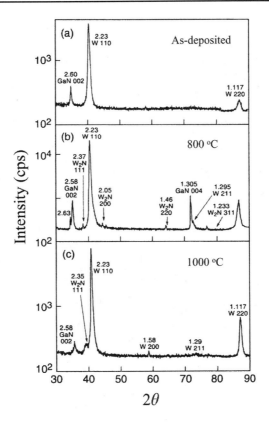

Figure 1.34. The x-ray diffraction results for the (a) as-deposited, (b) 800°C annealed, (c) 1000°C annealed W on n$^+$-GaN samples

The formation of the β-W$_2$N phase paralleled the low specific contact resistance, and is very likely to be critical for achieving the low resistance W ohmic contacts on GaN. If the N needed to form this interfacial compound has out-diffused from the GaN without decomposing the original structure, then there would exist an accumulation of N vacancies near the GaN surface, which act as donors. Thus, the surface GaN would be highly n-type and the carrier tunneling is enhanced. Another possibility is that the N out-diffusion aids in the formation of a Ga-enriched highly conductive layer at the GaN surface, and results in a lower barrier height.

The near-surface GaN crystal quality is of particular importance, and could directly influence the uniformity and stability of the W contacts. TEM results showed that the defect-laden regions were directly associated with contact spiking into the GaN epilayer, which in turn inhibited the formation and development of the β-W$_2$N interfacial phase with its associated smooth metal–semiconductor interface.[105] Protrusion of the metallization down the threading and misfit dislocations was observed at 800°C, extending >5000 Å in some cases, which is obviously a problem for multilayer structures. In contrast, contacting to a low-defect surface improved the lateral extent of the new phase needed for ohmic contact formation and impeded contact metal spiking. The excellent structural stability of the W on GaN is shown in the SEM micrographs of Figure 1.35, where a sharp interface is retained after 750°C annealing.

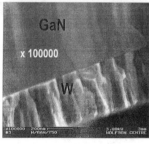

Figure 1.35. SEM micrographs of W contact on Si^+-implanted GaN after annealing at 750°C The GaN thinkess is 3 microns.

1.5.4 Behavior of Tungsten and Tungsten Silicide Contacts on p-Gallium Nitride

The 1 μm thickness GaN layers were grown on Al_2O_3 substrates by MBE using solid Ga and rf-activated N_2 from a plasma source. The Mg acceptor concentration was 10^{18} cm^{-3}. Cathodoluminescence revealed a strong peak at 383–385 nm, with very little deep–level emission. W or $WSi_{0.45}$ layers ~1000 Å thickness were deposited using an MRC501 sputtering system using an Ar discharge and an acceleration voltage of 90 V. Prior to sputtering, the samples were cleaned in HCl and buffered HF solutions. TLM patterns were defined by dry etching the exposed metal with SF_6–Ar, and forming mesas around the contact pads using BCl_3–N_2 dry etching to confine the current flow. For comparison Au(1000 Å)–Ni(500 Å) was deposited by e-beam evaporation, defined by lift-off and mesas formed by dry etching. The samples were annealed for 1 min at temperatures from 400–1100°C under a flowing N_2 ambient.

Figure 1.36 shows I–V characteristics for WSi, W and Ni–Au metallization on p-GaN, as a function of post-deposition annealing temperature. It is clear that annealing at low temperatures improves the intimate contact between the metals and GaN, and hence, the contact quality. However, even at the optimum annealing temperatures for each metal scheme (700°C for Ni–Au and W; 800°C for WSi_x) there is not true ohmic behavior, and the contacts are better described as leaky Schottky diodes. At annealing temperatures above these optimum conditions, the contact characteristics worsen and become more rectifying. Note that the formation of the W_2N phase at these temperatures could result in a decrease in hole concentration at the GaN surface, and degrade the contact properties.

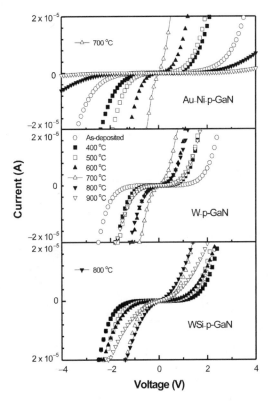

Figure 1.36. Annealing temperature dependence of I–V characteristics of WSi, W and Ni–Au contacts on p-GaN (60 s anneal times)

It is instructive to examine the contact morphology for different annealing temperatures. Figure 1.37 shows SEM micrographs of Ni–Au after annealing at 400°C (top left) and 700°C (top right), and of W after annealing at 400°C (bottom left) and 900°C (bottom right). In the former case the Ni–Au becomes strongly islanded, as reported by Venugopalan *et al.* [106], due to dissociation of the GaN by the reaction with the Ni.[56] The rate of the reaction becomes increasingly rapid above ~600°C, and in our contacts a 5 min anneal at 700°C produced extremely poor results both from an electrical and structural view point. The reacted nature of the Ni–Au contact will definitely be a problem in electronic devices such as heterojunction bipolar transistors or junction FETs, where the contact size is much smaller than in photonic devices, and need to be uniform for large numbers of devices. By sharp contrast, both the W and WSi did not show any loss of dimensional stability or surface morphology degradation even at >900°C. In addition, there is not a strong dependence of the electrical characteristics on annealing time, as shown in Figure 1.38. There is little change in the contact properties for 30 s–2 min, and contacts become slightly more rectifying for longer times, indicating that the interfacial reaction is very limited. In conjunction with the low defect density material grown by epitaxial overgrowth, the use of WSi and W metallization should prove useful in improving the thermal stability of p-ohmic contacts on GaN.

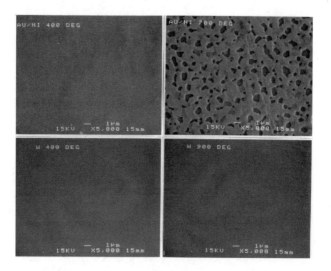

Figure 1.37. SEM micrographs of Ni–Au contacts on p-GaN after 60 s anneals at 400°C (top left) and 700°C (top right), and W contacts after annealing at 400°C (bottom left) and 900°C (bottom right)

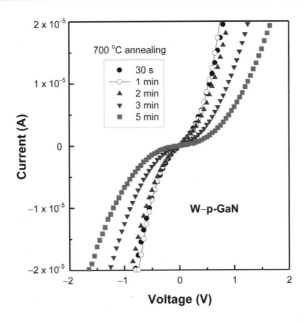

Figure 1.38. Annealing time dependence at 700°C of I–V characteristics from W contacts on p-GaN

From Fermi-Dirac statistics we can calculate the Fermi level position E_F for p-GaN containing 10^{18} acceptors as a function of absolute temperature T

$$N_A \frac{1}{1+2\exp[(E_a - E_F)/kT]} = N_V \exp(-(E_F-E_V)/kT) \quad (1.6)$$

where N_A is the acceptor concentration, $E_a = 171$ meV for Mg in GaN and N_V is the valence band density of states. Using this relation, we calculated the ionization efficiency for Mg as a function of sample temperature, as shown in Figure 1.39. At 25°C, only ~10% of the Mg is ionized, whereas the efficiency increases to ~57% at 300°C. Given the acceptor concentration in our films, this means the hole density rises from ~10^{17} cm^{-3} at 25°C to ~5.7×10^{17} cm^{-3} at 300°C. Past investigations have shown a combination of thermionic emission and thermionic field emission as the dominant conduction mechanisms in contacts to p-GaN.[91,107] At elevated operating temperatures, the hole density in the material increases rapidly, leading to a decrease in material sheet resistance. Concurrently, there should be more efficient thermionic emission and tunneling of holes across the metal-GaN barrier, leading to a decrease in specific contact resistance.

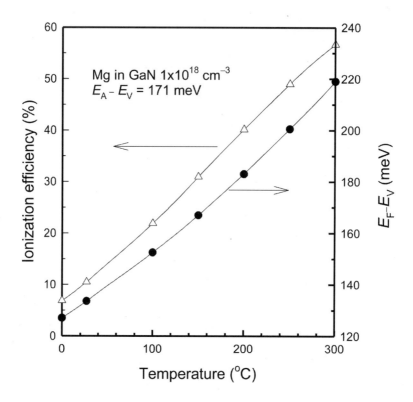

Figure 1.39. Ionization efficiency of Mg acceptors in GaN, and Fermi level position for GaN doped with 10^{18} cm^{-3} Mg acceptors as a function of temperature

Figure 1.40 shows I–V characteristics from the three different metallization schemes on p-GaN, as a function of measurement temperature. At 250°C the Ni–Au is truly ohmic, whereas the W and WSi$_x$ have linear characteristics at 300°C. Table 1.3 shows the r_c values at 300°C are 9.2 × 10^{-2} Ω-cm^2 (Ni–Au), 6.8 × 10^{-1} Ω-cm^2 (W) and 2.6 × 10^{-2} Ω-cm^2 (WSi). The substrate sheet resistance decreases with increasing temperature, indicating that the increased hole concentration plays a major role in decreasing r_c. Whilst device operation at elevated temperatures clearly would improve the p-contact resistance, it remains to be seen how much of a trade-off this entails in terms of degraded reliability.

1.5 Electrical Contacts to Gallium Nitride

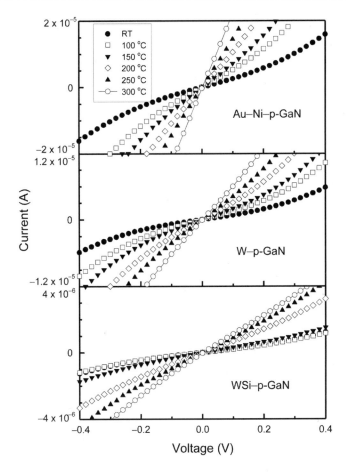

Figure 1.40. Measurement temperature dependence of I–V characteristics of Ni–Au, W and WSi contacts on p-GaN

Table 1.3. Temperature-dependent contact data for p-GaN

Contact	Measurement temperature (°C)	Specific contact resistance (Ω cm^2)	Contact resistivity (Ω mm)	Sheet resistance (Ω/)
Ni–Au	200	0.125	415.7	13900
Ni–Au	250	0.121	319.5	8470
Ni–Au	300	0.092	205.9	4600
W	300	0.682	758.4	–

1.6 Dry Etch Damage in Gallium Nitride

Dry etching has proven an effective technique for the formation of mesas in GaN devices. However, exposure to the energetic ions may result in significant damage, which often degrades the material properties and device performance. The ICP technique uses lower ion energies than RIE, and thus in general has lower damage levels as a result. At high source powers, however, ion flux incidence on the surface is so high that substantial damage can be detected to depths of a few hundred angstroms. Plasma-induced damage to GaN may take several forms, all of which lead to changes in its electrical and optical properties, as follows ion-induced creation of lattice defects, which generally behave as deep-level states and thus produce compensation, trapping or recombination in the material. Owing to channeling of the low energy ions that strike the sample, and rapid diffusion of the defects created, the effects can be measured as deep as 1000 Å from the surface, even though the projected range of the ions is only ≤10 Å.

1. Unintentional passivation of dopants by atomic hydrogen. The hydrogen may be a specific component of the plasma chemistry, or may be unintentionally present from residual water vapor in the chamber or from sources such as photoresist mask erosion. The effect of the hydrogen deactivation of the dopants is a strong function of substrate temperature, but may occur to depths of several thousand angstroms.
2. Creation of non-stoichiometric surfaces through preferential loss of one of the lattice elements. This can occur because of strong differences in the volatility of the respective etch products, leading to enrichment of the less volatile species, or by preferential sputtering of the lighter lattice element if there is a strong physical component to the etch mechanism. Typical depths of this non-stoichiometry are <100 Å.
3. Deposition of polymeric film from plasma chemistries involving CH_x radicals, or from reaction of photoresist masks with Cl_2-based plasma.

In GaN-based photonic devices, the constraints placed on dry etching are minimal. Usually, these structures are heavily doped and etch depths are comparatively large, and thus are fairly resistant to damage. Moreover, the etch proceeds to an n^+-GaN layer, onto which an ohmic contact is subsequently deposited. Preferential loss of N_2 from the near-surface region during the etch step is actually beneficial in this case because it leads to increased n-type doping levels, and hence lower contact resistances.[85,100,108,109] By sharp contrast, in electronic devices such as HBTs or Bipolar Junction Transistors (BJTs), the etching requirements are much more demanding. Low-damage processes are required to form mesas for the base and collector contacts without increasing recombination in the base–emitter junction or surface leakage in the base–collector junction.

To date, much less is known about the electrical effects of dry etch damage and its subsequent removal by chemical treatment or annealing in the GaN system than for other compound semiconductors. Most past work in this area has focused on n-type material. The sheet resistances of GaN, InGaN, InAlN and InN samples were found to increase in proportion to ion flux and ion energy in an ECR Ar plasma.[110,111] Ren et al.[112] examined the effect of ECR BCl_3–N_2 and CH_4–H_2 plasmas on the electrical performance of InAlN and GaN channel FET. They found that hydrogen passivation of the Si doping in the channel may occur if H_2 is a part of the plasma chemistry and that preferential loss of N_2 degraded the rectifying properties of Schottky contacts deposited on plasma-exposed surfaces. Saotume et al.[113] found pure Cl_2 plasma treatment decreased near band-edge (Photo luminescence) intensity of the GaN samples by a factor of approximately five through introduction of non-radiative levels, whereas subsequent photo-assisted wet etching restored this to about half of the original value.

There is very little information available on the electrical effects of plasma damage in p-type GaN. Shul et al.[114] reported that the sheet resistance of p-GaN increased upon exposure to pure ICP Ar discharge. The increases were almost linearly dependent on ion energy, but weakly dependent on ion flux.

In this section, we studied systematically the effects of ICP N_2, H_2, Ar or Cl_2–Ar discharge exposure under various conditions on the properties of n- and p-GaN Schottky diodes. The choice of these plasma chemistries enabled us to differentiate between ion mass effects and the role of physical versus chemical components of the etching. The depth and thermal stability of the damage have been determined. Wet etching and thermal annealing were employed to restore the electrical properties of the damaged materials.

1.6.1 Plasma Damage in n-Gallium Nitride

The layer structure and contact metals are shown schematically in Figure 1.41. The GaN was grown by rf plasma-assisted MBE on c-plane Al_2O_3 substrates. The Ti–Au ohmic contacts were patterned by lift-off and annealed at 750°C, producing contact resistances in the 10^{-4} $\Omega \cdot cm^{-2}$ region.

Samples were exposed to either pure N_2 or H_2 discharges in a Plasma-Therm 790 ICP system at a fixed pressure of 5 mTorr. The gases were injected into the ICP source at a flow rate of 15 standard cubic centimeters per minute (sccm). The experimentally varied parameters were source power (300–1000 W) and rf chuck power (40–250 W), which control ion flux and ion energy respectively. The Pt–Au Schottky metallization was then deposited through a stencil mask by e-beam evaporation. I–V characteristics were recorded on an HP4145A parameter analyzer, and the reverse breakdown voltage V_B was defined as the voltage at which the leakage current was 5×10^{-3} A. We found in all cases that plasma exposure caused significant increases in forward and reverse currents, with ideality factors increasing from typical values of 1.4 to 1.7 on control samples to >2. For this reason, we were unable to extract meaningful values of either ideality factor or barrier height.

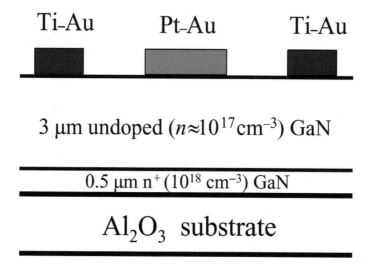

Fig.ure1.41. Schematic of GaN Schottky diode structure

Figure 1.42 shows a series of I–V characteristics from the GaN diodes fabricated on samples exposed to either H_2 or N_2 discharges at different source powers. It is clear that N_2 plasma exposure creates more degradation of the diode characteristics than does H_2 exposure. This implicates the ion mass ($^{28}N_2^+$, $^2H_2^+$ for the main positive ion species) as being more important in influencing the electrical properties of the GaN surface than a chemical effect, since H_2 would be likely to preferentially remove nitrogen from the GaN as NH_3.

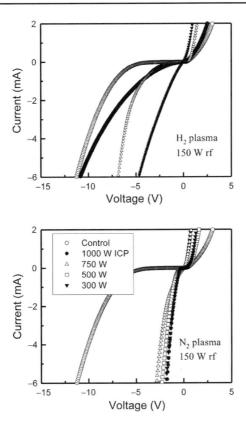

Figure 1.42. I–V characteristics from GaN diodes before and after H_2 (top) and N_2 (bottom) plasma exposure (150 W rf chuck power, 5 mTorr) at different ICP powers

The variations of V_B of the diodes with the source power during plasma exposure are shown in Figure 1.43. For any exposure to the N_2 discharges, V_B is severely reduced. By contrast, there is less degradation with the H_2 plasma exposures at higher source powers. This is likely related to the lower average ion energy at those conditions, as shown at the bottom of Figure 1.43. The average ion energy is approximately equal to the sum of dc self-bias and plasma potential, with the latter being in the range –22 to –28 V as determined by Langmuir probe measurements. Ion-induced lattice damage in GaN may display n-type conductivity, due to more N atoms being displaced. In addition, the heavy N_2^+ ions are also more effective in preferential sputtering of the N relative to Ga and creating N vacancies, compared with the H_2^+ ions. The net result is that N_2^+ ions will lead to more degradation of the surface electrical properties of GaN than do H_2^+ ions of similar energy.

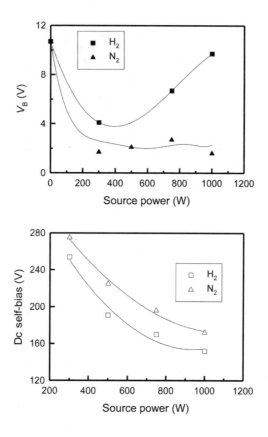

Figure 1.43. Variation of V_B in GaN diodes (top) and dc chuck self-bias (bottom) as a function of ICP source power in H_2 or N_2 plasmas (150 W rf chuck power, 5 mTorr)

Similar conclusions can be drawn from the data on the effect of increasing rf chuck power. Figure 1.44 shows the diode I–V characteristics from H_2 or N_2 plasma-exposed samples at fixed source power (500 W) but varying rf chuck power. There are once again very severe decreases in breakdown voltage and increases in leakage current. The dependence of V_B on rf chuck power during the plasma exposures is shown in Figure 1.45, along with the dc self-bias. The V_B values fall by more than a factor of two even for very low self-biases, and emphasize how sensitive the GaN surface is to degradation by energetic ion bombardment. The degradation saturates beyond ~100 W chuck power, corresponding to ion energies of ~175 eV. We assume that, once the immediate surface becomes sufficiently damaged, the contact properties basically cannot be made any worse and the issue is then whether the damage depth increases with the different plasma parameters. Since ion energy appears to be a critical factor in creating the near-surface damage, damage depth would be expected to increase with ion energy in a non-etching process. In the case of simultaneous etching and damage creation (*e.g.* in Cl_2–Ar etch processing), higher etch rates would lead to smaller depth of residual damage because the disordered region would be partially removed.

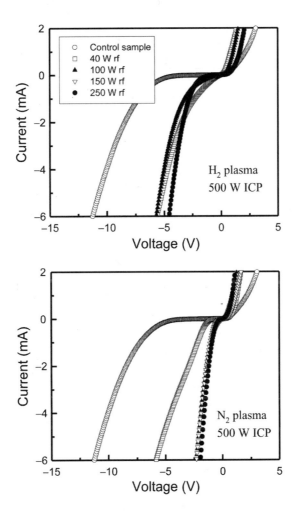

Figure 1.44. I–V characteristics from GaN diodes before and after H_2 (top) or N_2 (bottom) plasma exposure (500 W source power, 5 mTorr) at different rf chuck powers

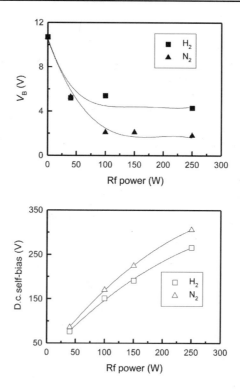

Figure 1.45. Variation of V_B in GaN diodes (top) and dc chuck self-bias (bottom) as a function of rf chuck power in H_2 or N_2 plasmas (500 W source power, 5 mTorr)

For completed n-GaN diodes exposed to ICP discharges we observed that the changes in the electrical properties were much less severe than those above. However, the low bias forward currents of the damaged devices were still increased by up to two orders of magnitude. Auger electron spectroscopy (AES) showed that plasma exposure created a N_2-deficient region around the periphery of the rectifying contact, which reduced the barrier to current conduction. By contrast, the reverse I–V is more strongly dependent on the bulk doping in the GaN under the contact and is less affected by the plasma damage.

1.6.2 Effect of Etching Chemistries on Damage

It would be expected that exposure to etching chemistries such as Cl_2–Ar create less and shallower damage, on the basis of the fast etching rate and hence improved damage removal. In this section we compare the effects of Cl_2–Ar and Ar ICP exposure on the electrical properties of n-GaN Schottky diodes. The layer structure consisted of 0.5 μm n^+ (10^{18} cm^{-3}) GaN grown on c-plane Al_2O_3 substrate by MBE, followed by 1 μm of nominally doped ($n\approx5\times10^{16}$ cm^{-3}) GaN. The mesas were etched with Ar–Cl_2 ICP discharges at low powers, and ohmic contacts were formed by e-beam evaporation and lift-off of Ti–Al–Pt–Au, followed by annealing at 800°C to remove dry etch damage and alloy the contacts. The samples were exposed to either $10Cl_2/5Ar$ or 15Ar (where the numbers denote the gas flow rate in standard cubic centimeters per minute) ICP discharges in a Plasma-Therm ICP reactor at a fixed pressure of 3 mTorr. We investigated a range of rf chuck powers (25–250 W) and etch times (4–100 s), with a fixed source power of 500 W. The Schottky metallization Pt–Au (ϕ= 50, 70 or 90 μm) was then deposited on the damaged surface by e-beam evaporation, followed by lift off. As schematic of the mesa diodes and an SEM micrograph are shown in Figure 1.46.

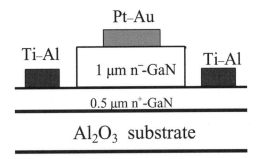

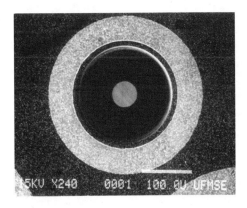

Figure 1.46. Schematic of the diode structure (top) and SEM of a complete device (bottom)

Figure 1.47 shows a series of I–V characteristics from n-type GaN diodes fabricated on samples exposed to either Cl_2–Ar (top) or Ar (bottom) discharges at different rf chuck powers. There is a significant reduction in V_B under all conditions, with Ar producing less damage at low chuck powers. This is probably related to two factors: the slightly higher chuck bias with Cl_2–Ar due to the lower positive ion density in the plasma (Cl is more electronegative than Ar) and the heavier mass of the Cl_2^+ ions compared to Ar^+. This is consistent with data on the relative effects of N_2 and H_2 plasma exposure, *i.e.* ion mass was more important in influencing the electrical properties of the GaN surface than any chemical effects.

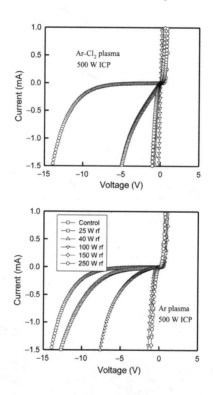

Figure 1.47. I–V characteristics from n-GaN samples exposed to ICP Cl_2–Ar (top) or Ar (bottom) discharges (500 W source power) as a function of rf chuck power prior to deposition of the rectifying contact

The variations of V_B and V_F with the rf chuck power during plasma exposure are shown in Figure 1.48 (top). At powers ≤100 W (this correspond to ion energies <~150 eV), the Cl_2–Ar creates more degradation of V_B, as discussed above, while at higher powers the damage saturates. This is also reflected in the variation of the forward on-voltage V_F with rf chuck power. Note that the etch rate in Cl_2–Ar discharges increases rapidly with the rf chuck power (shown at the bottom of the figure). The amount of the etch damage is the direct result of competition between defect introduction and etching processes.

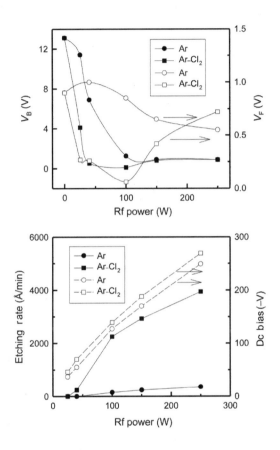

Figure 1.48. Variations of V_B and V_F (top), and n-GaN etching rate or dc self-bias (bottom) as a function of rf chuck power for n-GaN diodes exposed to ICP Ar and Cl_2–Ar discharges (500 W source power)

Figure 1.49 shows a series of I–V characteristics from n-type GaN diodes fabricated on samples exposed to the two different plasmas for different times at fixed rf chuck power (150 W) and source power (500 W). It is clear that the damage accumulates rapidly, with the I–V characteristics becoming linear at longer times. The breakdown voltages are <1 V for basically all plasma exposure times for both Cl_2–Ar and Ar. Figure 1.50 shows the variations in V_B and V_F in these diodes with plasma exposure, together with the etch depth versus etch time (bottom). As is readily apparent, V_B decreases dramatically after even short plasma exposures (4 s) and then tends to recover slightly up to ~25 s. It should be remembered that this is damage accumulating ahead of the etch front. The V_B values are saturated after some points due to the etching or sputtering effect. Since the etch rate with Cl_2–Ar is much faster (Figure 1.50, bottom), the damage accumulation in this chemistry should reach the saturation point much earlier, as shown in Figure 1.50 (top).

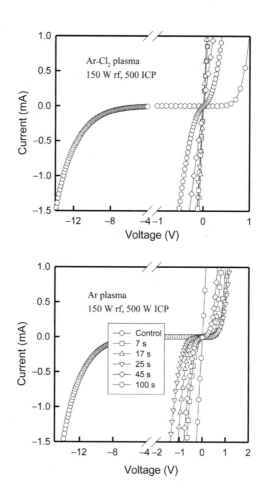

Figure 1.49. I–V characteristics from n-GaN samples exposed to ICP Cl_2–Ar (top) or Ar (bottom) discharges (150 W rf chuck power, 500 W source power) as a function of plasma exposure time prior to deposition of the rectifying contact

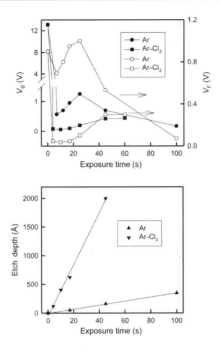

Figure 1.50. Variation of V_B and V_F (top) and of n-GaN etch depth (bottom) as a function of plasma exposure time for n-GaN diodes exposed to ICP Cl_2–Ar discharges (500 W source power 150 W rf chuck power)

The rapid accumulation could explain why Cl_2–Ar discharges produce more damage than pure Ar. Significant damage could be created in the near-surface region in this high-density plasma after even a very short exposure. The Schottky contact is most sensitive to the electrical properties of the immediate surface layer. For these reasons it is possible that Cl_2–Ar discharges, containing ions with higher energy and heavier mass, produce more degradation in the electrical properties of the GaN Schottky diodes than pure Ar discharges, even though the damage depth may be smaller due to high etch rate. Note that we have studied a worst-case scenario for damage introduction during ICP etching. In a real etch process of GaN device, damage would be less severe because much lower power conditions are employed.

1.6.3 Thermal Stability of Damage

To examine the thermal stability of the etch damage, the n-type GaN samples were exposed to the 500 W source power, 150 W rf chuck power (dc self-bias –221 V), 5 mTorr N_2 discharge, and then annealed in N_2 for 30 s at 300–850°C prior to deposition of the rectifying contact. Figure 1.51 (top) compares the AFM data from the control sample, the as-exposed sample and samples annealed at 550°C or 750°C. The fact that plasma exposure severely degraded the surface is clear, as the RMS surface roughness increases from 0.8 nm to 4.2 nm. Subsequent annealing essentially restored the initial morphology. I–V data from annealed samples are shown in Figure 1.51 (bottom). These samples show that increasing the annealing temperature up to 750°C brings a substantial improvement in V_B. However, for annealing at 850°C the diode began to degrade, and this is consistent with the temperature at which N_2 begins to be lost from the surface.[59]

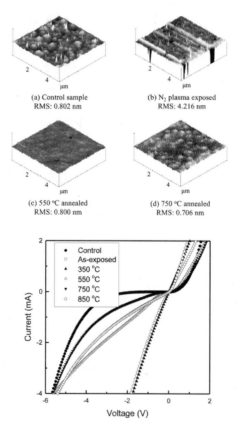

Figure 1.51. AFM scans of GaN surfaces (top) and I–V characteristics from GaN diodes (bottom) before and after N_2 plasma exposure (500 W source power, 150 W rf power, 5 mTorr) and subsequent annealing at different temperatures prior to deposition of the Schottky contacts

It is instructive to compare the thermal stability of Pt–Au contacts on the damaged and undamaged GaN. The metal was deposited on a control sample and a sample which was plasma exposed (N_2, 500 W source power, 150 W rf chuck power, 5 mTorr), then annealed at different temperatures for 30 s. The I–V characteristics are shown in Figure 1.52. The Pt–Au contact is stable to 700°C on unetched samples, whereas in the case where the samples were exposed to the N_2 plasma the I–V characteristics show continued worsening upon annealing. The poorer stability in etched samples could be related to the surface damage enhancing interfacial reaction between the Pt and GaN, and ohmic contact on the damaged materials can take advantage of this enhanced alloying process.

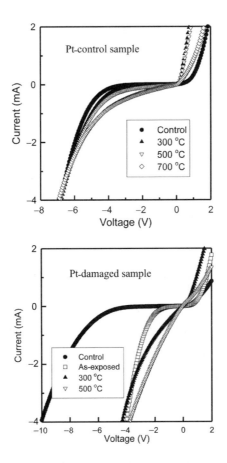

Figure 1.52. I–V characteristics from GaN diodes with Schottky metal deposited on control samples (top) and plasma-exposed samples (bottom) followed by annealing at different temperatures

Figure 1.53 shows I–V characteristics from control, samples which were exposed to Ar or Cl_2/Ar discharges at a fixed source power (500 W) and rf chuck power (150 W rf), and annealed in N_2. Again, the annealing produced a significant recovery of the electrical properties, and more restoration was achieved for samples exposed to the Ar plasma. The V_B values are shown in Figure 1.54, as a function of post-plasma exposure annealing temperature. Annealing temperatures between 700–800°C restored >70% of the original V_B value, but clearly annealing alone cannot remove all of the dry etch induced damage. Annealing temperatures above 800°C were found to lead to preferential loss of N_2 from the surface, with a concurrent degradation in V_B.

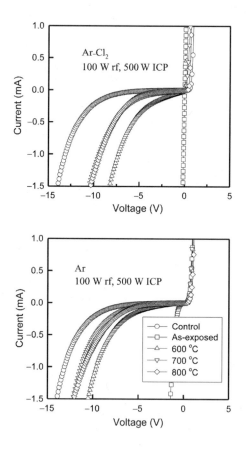

Figure 1.53. I–V characteristics from n-GaN samples exposed to ICP Cl_2–Ar (top) or Ar (bottom) discharges (500 W source power, 100 W rf chuck power) as a function of annealing temperature prior to deposition of the rectifying contact

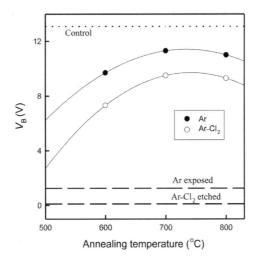

Figure 1.54. Variation of V_B in n-GaN diodes exposed to ICP Cl_2–Ar or Ar discharges (500 W source power, 100 W rf chuck power) with annealing temperature prior to deposition of the rectifying contact

1.6.4 Plasma Damage in p-Gallium Nitride

The layer structure consisted of 1 μm of undoped GaN ($n\approx5\times10^{16}$ cm^{-3}) grown on a c-plane Al_2O_3 substrate, followed by 0.3 μm of Mg doped ($p\approx5\times10^{17}$ cm^{-3}) GaN. The samples were grown by rf plasma-assisted MBE. Ohmic contacts were formed with Ni–Au deposited by e-beam evaporation, followed by lift-off and annealing at 750°C. The GaN surface was then exposed for 1 min to ICP H_2 or Ar plasmas in a Plasma-Therm 790 System. The 2 MHz ICP source power was varied from 300 to 1400 W, while the 13.56 MHz rf chuck power was varied from 20 to 250 W. The former parameter controls the ion flux incident on the sample, while the latter controls the average ion energy. Ti–Pt–Au contacts with 250 μm diameter were deposited through a stencil mask. The I–V characteristics of the diodes were recorded on an HP 4145A parameter analyzer. The unetched control diodes have reverse breakdown voltages of ~2.5–4 V depending on the wafer – these values were uniform (±12%) across a particular wafer.

Figure 1.55 shows the I–V characteristics from samples exposed to either H_2 (top) or Ar (bottom) ICP discharges (150 W rf chuck, 2 mTorr) as a function of source power. In both cases there is an increase in both the reverse breakdown voltage and the forward turn-on voltage, with these parameters increasing monotonically with the source power during plasma exposure.

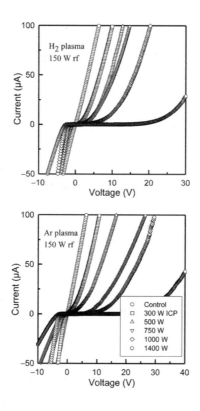

Figure 1.55. I–V characteristics from samples exposed to either H_2 (top) or Ar (bottom) ICP discharges (150 W rf chuck power) as a function of ICP source power prior to deposition of the Ti–Pt–Au contact

Figure 1.56 shows this increase in breakdown voltage as a function of source power, and also the variation of the chuck dc self-bias. As the source power increases, the ion density also increases and the higher plasma conductivity suppresses the dc bias developed. Note that the breakdown voltage of the diodes continues to increase even as this bias (and hence ion energy, which is the sum of this bias and the plasma potential) decreases. These results show that ion flux plays an important role in the change of diode electrical properties. The other key result is that Ar leads to consistently more of an increase in breakdown voltage, indicating that ion mass is important, rather than any chemical effect related to removal of N_2 or NH_3 in the H_2 discharges.

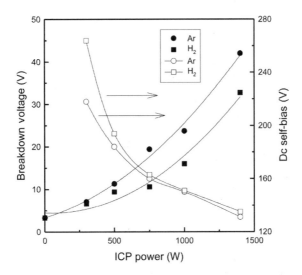

Figure 1.56. Variation of diode breakdown voltage in samples exposed to H_2 or Ar ICP discharges (150 W rf chuck power) at different ICP source powers prior to deposition of the Ti–Pt–Au contact. The dc chuck self-bias during plasma exposure is also shown

The increase in breakdown voltage on the p-GaN is due to a decrease in hole concentration in the near-surface region through the creation of shallow donor states. The key question is whether there is actually conversion to an n-type surface under any of the plasma conditions. Figure 1.57 (top) shows the forward turn-on characteristics of the p-GaN diodes exposed to different source power Ar discharges. At low source power (300 W), the turn-on remains close to that of the unexposed control sample, *i.e.* the surface remains p-type with no significant reduction in conductivity. However, there is a clear increase in the turn-on voltage at higher source powers, and in fact at ≥750 W the characteristics are those of an n–p junction, *i.e.* the surface has converted to n-type. The turn-on voltage at the highest flux conditions is ~3.3 V (Figure 1.57 (bottom)), which is close to the build-in potential of a GaN n–p diode. Since the hole concentration in the GaN is ~10^{17} cm^{-3}, the Mg acceptor concentration is ~10^{19} cm^{-3} based on the ionization level of ~170 meV. This means the plasma exposure at high flux conditions produces >10^{19} cm^{-3} shallow donor states and there is surface conversion. According to our previous study [111], the obvious conclusion is that nitrogen vacancies create these shallow donor levels. We found that N_2 loss from the p-GaN surface during thermal annealing at temperatures >900°C also produced significantly decreased p-type conduction.

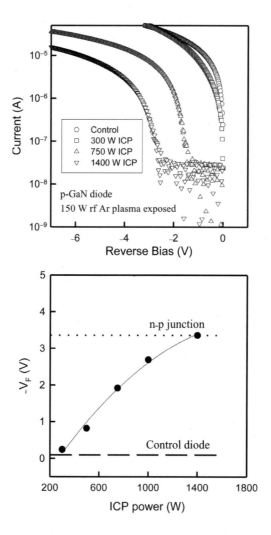

Figure 1.57. Forward turn-on characteristics (top) and turn-on voltage (bottom) of diodes exposed to ICP Ar discharges (150 W rf chuck power) at different ICP source powers prior to deposition of the Ti–Pt–Au contact

The influence of rf chuck power on the diode I–V characteristics is shown in Figure 1.58 for both H_2 and Ar discharges at fixed source power (500 W). A similar trend is observed as for the source power experiments, namely the reverse breakdown voltage increases, consistent with a reduction in p-doping level near the GaN surface. Surface type conversion also occurred at high powers.

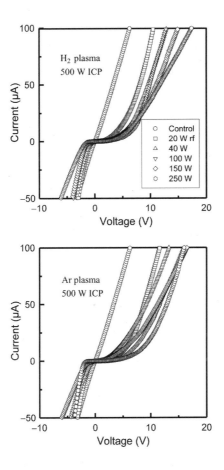

Figure 1.58. I–V characteristics from samples exposed to either H_2 (top) or Ar (bottom) ICP discharges (500 W source power) as a function of rf chuck power prior to deposition of the Ti–Pt–Au contact

Figure 1.59 plots breakdown voltage and dc chuck self-bias as a function of the applied rf chuck power. The breakdown voltage initially increases rapidly with ion energy (the self-bias plus ~25 V plasma potential) and saturates above ~100 W probably due to the fact that sputtering yield increases and some of the damaged region is removed. Note that there are very large changes in breakdown voltage even for low ion energies, emphasizing the need to control both flux and energy carefully.

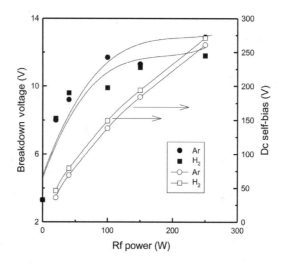

Figure 1.59. Variation of diode breakdown voltage in samples exposed to H_2 or Ar ICP discharges (500 W source power) at different rf chuck powers prior to deposition of the Ti–Pt–Au contact. The dc chuck self-bias during plasma exposure is also shown

1.6.5 Thermal Stability of Damage

One important method of removing plasma-induced damage is annealing. In these experiments we exposed the samples to the same type of plasma (Ar, 750 W source power, 150 W rf chuck power) and then annealed under N_2 at different temperatures. Figure 1.60 (top) shows the I–V characteristics of these different samples, and Figure 1.60 (bottom) shows the resulting breakdown voltages as a function of annealing temperature. On this wafer, plasma exposure caused an increase in breakdown voltage from ~2.5 to ~18 V. Subsequent annealing at 400°C initially decreased the breakdown voltage, but higher temperatures produced a large increase. In some cases, the samples were found to be highly compensated after annealing at 600–700°C. At temperatures above 700°C, the diode characteristics returned toward their initial values and were back to the control values by 900°C. This behavior is similar to that observed in implant-isolated compound semiconductors, where ion damage compensates the initial doping in the material, producing higher sheet resistances. In many instances the damage site density is larger than that needed to trap all of the free carriers, and trapped electrons or holes may move by hopping conduction. Annealing at higher temperatures removes some of the damage sites, but there are still enough to trap all the conduction electrons/holes. Under these conditions the hopping conduction is reduced and the sample sheet resistance actually increases. At still higher annealing temperatures, the trap density falls below the conduction electron or hole concentration and the latter are returned to their respective bands. Under these conditions the sample sheet resistance returns to its pre-implanted value. The difference in the plasma-exposed samples is that the incident ion energy is a few hundred electron-volts compared to a few hundred kilo-electron-volts in implant-isolated material. In the former case the main electrically active defects produced are nitrogen vacancies near the surface, whereas in the latter case there will be vacancy and interstitial complexes produced in far greater numbers to far greater depths. We did not examine the time dependence of the damage removal, but expect it would show a square-root power of annealing time.

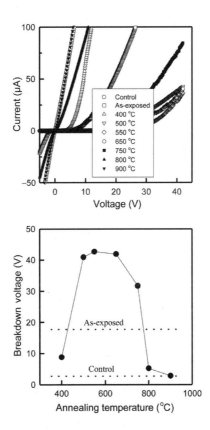

Figure 1.60. I–V characteristics from samples exposed to ICP Ar discharges (750 W source power, 150 W rf chuck power) and subsequently annealed at different temperatures prior to deposition of the Ti–Pt–Au contact (top) and breakdown voltage as a function of annealing temperature (bottom)

In our previous work on plasma damage in n-GaN [111] we found that annealing at ~750°C almost returned the electrical properties to their initial values. If the same defects are present in both n- and p-type materials after plasma exposure, then this difference in annealing temperature may be a result of a Fermi level dependence on the annealing mechanism.

1.6.6 Determination of Damage Profile in Gallium Nitride

An important concern is the depth of the plasma-induced damage. We found we were able to etch p-GaN very slowly in boiling NaOH solutions, at rates that depended on the solution molarity (Figure 1.61) even without any plasma exposure of the material. This enabled us to directly measure the damage depth in plasma exposed samples in two different ways.

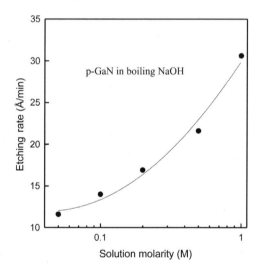

Figure 1.61. Wet etching rate of p-GaN in boiling NaOH solutions as a function of solution molarity

The first method involved measuring the etch rate as a function of depth from the surface. Defective GaN resulting from plasma, thermal or implant damage can be wet chemically etched at rates much faster than undamaged material because the acid or base solutions are able to attack the broken or strained bonds present. Kim et al.[115] reported that the wet etch depth on thermally or ion-damaged GaN was self-limiting. Figure 1.62 shows the GaN etch rate as a function of depth in samples exposed to a 750 W source power, 150 W rf chuck power Ar discharges (−158 V dc bias). The etch rate is a strong function of the depth from the surface and saturates between ~425 and 550 Å. Within this depth range the etch rate is returned to the "bulk" value characteristic of undamaged p-GaN.

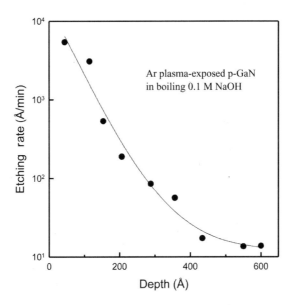

Figure 1.62. Wet etching rate of Ar plasma-exposed (750 W source power, 150 W rf chuck power) p-GaN as a function of depth into the sample

The second method to establish damage depth, of course, is simply to measure the I–V characteristics after removing different amounts of material by wet etching prior to deposition of the rectifying contact. Figure 1.63 (top) shows the I–V characteristics from samples exposed to 750 W source power, 150 W rf chuck power (–160 V dc chuck bias) Ar discharges and subsequently wet etched to different depths using 0.1 M NaOH solutions before deposition of the Ti–Pt–Au contact. Figure 1.63 (bottom) shows the effect of the amount of material removed on the diode breakdown voltage. Within the experimental error of ±12%, the initial breakdown voltage is re-established in the range 400–450 Å. This is consistent with the depth obtained from the etch-rate experiments described above, and corresponds to an ion energy of ~180 eV.

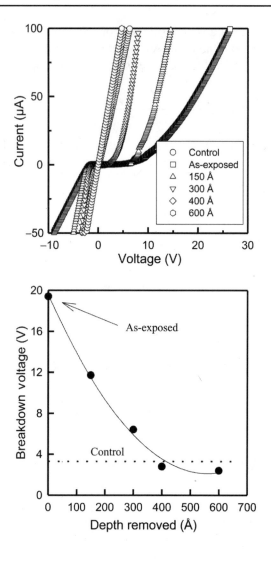

Figure 1.63. I–V characteristics from samples exposed to ICP Ar discharges (750 W source power, 150 W rf chuck power) and subsequently wet etched to different depths prior to deposition of the Ti/Pt/Au contact (top) and breakdown voltage as a function of depth removed (bottom).

It would be instructive to determine and compare the damage depth created after exposure to Cl_2–Ar and Ar discharges (500 W source power, 150 W rf chuck power, 1 min, –192 V or –171 V dc chuck bias). For these plasma conditions, we did not observe type conversion of the surface. The values of forward turn-on voltage increased after plasma exposure, due to the lowered net hole concentration. The electrical properties are almost restored after depths of 500–600 Å were removed by NaOH etching, as can be seen in Figure 1.64. This data can also be obtained from Figure 1.65, where the wet etch depth in plasma-damaged p-GaN is plot as a function of etching time. What is clear from this data is that the damage depth produced in Cl_2–Ar discharges is only slightly smaller than that in Ar discharges, even though the etch rate of GaN in the former plasma is almost 30 times higher.

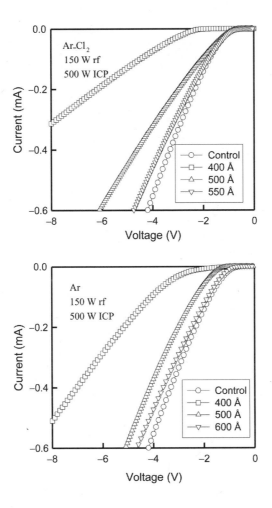

Figure 1.64. Forward I–V characteristics from p-GaN samples exposed to ICP Cl_2–Ar (top) or Ar (bottom) discharges (500 W source power, 150 W rf chuck power) and wet etched in boiling NaOH to different depths prior to deposition of the rectifying contact

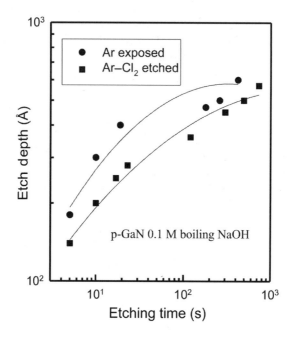

Figure 1.65. Wet etch depth versus etch time in boiling NaOH solutions for plasma-damaged p-GaN

The damage depth in n-type GaN can be established by photoelectrochemically (PEC)[116] wet etching different amounts of the plasma-exposed material, and then measuring the electrical properties of the GaN surfaces. For samples exposed to a 500 W source power, 150 W rf chuck power N_2 discharge, the reverse breakdown voltage was restored to about half of its initial value after removal of ~260 Å damaged layer. However, increasingly rough surfaces of PEC etching for larger removal depths were usually observed. The estimated damage depth for the N_2 plasma conditions mentioned above could be similar to that established in p-type GaN samples exposed to similar plasma conditions.

1.7 Conclusions and Future Trends

Recognized for their stability in harsh environments and their unique optical and electronic properties, the III–V nitrides have become one of the few select classes of materials being engineered into optoelectronics and microelectronic devices demanded by today's technologically advancing society. Although early progress was slowed by the lack of ideal substrates and p-type doping ability, the pace of successful GaN-based device fabrication has accelerated within the last few years. The commercialization of blue and green LEDs and the achievement of long lifetime LDs have been followed by the realization of UV photodetectors and numerous different electronic devices. However, the scientific understanding necessary to routinely reproduce these results and advance device performance is still lacking in many critical areas. While further improvement in heteroepitaxial growth is currently gaining considerable attention, a reliable processing procedure is another critical factor for GaN to reach its full potential, especially for microwave power electronics. This dissertation has focused on several important aspects of GaN processing, with efforts directed towards optimizing GaN technology and improving GaN-based device performance.

A novel RTP at temperatures up to 1500°C has been developed for GaN and related materials. High quality AlN has proven to be an efficient encapsulant at these temperatures, and can be removed selectively in KOH solutions. High-dose Si implantation followed by 1400°C annealing produced metallic doping levels ($\sim 5 \times 10^{20}$ cm^{-3}) in GaN. Post-implant annealing at 1100°C was found to be insufficient to remove the lattice damage, while 1400°C produced much lower defect densities. The implantation and activation of other common donors and acceptors in GaN have also been examined. The Group VI donors, though free of the potential drawback of being amphoteric, did not show any advantages over Si for creation of n-type doping in GaN. Only Mg was found to produce p-type conductivity, while C- and Be-implanted samples remained n-type due to strong compensation effects. None of the implanted species showed measurable diffusion at high temperatures, but Be did display an apparent defect-assisted redistribution at 900°C. A summary of these results is shown in Table 1.4.

To fully develop the potential of ion implantation in advanced electronics, there is still a need to examine further the possible implanted dopants in GaN, especially to improve p-type doping efficiency in GaN or AlGaN. One attractive technique is to improve the occupation of dopants on the desirable lattice sites through co-implantation, such as co-implantation of C, Ge or Sn with Group III species for more conducting p-type nitrides, and co-implantation of Si and Group VI dopants to get even better n-type conductivity.

Table 1.4. Ion implantation doping in GaN

	Max achievable doping level (cm^{-3})	Diffusivity (cm$^2\cdot$s^{-1})	Ionization level (meV)
Donors			
Si	5×10^{20}	<2×10^{-13} (1500°C)	28
S	5×10^{18}	<2×10^{-13} (1400°C)	48
Se	2×10^{18}	<2×10^{-13} (1450°C)	–
Te	1×10^{18}	<2×10^{-13} (1450°C)	50
Si	5×10^{20}	<2×10^{-13} (1500°C)	28
S	5×10^{18}	<2×10^{-13} (1400°C)	48
Se	2×10^{18}	<2×10^{-13} (1450°C)	–
O	3×10^{18}	<2×10^{-13} (1200°C)	30
Acceptors			
Mg	~5×10^{17}	<2×10^{-13} (1450°C)	170
Be	n-type	Defect-assisted	165
C	n-type	<2×10^{-13} (1400°C)	–

High sheet resistances of ~10^{12} Ω/square in n-GaN and 10^{10} Ω/square in p-GaN have been achieved by implantation of O, Cr, Fe and Ti. The concentration of electrically active deep energy levels related to the chemical nature of these species was found to be <7×10^{17} cm^{-3}, and the implanted GaN displayed typical damage-related isolation behavior. Temperature-dependent measurements showed that the defect levels are in the range 0.2–0.49 eV in n-GaN and are ~0.44 eV in p-GaN, which are far from midgap, but sufficient to produce high resistivity materials. Future work in the area of implant isolation includes searching for high compensation in InGaN and thermally stable isolation for use at elevated temperatures.

W-based contacts deposited on Si-implanted GaN produced ohmic contacts with low resistance of 10^{-6} Ω-cm^2. Minimal reaction at the W–GaN interface was observed even at 1000°C. The behavior of W and WSi on p-type GaN has been compared with that of Ni–Au. True ohmic characteristics were obtained only at elevated temperatures. The refractory contact schemes offer superior thermal stability to the standard metallization used in photonic devices, *i.e.* Ti–Al and Ni–Au, and, therefore, are very promising for GaN-based ultra-high-power electronics. At this point, difficulty in achieving low resistance p-type ohmic behavior is still one of the major obstacles in fabricating efficient photonic devices. Further improvements in this area must be achieved at the next stage, including better understanding of metal–GaN reactions and use of these reactions for tunneling and/or lowering barrier height, exploring and incorporating a graded structure or superlattice layer, or highly doped material with a narrower band gap for ohmic contacts.

Dry etch damage is another key issue for optimization of GaN processing. In this work, plasma damage in n- and p- GaN has been studied systematically using Schottky diode measurements. ICP discharges (including real etching chemistries) produced large changes in diode characteristics. The results are consistent with creation of nitrogen vacancy – related shallow donors in the near-surface region (<600 Å), which decrease the doping concentration in p-GaN and increase electron density in n-GaN. Surface type conversion was observed in p-GaN at high ion fluxes and high ion energies. The damage was found to accumulate rapidly, and ion mass appeared to be a key parameter in determining the magnitude in degradation of material properties. Annealing at ~750°C for n-GaN and 850°C for p-GaN, or wet chemical etching have proven to be efficient for damage removal. For GaN-based microelectronics, other promising low-damage, etching such as low energy electron-enhanced etching [117–122] or photo-assisted vapor etching,[124] could become alternative techniques and attract more and more attention in the future.

GaN and related materials provide exciting opportunities in material research and device engineering, and are perhaps a model system for the interplay between science and technology. In the photonics arena, device applications have appeared before there was a complete understanding of the growth and defect issues. Today, research regarding the III–V nitride materials is focused on the optimization of these and other devices, and realization of their theoretical potential; thus, methods to improve heteroepitaxial growth and critical advances in processing technology are once again at the forefront.

References

1. Morkoc H, Strite S, Gao GB, Lin ME, Sverdlov B, Burns M (1994) J Appl Phys 76: 1363
2. Chow TP, Tyagi R (1994) IEEE Trans Electron Dev 41: 1481
3. Bandic ZZ, Bridger PM, Piquette EC, McGill TC, Vaudo RP, Phanse VM, Redwing JM (1998) Appl Phys Lett 74: 1266
4. Gaska R, Yang JW, Osinsky A, Chen Q, Khan MA, Orlov AO, Snider GL, Shur MS (1998) Appl Phys Lett 72: 707
5. Uzawa Y, Wang Z, Kawakami A, Komiyama B (1995) Appl Phys Lett 66: 1992
6. Ambacher O (1998) J Phys D 31: 2653
7. Fischer S, Gisbertz A, Meyer BK, Topf M, Koynov S, Dirnstorfer I, Volm D, Ueckev R, Reiche P (1996) Symp Proc Electrochem. Soci-

ety, Los Angeles, CA Oct 1996 (Electrochem. Soc., Pennington, NJ 1996).
8. Kung P, Saxler A, Zhang X, Walker D, Lavado R, Razaghi M (1996) Appl Phys Lett 69: 2116
9. Gotz W, Romano LT, Walker J, Johnson NM, Molnar RJ (1998) Appl Phys Lett 72: 1214
10. Wetzel C, Volm D, Meyer BK, Pressel K, Nilsson S, Mokhov EN, Baranov PG (1994) Appl Phys Lett 65: 1033
11. Vodakov YA, Mokhov EN, Roenkov AD, Boiko ME, Baranov PG, (1998) J Cryst Growth 10: 183
12. Moustakas TD, Lei T, Molnar RJ (1993) Physica B 185: 39
13. Van Hove JM, Hickman R, Klaassen JJ, Chow PP, Ruden PP (1997) Appl Phys Lett 70: 282
14. Ilegems M (1972) J Cryst Growth 13–14: 360
15. Morita M, Isogai S, Shimizu N, Tsubouchi K, Mikohiba N (1981) Jpn J Appl Phys 19: L173
16. Amano H, Sawaki N, Akasaki I, Toyada Y (1986) Appl Phys Lett 48: 353
17. Keller S, Keller PB, Wu YF, Heying B, Kapolnekv D, Speck JS, Mishra UK, DenBaars SP (1996) Appl Phys Lett 68: 1525
18. Nakamura S, Mukai T, Senoh M (1994) Appl Phys Lett 64: 1687
19. Kato Y, Kitamura S, Hiramatsu K, Sawaki N (1994) J Cryst Growth 144: 133
20. Marchand H, Ibbetson JP, Fini PT, Kozodoy P, Keller S, DenBaars S, Speck, Mishra UK (1998) Mater Res Soc Int J Nitride Semi Res 3: 3
21. Madar R, Jacob G, Hallais J, Frucgart R (1975) J Cryst Growth 31: 197
22. Karpinski J, Porowski S, Miotkowska S (1982) J Cryst Growth 56: 77
23. Burm J, Chu K, Davis W, Schaff WJ, Eastman LF, Eustis TJ (1997) Appl Phys Lett 70: 464
24. Rode DL, Gaskill DK (1995) Appl Phys Lett 66: 1972
25. Albrecht JD, Wang RP, Ruden PP, Farahmand M, Brennan KF (1998) J Appl Phys 83: 4777
26. Molnar RJ, Moustakas TD (1993) Bull Am Phys Soc 38: 445
27. Gotz W, Johnson NM, Walker J, Bour DP, Amano H, Akasaki I (1995) Appl Phys Lett 67: 2606
28. Amano H, Kito M, Hiramatsu K, Akaski I (1989) Jpn J Appl Phys 28: L2112
29. Nakamura S, Muskai T, Senoh M, Iwasa N (1992) Jpn J Appl Phys 31: L139

30. Pearton SJ, Abernathy CR, Vartuli CB, Zolper JC, Yuan C, Stall RA (1995) Appl Phys Lett 67: 1435
31. Zolper JC, Wilson RG, Pearton SJ, Stall RA (1996) Appl Phys Lett 68: 1945
32. Brandt O, Yang H, Kostal H, Ploog K (1996) Appl Phys Lett 69: 2707
33. Nakamura S, Muskai T, Senoh M (1995) Appl Phys Lett 67: 1868
34. Nakamura S, Senoh M, Iwasa N, Nagahama S, Yamada T, Muskai T (1995) Jpn J Appl Phys B10: L1332
35. Nakamura S, Senoh M, Nagahama SI, Iwasa N, Yamada T, Matsushita T, Kiyoku H, Ugimoto YS (1996) Jpn J Appl Phys 35: L217
36. Nakamura S, Senoh M, Nagahama SI, Iwasa N, Yamada T, Matsushita T, Kiyoku H, Ugimoto YS (1996) Appl Phys Lett 68: 2105
37. Nakamura S, Senoh M, Nagahama SI, Iwasa N, Yamada T, Matsushita T, Kiyoku H, Ugimoto YS, Kozaki T, Umemoto H, Sano M, Chocho K (1998) Appl Phys Lett 72: 211
38. Khan MA, Bhattarai AR, Kuznia JN, Olson DT (1993) Appl Phys Lett 63: 1214
39. Khan MA, Kuznia JN, Olson DT, Schatt W, Burm J, Shur MS (1994) Appl Phys Lett 65: 1121
40. Sheppard ST, Doverspike K, Pribble WL, Allen ST, Palmour JW, Kehias LT, Jenkins TJ (1999) IEEE Electron Dev Lett 20: 161
41. Daumiller L, Kirchner C, Kamp M, Ebeling KJ, Pond L, Weitzel CE, Kohn E (1998) presented of 56th Dev Res Conf, Charlottesville, VA Dec1998.
42. Asbeck PM, Yu ET, Lau SS, Sulliran GJ, Van Hove JM, Redwing J (1997) Electron Lett 33: 1230
43. Foutz BE, Eastman LF, Bhapkar UV, Shur MS (1997) Appl Phys Lett 70: 2849
44. Khan MA, Kuznia TN, Bhattaraia AR, Olson DT (1993) Appl Phys Lett 62: 1786
45. Binari SC, Rowland LB, Kruppa W, Kelner G, Doverspike K, Gaskill DK (1994) Electron Lett 30: 1248
46. Zolper JC, Shul RJ, Baca AG, Wilson RG, Pearton SJ, Stall RA (1996) Appl Phys Lett 68: 2273
47. Binari SC (1995) Electrochem Soc Proc 95(21): 136
48. Ren F, Hong M, Chu SNG, Marcus MA, Schurman MJ, Baca AG, Pearton SJ, Abernathy CR (1999) Appl Phys Lett 73: 3893
49. McCarthy LS, Kozodoy P, Rodwell MJW, DenBaars SP (1999) IEEE Electron Dev Lett 20: 277

50. Han J, Baca AG, Shul RJ, Willison CG, Zhang L, Ren F, Zhang A, Dang G, Donovan SM, Cao XA, Cho H, Jung KB, Abernathy CR, Pearton SJ, Wilson RG (1999) Appl Phys Lett 74: 2702
51. Zolper JC, Han J, Biefeld RM, Van Deusen SB, Wampler WR, Reiger DJ, Pearton SJ, Williams JS, Tan HH, Stall R (1998) J Electron Mater 27: 179
52. Binari SC, Dietrich HB, Kelner G, Rowland LB, Doverspike K, Wickenden DK (1995) J Appl Phys 78: 3008
53. Zolper JC, Pearton SJ, Abernathy CR, Vartuli CB (1995) Appl Phys Lett 66: 3042
54. Liu QZ, Lau SS (1998) Solid-State Electron 42: 677
55. Lin ME, Ma Z, Huang FY, Fan ZF, Allen HL, Morkoc H (1994) Appl Phys Lett 61: 1003
56. Bermudez VM, Kaplan R, Khan MA. Kuznia JN (1993) Phys Rev B 48: 2436
57. Schmitz AC, Ping AT, Khan MA, Chen Q, Yang JW, Adesida I, Electron Lett 32: 1832
58. Liu QZ, Ya LS, Lau SS, Redwing JM, Rerkins NR, Kuech TF (1997) Appl Phys Lett 70: 1275
59. Foresi JS, Moustakas TD (1993) Appl Phys Lett 62: 2859
60. Adesida I, Mahajan A, Andideh E, Khan MA, Olsen DT, Kuznia JN (1993) Appl Phys Lett 63: 2777
61. Shul RJ, McClellan GB, Pearton SJ, Abernathy CR, Constantine C, Barratt C (1996) Electron Lett 32: 1408
62. Shul RJ, McClellan GB, Casalnnoro SA, Rieger DJ, Pearton SJ, Constantine C, Barratt C, Karlicek RK (1996) Appl Phys Lett 69: 1119
63. Pearton SJ, Abernathy CR, Ren F (1994) Appl Phys Lett 64: 2294
64. Pankove JI, Hutchby JA (1974) Appl Phys Lett 24: 281
65. Maruska HP, Lioubtchenko M, Tetreault TG, Osinski M, Pearton SJ, Schurman M, Vando R, Sakai S, Chen Q, Shul RJ (1998) Mat Res Soc Symp 483: 345 (MRS, Pittsburgh, PA, 1998)
66. Tan HH, Williams JS, Zou J, Cockayne DJH, Pearton SJ, Zolper JC, Stall RA (1998) Appl Phys Lett 72: 1190
67. Abernathy CR (1995) Mater Sci Eng R 14: 203
68. Yi CC, Wessels BW (1996) Appl Phys Lett 69: 3026
69. Bogulawski P, Briggs EL, Bernholc J (1995) Phys Rev B 51: 17255
70. Abernathy CR, MacKenzie JD, Pearton SJ, Hobson WS (1995) Appl Phys Lett 66: 1969
71. Wilson RG, Pearton SJ, Abernathy CR, Zavada JM (1995) Appl Phys Lett 66: 2238
72. Deal MD, Robinson HG (1989) Appl Phys Lett 55: 1990

73. Robinson HG, Deal MD, Stevenson DA (1991) Appl Phys Lett 58: 2000
74. Robinson HG, Deal MD, Griffin PB, Amaratunga G, Griffin PB, Stevenson DA, Plummer JD (1992) J Appl Phys 71: 2615
75. Pearton SJ (1993) Int J Mod Phys 7: 4687
76. Deal MD, Hu CJ, Lee CC, Robinson HG (1993) Mat Res Soc Symp 300: 365 (MRS, Pittsburgh, PA, 1993)
77. Zolper JC, Han J, Van Deusen SB, Biefeld RM, Crawford MH, Han J, Suski T, Baranowski JM, Pearton SJ (1998) Mat Res Soc Symp Proc 482: 609 (MRS, Pittsburgh, PA, 1998)
78. Harrington G, Hsin Y, Liu QZ, Asbeck PM, Lau SS, Khan MA, Yang JW, Chen Q (1998) Electron Lett 34: 193
79. Pearton SJ, Iannuzzi MP, Reynolds CL, Peticolas L (1988) Appl Phys Lett 52: 395
80. Zolper JC, Sherwin ME, Baca AG, Schneider RP (1995) J Electron Mater 24: 21
81. Cohen M, Fritsche H, Ovshinsky S (1969) Phys Rev Lett 22: 1065
82. Pearton SJ (1990) Mater Sci Rep 4: 313
83. Stavola M (ed) (1998) Identification of defects in semiconductors, In: *Semiconductors and Semimetals*, Vols 51A and 51B, Academic Press, San Diego
84. Pressel K, Thurian P (1999) In: *Properties, Processing and Applications of GaN and Related Semiconductor,* Edgar JH, Strite A, Akasaki I, Amano H, Wetzel C (eds) EMIS Data Review 23, INSPEC, IEEE, London
85. Fan Z, Mohammad SN, Kim W, Aktas O, Botchkarev AE, Morkoc H (1996) Appl Phys Lett 68: 1672
86. Fomenko VS (1981) *Emission Properties of Materials*, Naukora Dumka, Kiev
87. Ishikawa H, Kobayashi S, Koide Y, Yamasaki S, Nagai S, Umezaki J, Koike M, Murakami M (1997) J Appl Phys 81: 1315
88. Vassilevski KV, Rastegaeva MG, Babanin AI, Nikitina IP, Dmitriev VA (1996) MRS Internet J Nitride Semicond. Res 1: 38
89. Mori T, Kozawa T, Ohwaki T, Taga Y, Nagai S, Yamasaki S, Asami A, Shibata N, Koike M (1996) Appl Phys Lett 69: 3537
90. Kim T, Khim J, Chae S, Kim T (1997) Mat Res Soc Symp Proc 468: 427
91. King DJ, Zhang L, Ramer JC, Hersee SD, Lester LF (1997) Mat Res Soc Symp Proc 468: 421 (MRS, Pittsburgh, PA, 1997)
92. Smith LL, Davis RF, Kim MJ, Carpenter RW, Huang Y (1997) J Mater Res 12: 2249

93. Koide Y, Maeda T, Kawakami T, Fujita S, Uemura T, Shibata N, Murakami M (1999) J Electron Mater 28: 341
94. Ho J-K, Jong C-S, Chiu CC, Huang C-N, Chen C-Y, Shih K-K (1999) Appl Phys Lett 74: 1275
95. Kim JK, Lee J-L, Lee JW, Shin HE, Park YJ, Kim T (1998) Appl Phys Lett 73: 2953
96. Lee J-L, Weber M, Kim JK , Lee JW, Park YJ, Kim T, Lynn K (1999) Appl Phys Lett 74: 2289
97. Suzuki M, Kawakami T, Arai T,Koide Y, Uemura T, Shibata N, Murakami M (1999) Appl Phys Lett 74: 275
98. Schubert EF, Grieshaber W, Goepfert LD (1996) Appl Phys Lett 69: 37371
99. Zhou L, Ping AT, Khan F, Osinsky A, Adesida I (2000) Electron Lett 36: 91
100. Ewards NV, Bremser MD, Weeks TW, Kerm RS, Davis RF, Aspnes DS (1996) Appl Phys Lett 69: 2065
101. Bermudez VM (1996) J Appl Phys 80: 1190
102. Smith LL, King SW, Nemanich RJ, Davis RF (1996) J Electron Mater 25: 805
103. Jang JS, Park SJ, Seong TY (1999) J Vac Sci Technol B 17: 2667
104. Cole MW, Eckart EW, Han WY, Pfeffer RL, Monahan T, Ren, Yuan C, Stall RA, Pearton SJ, Li Y, Lu Y (1996) J App Phys 80: 278
105. Cole MW, Ren F, Pearton SJ (1997) J Electron Soc 144: L275; Appl Phys Lett 71: 3004
106. Venugopalan HS, Mohney SE, Luther BP, DeLucca JM, Wolter SD, Redwing JM, Bulman GE (1997) Mat Res Soc Symp Proc 468: 431
107. Trexler JT, Miller SJ, Holloway PH, Khan MA (1996) Mat Res Soc Symp Proc 395: 819 (MRS, Pittsburgh, PA, 1996)
108. Chen JY, Pan CJ, Chi GC (1999) Solid-State Electron 43: 649
109. Ping AT, Chen Q, Yang JW, Khan MA, Adesida I (1998) J Electron Mater 27: 261
110. Eddy CR Jr, Molnar B (1996) Mat Res Soc Symp Proc 395: 745; (1999) J Electron Mater 28: 314
111. Pearton SJ, Lee JW, MacKenzie JD, Abernathy CR, Shul RJ (1995) Appl Phys Lett 67: 2329
112. Ren F, Lothian JR, Pearton SJ, Abernathy CR, Vartuli CB, MacKenzie JD, Wilson RG, Karlicek RF (1997) J Electron Mater 26: 1287
113. Saotume K, Matsutani A, Shirasawa T, Mori M, Honda T, Sakaguchi T, Koyama F, Iga K (1997) Mat Res Soc Symp Proc 449: 1029 (MRS, Pittsburgh, PA, 1997)

114. Shul RJ, Zhang L, Baca AG, Willison CG, Han J, Pearton SJ, Ren F, Zolper JC, Lester LF (1999) Mat Res Soc Symp Proc Vol 573: 161 (MRS, Pittsburgh, PA, 1999)
115. Kim BJ, Lee JW, Park HS, Kim TL (1998) J Electron Mat 27: L32
116. Youtsey C, Bulman G, Adesida I (1998) J Electron Mater 27: 282
117. Ohba Y, Hatano A (1994) J Cryst Growth 145: 214
118. Pearton SJ, Zolper JC, Shul RJ, Ren D (1999) J Appl Phys 86: 1
119. Bridger PM, Bandic ZZ, Piquette EC, McGill TC (1998) Appl Phys Lett 73: 3438
120. Bandic ZZ, Bridger PM, Piquette EC, McGill TC (1998) Appl Phys Lett 72: 3166
121. *Atlas User's Manual* (1998) Silvaco International, Santa Clara, CA
122. *Properties, Processing and Applications of GaN and Related Semiconductors* (1999) Edgar JH, Strife S, Akasaki I, Amano H, Wetzel C (eds) EMIS DataReview No. 23 IEEE, London
123. Gillis HP, Choutov DA, Martin KP, Pearton SJ, Abernathy CR (1996) J Electrochem Soc 143: L251
124. Leonard RT, Bedair SM (1996) Appl Phys Lett 68: 794

2 Dry Etching of Gallium Nitride and Related Materials

2.1 Abstract

In this chapter, the characteristics of dry etching of the AlGaInN materials system in different reactor types and plasma chemistries are reviewed, along with the depth and thermal stability of etch-induced damage. The application to device processing for both electronics and photonics is also discussed.

2.2 Introduction

Owing to limited wet chemical etch results for the Group III nitrides, a significant amount of effort has been devoted to the development of dry etch processing.[1–19] Dry etch development was initially focused on mesa structures where high etch rates, anisotropic profiles, smooth sidewalls and equirate etching of dissimilar materials were required. For example, commercially available LEDs and laser facets for GaN-based LDs were patterned using RIE. However, as interest in high power, high temperature electronics [20–23] increased, etch requirements expanded to include smooth surface morphology, low plasma-induced damage and selective etching became important. Dry etch development is further complicated by the inert chemical nature and strong bond energies of the Group III nitrides compared with other compound semiconductors. GaN has a bond energy of 8.92 eV/atom, InN 7.72 eV/atom and AlN 11.52 eV/atom.[24]

2.3 Plasma Reactors

Dry plasma etching has become the dominant patterning technique for the Group III nitrides, due to the shortcomings in wet chemical etching.

Plasma etching proceeds by either physical sputtering, chemical reaction, or a combination of the two often referred to as ion-assisted plasma etching, Physical sputtering is dominated by the acceleration of energetic ions formed in the plasma to the substrate surface at relatively high energies, typically >200 eV. Owing to the transfer of energy and momentum to the substrate, material is ejected from the surface. This sputter mechanism tends to yield anisotropic profiles; however, it can result in significant damage, rough surface, morphology, trenching, poor selectivity and nonstoichiometric surfaces, thus minimizing device performance. The measured sputter rates for GaN, InN, AlN and InGaN as a function of Ar^+ ion energy increased with ion energy but were quite slow, <600 Å/min, due to the high bond energies of the Group III–N bond.[25]

Chemically dominated etch mechanisms rely on the formation of reactive species in the plasma which absorb to the surface, form volatile etch products and then desorb from the surface. Since ion energies are relatively low, etch rates in the vertical and lateral directions are often similar, thus resulting in isotropic etch profiles and loss of critical dimensions. However, owing to the low ion energies used, plasma-induced damage is minimized. Alternatively, ion-assisted plasma etching relies on both chemical reactions and physical sputtering to yield anisotropic profiles at reasonably high etch rates. Provided the chemical and physical components of the etch mechanism are balanced, high-resolution features with minimal damage can be realized and optimum device performance can be obtained.

2.3.1 Reactive Ion Etching

RIE utilizes both the chemical and physical components of an etch mechanism to achieve anisotropic profiles, fast etch rates and dimensional control. RIE plasmas are typically generated by applying rf power of 13.56 MHz between two parallel electrodes in a reactive gas (see Figure 2.1(a)). The substrate is placed on the powered electrode, where a potential is induced and ion energies, defined as they cross the plasma sheath, are typically a few hundred electron-volts. RIE is operated at low pressures, ranging from a few millitorr up to 200 mTorr, which promotes anisotropic etching due to increased mean free paths and reduced collisional scattering of ions during acceleration in the sheath.

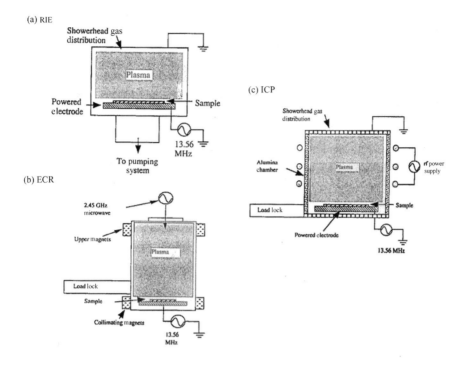

Figure 2.1. Schematic diagram of (a) RIE, (b) ECR, and (c) ICP etch platforms

Adesida et al. [18] were the first to report RIE of GaN in $SiCl_4$-based plasmas. Etch rates increased with increasing dc bias, and were >500 Å/min at −400 V. Lin et al. reported similar results for GaN in BCl_3 and $SiCl_4$ plasmas with etch rates of 1050 Å/min in BCl_3 at 150 W cathode (area 250 in^2) rf power. Additional RIE results have been reported for HBr- [27], CHF_3- and CCl_2F_2-based [28] plasmas, with etch rates typically <600 Å/min. The best RIE results for the Group III nitrides have been obtained in chorine-based plasmas under high ion energy conditions where the Group III–N bond breaking and the sputter desorption of etch products from the surface are most efficient. Under these conditions, plasma damage can occur and degrade both electrical and optical device performance. Lowering the ion energy or increasing the chemical activity in the plasma to minimize the damage often results in slower etch rates or less anisotropic profiles, which significantly limits critical dimension. Therefore, it is necessary to pursue alternative etch platforms which combine high quality etch characteristics with low damage.

2.3.2 High-Density Plasmas

The use of high-density plasma etch systems, including ECR, ICP and magnetron RIE (MRIE), has resulted in improved etch characteristics for the group-III nitrides as compared to RIE. This observation is attributed to plasma densities which are two to four orders of magnitude higher than RIE, thus improving the Group III–N bond breaking efficiency and the sputter desorption of etch products formed on the surface. Additionally, since ion energy and ion density can be more effectively decoupled compared with RIE, plasma-induced damage is more readily controlled. Figure 2.1(b) shows a schematic diagram of a typical low-profile ECR etch system. High-density ECR plasmas are formed at low pressures with low plasma potentials and ion energies due to magnetic confinement of electrons in the source region. The sample is located downstream from the source to minimize exposure to the plasma and to reduce the physical component of the etch mechanism. Anisotropic etching can be achieved by superimposing an rf bias (13.56 MHz) on the sample and operating at low pressure (<5 mTorr) to minimize ion scattering and lateral etching. However, as the rf biasing is increased, the potential for damage to the surface increases. Figure 2.2 shows a schematic of the plasma parameters and sample position in a typical high-density plasma reactor.

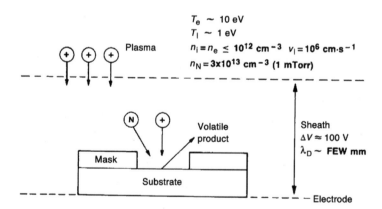

Figure 2.2. Schematic diagram of high-density plasma etching process

ECR etch rates for GaN, InN and AlN increased as either the ion energy (dc bias) or ion flux (ECR source power) increased.[29,30] Etch rates of 1100 Å/min for AlN and 700 Å/min for GaN at −150 V dc bias in a Cl_2–H_2 plasma and 350 Å/min for InN in a CH_4–H_2–Ar plasma at −250 V dc bias were reported. The etched features were anisotropic and the surface remained stoichiometric over a wide range of plasma conditions. GaN ECR etch data have been reported by several authors, with etch rates as high as 1.3 µm/min.[31–42]

ICP offers another high-density plasma etch platform to pattern Group III nitrides. ICP plasmas are formed in a dielectric vessel encircled by an inductive coil into which rf power is applied (see Figure 2.1(c)). The alternating electric field between the coils induces a strong alternating magnetic field trapping electrons in the center of the chamber and generating a high-density plasma. Since ion energy and plasma density can be effectively decoupled, uniform density and energy distributions are transferred to the sample while keeping ion and electron energy low. Thus, ICP etching can produce low damage while maintaining fast etch rates. Anisotropy is achieved by superimposing an rf bias on the sample. ICP etching is generally believed to have several advantages over ECR, including easier scale-up for production, improved plasma uniformity over a wider area and lower cost of operation. The first ICP etch results for GaN were reported in a Cl_2–H_2–Ar ICP plasma with etch rates as high as ~6875 Å/min.[43,44] Etch rates increased with increasing dc bias, and etch profiles were highly anisotropic with smooth etch morphologies over a wide range of plasma conditions. GaN etching has also been reported in a variety of halogen- and methane-based ICP discharges.[45–54] Use of a Cl_2–Ar–O_2 chemistry produced good selectivity for GaN and InGaN over AlGaN (up to ~50), due to formation of an oxide on the AlGaN.[52,53]

MRIE is another high-density etch platform and is comparable to RIE. In MRIE, a magnetic field is used to confine electrons close to the sample and minimize electron loss to the wall.[55–57] Under these conditions, ionization efficiencies are increased and high plasma densities and fast etch rates are achieved at much lower dc biases (less damage) compared with RIE. GaN etch rates of ~3500 Å/min were reported in BCl_3-based plasmas at dc biases <–100 V.[58] The etch was fairly smooth and anisotropic.

2.3.3 Chemically Assisted Ion Beam Etching

Chemically assisted ion beam etching (CAIBE) and reactive ion beam etching (RIBE) have also been used to etch Group III nitride films.[19,59–62] In these processes, ions are generated in a high-density plasma source and accelerated by one or more grids to the substrate. In CAIBE the reactive gases are added to the plasma downstream of the acceleration grids, thus enhancing the chemical component of the etch mechanism, whereas in RIBE the reactive gases are introduced in the ion source. Both etch platforms rely on relatively energetic ions (200–2000 eV) and low chamber pressures (<5 mTorr) to achieve anisotropic etch profiles. However, with such high ion energies, the potential for plasma-induced damage exists. Adesida and co-workers reported CAIBE etch rates for GaN as high as 2100 Å/min with 500 eV Ar^+ ions and Cl_2 or HCl ambients.[18,59,60] Rates increased with beam current, reactive gas flow rate and substrate temperature. Anisotropic profiles with smooth etch morphologies were observed.

2.3.4 Reactive Ion Beam Etching

The RIBE removal rates for GaN, AlN and InN are shown in Figure 2.3 as a function of Cl_2 percentage in Cl_2–Ar beams at 400 eV and 100 mA current. The trend in removal rates basically follows the bond energies of these materials. At fixed Cl_2/Ar ratio, the rates increased with beam energy. At very high voltages, one would expect the rates to saturate or even

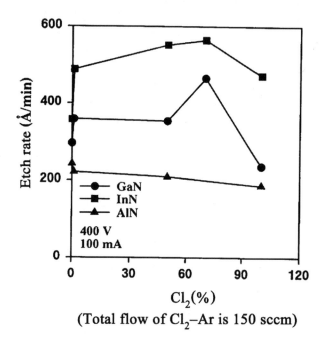

Figure 2.3. RIBE nitride removal rates as a function of Cl_2 percentage in Cl_2–Ar beams

decrease due to ion-assisted desorption of the reactive chlorine from the surface of the nitride sample before it can react to form the chloride etch products.

There was relatively little effect of either beam current or sample temperature on the RIBE removal rates of the nitride. The etch profiles are anisotropic, with light trenching at the base of the features. This is generally ascribed to ion deflection from the sidewalls causing an increased ion flux at the base of the etched features.

2.3.5 Low-Energy Electron-Enhanced Etching

Low-energy electron-enhanced etching (LE4) of GaN has been reported by Gillis and co-workers.[16,63–65] LE4 is an etch technique which depends on the interaction of low-energy electrons (<15 eV) and reactive species at the substrate surface. The etch process results in minimal surface damage since there is negligible momentum transferred from the electrons to the substrate. GaN etch rates of ~500 Å/min in an H_2-based LE4 plasma and ~2500 Å/min in a pure Cl_2 LE4 plasma have been reported.[16,65] GaN has also been etched using photo-assisted dry etch processes where the substrate is exposed to a reactive gas and uv laser radiation simultaneously. Vibrational and electronic excitations lead to improved bond breaking and desorption of reactant products. Leonard and Bedair [66] reported GaN etch rates of <80Å/min in HCl using a 193 nm ArF excimer laser.

GaN etch rates are compared in Figure 2.4 for RIE, ECR and ICP Cl_2–H_2–CH_4–Ar plasmas and an RIBE Cl_2–Ar plasma. CH_4 and H_2 were removed from the plasma chemistry to eliminate polymer deposition in the RIBE chamber. Etch rates increased as a function of dc bias independent of etch technique. GaN etch rates obtained in the ICP and ECR plasmas were much faster than those obtained in RIE and RIBE. This was attributed to higher plasma densities (one to four orders of magnitude higher) which resulted in more efficient breaking of the Group III–N bond and sputter desorption of the etch products. Slower rates observed in the RIBE may also be due to lower operational pressures (0.3 mTorr compared with 2 mTorr for the ICP and ECR) and/or lower ion and reactive neutral flux at the GaN surface due to high source-to-sample separation.

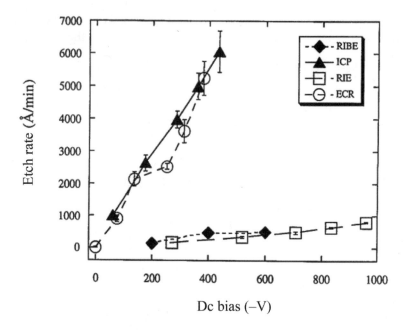

Figure 2.4. GaN etch rates in RIE, ECR, ICP and RIBE Cl_2-based plasmas as a function of dc bias

2.4 Plasma Chemistries

2.4.1 Chlorine-Based Plasmas

Etch characteristics are often dependent upon plasma parameters, including pressure, ion energy and plasma density. As a function of pressure, plasma conditions, including the mean free path and the collisional frequency, can change, resulting in changes in both ion energy and plasma density. GaN etch rates are shown as a function of pressure for an ICP-generated BCl_3–Cl_2 plasma in Figure 2.5. Etch rates increased as the pressure was increased from 1 to 2 mTorr and then decreased at higher pressures. The initial increase in etch rate suggested a reactant-limited regime at low pressure; however, at higher pressures the etch rates decreased due either to lower plasma densities (ions or radical neutrals), redeposition or polymer formation on the substrate surface. At pressures <10 mTorr, GaN etches were anisotropic and smooth, while at pressures >10 mTorr the etch profile undercut and poorly defined due to a lower mean free path, collisional scattering of the ions and increased lateral etching of the GaN.

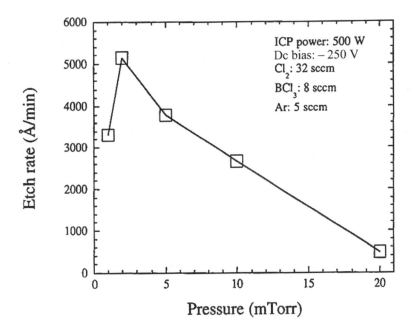

Figure 2.5. GaN etch rates as a function of pressure in an ICP-generated BCl_3–Cl_2–Ar plasma at 32 sccm Cl_2, 8 sccm BCl_3, 5 sccm Ar, 500 W IPC source power, dc bias –150 V and 10°C electrode temperature

GaN etch rates are plotted as a function of dc bias (which correlates to ion energy) for an ICP-generated BCl_3–Cl_2 plasma in Figure 2.6. The GaN etch rates increased monotonically as the dc bias or ion energy increased. Etch rates increased due to improved sputter desorption of etch products from the surface, as well as from more efficient breaking of the Ga–N bonds. Etch rates have also been observed to decrease under high ion bombardment energies due to sputter desorption of reactive species from the surface before the reactions occur. This is often referred to as an adsorption limited etch regime. In Figure 2.7, SEM micrographs are shown for (a) –50, (b) –150 and (c) –300 V dc bias. The etch profile became more anisotropic as the dc bias increased from –50 to –150 V dc bias, due to the perpendicular path of the ions relative to the substrate surface, maintaining straight wall profiles. However, as the dc bias was increased to –300 V, a tiered etch profile with vertical striations in the sidewall was observed due to erosion of the resist mask edge. The GaN may become rougher at these conditions due to mask redeposition and preferential loss of N_2.

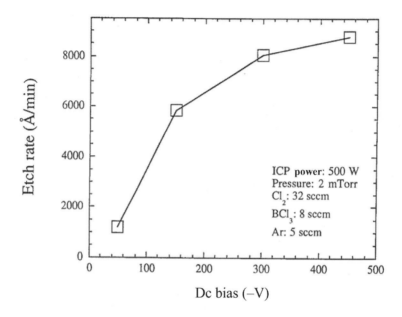

Figure 2.6. GaN etch rates as a function of dc bias in an ICP-generated BCl_3–Cl_2–Ar plasma at 32 sccm Cl_2, 8 sccm BCl_3, 5 sccm Ar, 500 W ICP source power, 2 mTorr pressure and 10°C electrode temperature

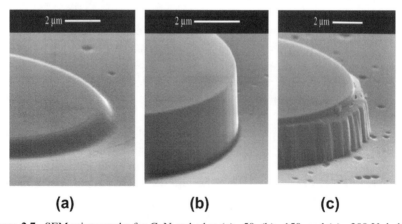

Figure 2.7. SEM micrographs for GaN etched at (a) –50, (b) –150, and (c) –300 V dc bias. ICP etch conditions were 32 sccm Cl_2, 8 sccm BCl_3, 5 sccm Ar, 500 W ICP source power, 2 mTorr pressure and 10°C electrode temperature

In Figure 2.8, GaN etch rates are shown as a function of ICP source power while the dc bias was held constant at −250 V. GaN etch rates increased as the ICP source power increased, due to higher concentrations of reactive species (which increases the chemical component of the etch mechanism) and/or higher ion flux (which increases the bond breaking and sputter desorption efficiency of the etch). Etch rates have also been observed to stabilize or decrease under high plasma flux conditions due either to saturation of reactive species at the surface or sputter desorption of reactive species from the surface before the reactions occur. The etch profile was anisotropic and smooth up to 1000 W ICP power, where the feature dimensions were lost and sidewall morphology was rough due to erosion of the mask edge under high plasma flux conditions.

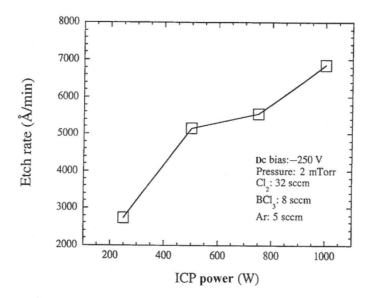

Figure 2.8. GaN etch rates as a function of ICP source power in an ICP-generated $BCl_3/Cl_2/Ar$ plasma at 32 sccm Cl_2, 8 sccm BCl_3, 5 sccm Ar, −250 V dc bias, 2 mTorr pressure and 10°C electrode temperature

In addition to etch rates, the etch selectivity or the ability to etch one film at higher rates than another can be very important in device fabrication. For example, optimization of etch selectivity is critical to control threshold voltage uniformity for high electron mobility transistors (HEMTs), to accurately stop on either the emitter or collector regions for metal contacts for HBTs, and for low resistivity *n*-ohmic contacts on InN layers. Several studies have recently reported etch selectivity for the Group III nitrides.[50,53,67–71] For example, Figure 2.9 shows GaN, InN and AlN etch rates and etch selectivities as a function of cathode rf power in an ICP-generated Cl_2–Ar plasma. Etch rates for all three films increased with increasing cathode rf power or dc bias due to improved breaking of the Group III–N bonds and more efficient sputter desorption of the etch products. Increasing InN etch rates were especially significant, since $InCl_3$, the primary In etch product in a Cl-based plasma, has a relatively low volatility. However, under high dc bias conditions, desorption of the $InCl_3$ etch products occurred prior to coverage of the etch surface.[68] The GaN:InN and GaN:AlN etch selectivities were <8:1 and decreased as the cathode rf power or ion energy increased. Smith and co-workers [53] reported similar results in a Cl_2–Ar ICP plasma where GaN:AlN and GaN:AlGaN selectivities decreased as dc bias increased. 53 At –20 V dc bias, etch selectivities of ~39:1 were reported for GaN:AlN and ~10:1 for GaN:AlGaN.

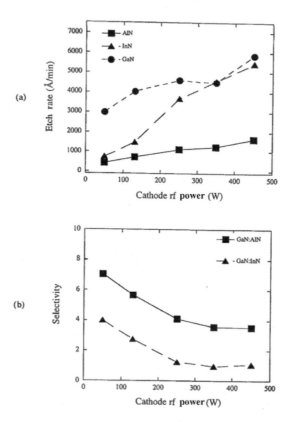

Figure 2.9. GaN, InN and AlN (a) etch rates and (b) GaN:AlN and GaN:InN etch selectivities as a function of dc bias in a Cl_2–Ar ICP. Plasma conditions were: 25 sccm Cl_2, 5 sccm Ar, 2 mTorr chamber pressure, 500 W ICP-source power and 25°C cathode temperature

Temperature-dependent etching of the Group III nitrides has been reported in ECR and ICP etch systems [32,42,54]. Etch rates are often influenced by the substrate temperature, which can affect the desorption rate of etch produced, the gas–surface reaction kinetics and the surface mobility of the reactants. Substrate temperature can be controlled and maintained during the etch process by a variety of clamping and backside heating or cooling procedures. GaN and InN etch rates are shown in Figure 2.10 as a function of temperature in a Cl_2–H_2–Ar ICP. GaN etch rates were much faster than InN due to higher volatility of the $GaCl_3$ etch products compared with $InCl_3$ and showed little dependence on temperature. However, the InN etch rates showed a considerable temperature dependence, increasing at 150°C due to higher volatilities of the $InCl_3$ etch products at higher substrate temperatures.

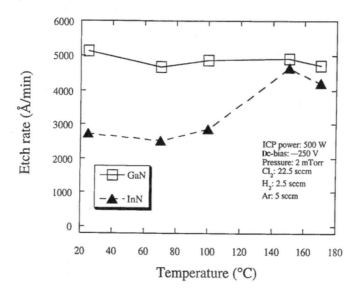

Figure 2.10. GaN and InN etch rates as a function of temperature for an ICP-generated Cl_2–H_2–Ar plasmas. ICP etch conditions were 22.5 sccm Cl_2, 2.5 sccm H_2, 5 sccm Ar, 500 W ICP source power, −250 V dc bias and 2 mTorr pressure

Several different plasma chemistries have been used to etch the Group III nitrides. As established above, etch rates and profiles can be strongly affected by the volatility of the etch products formed. Table 2.1 shows the boiling points of possible etch products for the Group III nitrides exposed to halogen- and hydrocarbon-based plasmas. For halogen-based plasmas, etch rates are often limited by the volatility of the Group III halogen etch product. For Ga- and Al-containing films, chlorine-based plasmas typically yield fast rates with anisotropic, smooth etch profiles. CH_4–H_2-based plasma chemistries have also yielded smooth, anisotropic profiles for Ga-containing films, although at much slower rates. Based only on a comparison of etch product volatility, slower etch rates in CH_4-based plasmas is unexpected, since the $(CH_3)_3Ga$ etch product has a much lower boiling point than $GaCl_3$. This observation demonstrates the complexity of the etch process, where redeposition, polymer formation, and gas-phase kinetics can influence the results. As shown above, etch rates for In-containing films obtained in room temperature chlorine-based plasmas tend to be slow with rough surface morphology and overcut profiles due to the low volatility of the $InCl_3$ and preferential loss of the Group V etch products. However, at elevated temperatures (>130°C), the $InCl_3$ volatility increases and the etch

Table 2.1. Boiling points for possible etch products of Group III nitride films etched in halogen- or CH_4–H_2-based plasmas

Etch products	Boiling points (°C)
$AlCl_3$	183
AlF_3	Na
AlI_3	360
$AlBr_3$	263
$(CH_3)_3Al$	126
$GaCl_3$	201
GaF_3	1000
GaI_3	Sublimes 345
$GaBr_3$	279
$(CH_3)_3Ga$	55.7
$InCl_3$	600
InF_3	>1200
InI_3	Na
$InBr_3$	Sublimes
$(CH_3)_3In$	134
NCl_3	<71
NF_3	−129
NBr_3	Na
NI_3	Explodes
NH_3	−33
N_2	−196
$(CH_3)_3N$	−33

2.4 Plasma Chemistries

rates and surface morphology improve.[32,42,54,72–74] Significantly better room temperature etch results are obtained in CH_4–H_2-based plasmas due to the formation of more volatile $(CH_3)_3In$ etch products.[30,75]

The source of reactive plasma species and the addition of secondary gases to the plasma can vary etch rates, anisotropy, selectivity and morphology. The fragmentation pattern and gas-phase kinetics associated with the source gas can have a significant effect on the concentration of reactive neutrals and ions generated in the plasma, thus affecting the etch characteristics. Secondary gas additions and variations in gas ratios can change the chemical:physical ratio of the etch mechanism. The effect of Ar, SF_6, N_2 and H_2 additions to Cl_2- and BCl_3-based ICPs and ECR plasmas for GaN etching has been reported.[76] In general, GaN etch rates were faster in Cl_2-based plasmas compared with BCl_3 due to a higher concentration of reactive Cl. The addition of H_2, N_2 or SF_6 to either Cl_2- or BCl_3-based plasmas changed the relative concentration of reactive Cl in the plasma, which directly correlated to the GaN etch rate. For example, in Figure 2.11, GaN etch rates are shown as a function of H_2 concentration for ECR- and ICP-generated Cl_2–H_2–Ar plasmas. GaN etch rates in the ECR and ICP increased slightly as H_2 was initially added to the Cl_2–Ar plasma, indicating a reactant-limited regime. Monitoring the ECR plasma with quadrupole mass spectrometry (QMS) showed that the Cl concentration (indicated by $m/e=35$) remained relatively constant at 10% H_2. As the H_2 concentration was increased above 10%, the Cl concentration decreased and a peak corresponding to HCl increased. GaN etch rates decreased at H_2 concentrations >10% in both the ECR and ICP, presumably due to the consumption of reactive Cl by hydrogen. In Figure 2.12, H_2 was added to BCl_3-based ECR and ICP plasmas. In the ECR plasma, the GaN etch rate increased at 10% H_2, corresponding with an increase in the reactive Cl concentration as observed by Optical Emission Spectroscopy (OES). As the H_2 concentration was increased further, GaN etch rates decreased, the Cl concentration decreased, and the HCl concentration increased, presumably due to the consumption of reactive Cl by hydrogen. In the ICP reactor, GaN etch rates were slow and decreased as up to 80% hydrogen was added to the plasma, where a slight increase was observed.

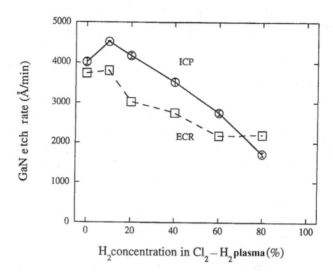

Figure 2.11. GaN etch rates in an ICP and ECR Cl$_2$H$_2$–Ar plasma as a function of hydrogen conservation

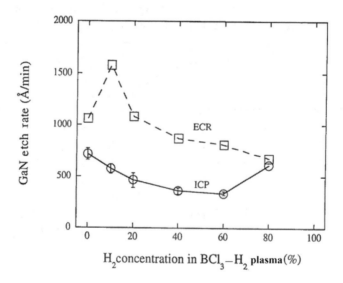

Figure 2.12. GaN etch rates in an ICP and ECR BCl$_3$–H$_2$–Ar plasma as a function of hydrogen conservation

Another example of plasma chemistry-dependent etching of GaN is shown in Figure 2.13 for Cl_2–N_2–Ar and BCl_3–N_2–Ar ICP-generated plasmas. In the Cl_2-based plasma, GaN etch rates decreased as the percentage of N_2 increased, presumably due to a reduction in reactive Cl. In the BCl-based plasma, GaN etch rates increased up to 40% N_2 and then decreased at higher N_2 concentration. This observation has also been reported for ECR and ICP etching of GaAs, GaP and In-containing films.[77–79] Ren *et. al.* [77] first observed maximum etch rates for In-containing films (InGaN and InGaP) in an ECR discharge at a gas ratio of 75/25 for BCl_3/N_2. Using optical emission spectroscopy (OES), Ren reported maximum emission intensity for atomic and molecular Cl at 75% BCl_3 as well as a decrease in the BCl_3 intensity and the appearance of a BN emission line. The authors speculated that N_2 enhanced the dissociation of BCl_3, resulting in higher concentrations of reactive Cl and Cl ions and thus higher etch rates. Additionally, the observation of BN emission suggested that less B was available to recombine with reactive Cl. This explanation may also be applied to the peak GaN etch rates observed at 40% N_2 in the ICP BCl_3–N_2–Ar plasmas. However, OES of the BCl_3–N_2–Ar ICP discharge did not reveal higher concentrations of reactive Cl nor a BN peak emission. In Figure 2.14, OES spectra are shown for (a) 100% BCl_3 (b) 75% BCl_3–25% N_2, (c) 25% BCl_3–75% N_2 and (d) 100% N_2 ICPs. As N_2 was added to the BCl_3 plasma, the BCl_3 emission (2710 Å) and Cl emission (5443 and 5560 Å) decreased but the BN emission (3856 Å) was not obvious.

BCl_3–Cl_2 plasmas have shown encouraging results in the etching of GaN films.[45,52] The addition of BCl_3 to a Cl_2 plasma can improve sputter desorption due to higher mass ions and reduce surface oxidation by gettering H_2O from the chamber. In Figure 2.15, GaN etch rates are shown as a function of percentage of Cl_2 in a BCl_3–Cl_2–Ar ICP. As the percentage Cl_2 increased, GaN etch rates increased up to 80% due to higher concentrations of reactive Cl. OES showed that the Cl emission intensity increased and the BCl emission intensity decreased as the percentage Cl_2 increased. Slower GaN etch rates in a pure Cl_2 plasma were attributed to less efficient sputter desorption of etch products in the absence of BCl_3. Similar results were reported by Lee *et. al.*[45] The fastest GaN etch rates were observed at 10% BCl_3, where the ion current density and Cl radical density were the greatest as measured by OES and a Langmuir probe.

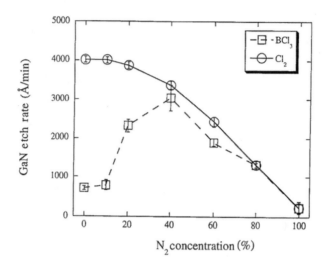

Figure 2.13. GaN etch rates as a function of percentage N_2 for ICP-generated Cl_2- and BCl_3-based plasmas

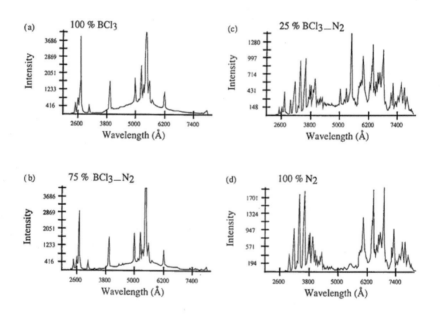

Figure 2.14. OES spectra for an ICP-generated BCl_3–N_2 plasma as a function of percentage BCl_3

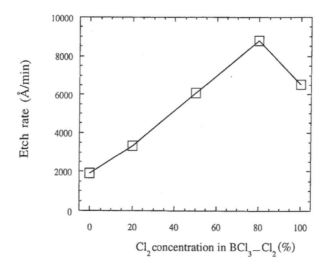

Figure 2.15. GaN etch rates in an ICP BCl_3-Cl_2 plasmas as a function of Cl_2

In general, GaN:AlN and GaN:InN etch selectivities are <10:1 as a function of plasma chemistry for Cl_2- or BCl_3-based plasmas. GaN:AlN and GaN:InN etch selectivities were higher for Cl_2-based ICPs compared with BCl_3-based ICP due to the higher concentration of reactive Cl produced in the Cl_2-based plasmas, thus resulting in faster GaN etch rates.[71] Alternatively, InN and AlN etch rates showed much less dependence on plasma chemistry and were fairly comparable in Cl_2- and BCl_3-based plasmas. An example of etch selectivity dependence on plasma chemistry is shown in Figure 2.16. GaN, AlN and InN etch rates and etch selectivities are plotted as a function of percentage SF_6 for a Cl_2-SF_6-Ar ICP. GaN and InN etch rates decreased as SF_6 was added to the plasma due to the consumption of Cl by S and, therefore, lower concentrations of reactive Cl. The AlN etch rates increased with the addition of SF_6 and reached a maximum at 20% SF_6. As SF_6 was added to the Cl_2 plasma, slower AlN etch rates were expected due to the formation of low-volatility AlF_3 etch products. However, due to the high ion flux in the ICP, the sputter desorption of the AlF_3 may occur prior to passivation of the surface.[68] Therefore, the GaN:AlN selectivity decreased rapidly from ~6:1 to <1:1 with the addition of SF_6. The GaN:InN selectivity reached a maximum of 4:1 at 20% SF_6.

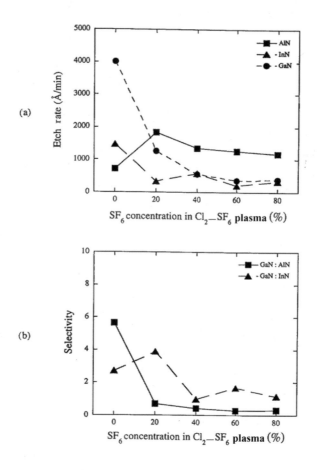

Figure 2.16. GaN, InN and AlN (a) etch rates and (b) GaN:AlN and GaN:InN etch selectivities as a function of SF_6 conservation in a Cl_2–SF_6–Ar ICP

The simple Cl_2–Ar chemistry works very well for most device fabrication processes, providing controllable etch rates. Even at biases <90 V, the GaN etch rate is still typically ~1000Å·min^{-1}.

2.4.2 Iodine- and Bromine-Based Plasmas

Other halogen-containing plasmas including ICl–Ar, IBr–Ar, BBr$_3$–Ar and BI$_3$–Ar have been used to etch GaN with promising results.[38,67,80–82]

Vartuli *et.al.* [83] reported GaN, InN, AlN, InN, InAlN and InGaN etch rates and selectivities in ECR ICl–Ar and IBr–Ar plasmas. In general, etch rates increased for all films as a function of dc bias due to improved Group III–N bond breaking and sputter desorption of etch products from the surface. GaN etch rates >1.3 μm/min were obtained in the ICl–Ar plasma at a rf power of 250 W (bias of –200 V), while GaN etch rates were typically <4000 Å/min in IBr–Ar. Cho *et al.* [82] reported GaN etch rates typically <2000 Å/min in ICP-generated BI_3–Ar and BBR_3–Ar plasmas. ICl–Ar and IBr–Ar ECR plasmas yielded GaN:InN, GaN:AlN, GaN:InGaN and GaN:InAlN selectivities <6:1; however, etch selectivities >100:1 were obtained for InN:GaN and InN:AlN in BI_3–Ar plasmas.[67,68,80–82] Fast etch rates obtained for InN were attributed to the high volatility of the InI_3 etch products compared with the GaI_3 and AlI_3 etch products, which can form passivation layers on the surface. Maximum selectivities of ~100:1 for InN:AlN and ~7.5 for InN:GaN were reported in the BBr_3–Ar plasma.[82] InI_x products have higher volatility than corresponding $InGl_x$ species, making iodine an attractive enchant for InGaN alloys. The interhalogen compounds are weakly bonded and, therefore, should easily break apart under plasma excitation to form reactive iodine, bromine and chlorine.

Figure 2.17 shows etch rates for the binary nitrides and selectivities for InN over both GaN and AlN as a function of the boron halide percentage by flow in the gas load. The dc chuck self-bias decreases as the BI_3 content increases, suggesting that the ion density in the plasma is increasing. The InN etch rate is proportional to the BI_3 content, indicating the presence of a strong chemical component in its etching. In comparison, AlN and GaN show very low rates until ~50% BI_3 (~500 Å/min^{-1} for AlN and ~1700 Å/min^{-1} for GaN). An increase in the BI_3 content in the discharges actually produces a fall off in the etch rate for both AlN and GaN. We expect there are several possible mechanisms by which to explain these data. First, the decrease in chuck self-bias, and hence ion energy, under these conditions may more than compensate for the higher active iodine neutral flux. Second, the formation of the less volatile GaI_x and AlI_x etch products may create a selvedge layer which suppresses the etch rate. This mechanism occurs in the Cl_2 reactive ion etching of InP. In this system, etching does not occur unless elevated sample temperatures or higher dc biases are used to facilitate removal of the $InCl_3$ etch product. In InN, etch selectivity to both materials initially increases but also goes through a minimum. Note, however, that selectivities of >100 can be achieved for both InN:AlN and InN:GaN.

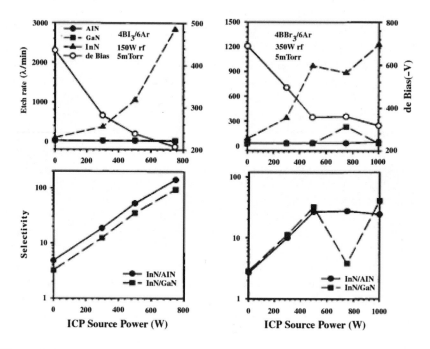

Figure 2.17. Nitride etch rates (top) and etch selectivities for InN:AlN and InN:GaN (bottom in BI_3–Ar or BBr_3–Ar discharges (750 W source power, 5 mTorr) as a function of the boron halide content

Data for BBr_3–Ar discharges are also shown in Figure 2.17 for fixed source power (750 W) and rf chuck power (350 W). Higher rf powers were required to initiate etching with BBr_3 compared with BI_3, and the dc self-bias increased with the BBr_3 content. The etch rate of InN is again a strong function of the boron halide content, GaN shows significant rates (~1800 Å/min^{-1}) only for pure BBr_3 discharges, and AlN shows very low etch rates over the whole range of conditions investigated. Maximum selectivities of ~100:1 for InN:AlN and ~7.5:1 for InN:GaN are obtained.

Based on the results in Figure 2.17, we chose fixed plasma compositions, and varied the ion energy and flux through control of the source and chuck powers. Figure 2.18 shows that source power had a significant effect only on the InN etch rate for both $4BI_3/6Ar$ and $4BBr_3/6Ar$ discharges at fixed rf power (150 W). The etch rate of InN continues to increase with source power, which controls the ion flux and dissociation of the discharge, whereas the GaN and AlN rates are low for both plasma chemistries. The InN etch rates are approximately a factor of two faster in BI_3–Ar than in BBr_3–Ar, even for lower rf chuck powers. This is expected from taking into consideration the relative stabilities of the respective In etch products (the InI₃ melting point is 210°C; InBr₃ sublimes at <600°C). The resultant selectivities are shown at the bottom of Figure 2.18; once again, a value of ~100:1 for InN over GaN is achieved with BI₃, whereas BBr₃ produced somewhat lower values.

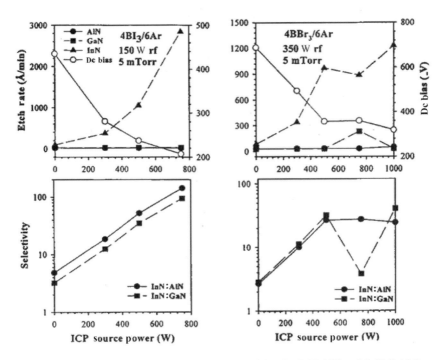

Figure 2.18. Nitride etch rates (top) and etch selectivities for InN:AlN and InN:GaN (bottom) in BI₃–Ar or BBr3–Ar discharges as a function of source power

The dependence of the etch rate and InN:AlN and InN:GaN selectivities on rf chuck power for both plasma chemistries at fixed source power (750 W) is shown in Figure 2.19. Whereas the GaN and AlN etch rates (top left) increase only at the highest chuck powers investigated for 4BI₃/6Ar discharges, the InN etch rate increases rapidly to 250 W. This is consistent with a strong ion-assisted component for the latter under these conditions. The subsequent decrease in the etch rate at higher power produces corresponding maxima (≥100) in etch selectivity for chuck powers in the range 150–250 W. This type of behavior is quite common to high-density plasma etching of III–V materials, where the etching is predominantly ion-assisted desorption of somewhat volatile products, with insignificant rates under ion-free conditions. In this scenario, at very high ion energies, the active etching species (iodine neutral in this case) can be removed by sputtering before they have a chance to complete the reaction with substrate atoms. Similar data for BBr$_3$–Ar mixtures are also shown in Figure 2.19. For this chemistry the InN etch rate saturates and we did not observe any reduction in etch rate, although this might be expected to occur if higher powers could be applied (our power supply is limited to 450 W). GaN does show an etch rate maximum at ~350 W, producing a minimum in the resultant InN:GaN selectivity. The etch selectivity of InN over the other two nitrides for BI$_3$–Ar is again much higher than for BBr$_3$–Ar.

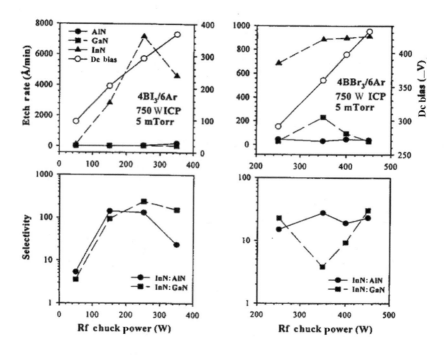

Figure 2.19. Nitride etch rates (top) and etch selectivities for InN:AlN and InN:GaN (bottom) in BI$_3$–Ar or BBr$_3$–Ar discharges as a function of rf chuck power

The effect of plasma composition on etch rates and selectivities if GaN, AlN and InN in ICl–Ar and IBr–Ar discharges at 750 W source power, 250 W rf chuck power and 5 mTorr is shown in Figure 2.20. The etch rates of InN and AlN are relatively independent of the plasma composition for both chemistries over a broad composition range, indicating the etch mechanism is dominated by physical sputtering. The dc bias voltage increased with increasing interhalogen concentrations. The decrease in ion flux also implies an increase in the concentrations of neutral species such as Cl, Br and I. The etch rate of GaN steadily increased with increasing ICl concentration.

2.4 Plasma Chemistries 121

By contrast, the etch rate of GaN saturated beyond 66.7% IBr. These results indicate that etching of GaN in both chemistries can be attributed more to chemical etching by increased concentrations of reactive neutrals than to ion-assisted sputtering. The effect of plasma composition showed an overall trend of decrease in selectivities for InN over both AlN and GaN as the concentration of ICl and IBr increased.

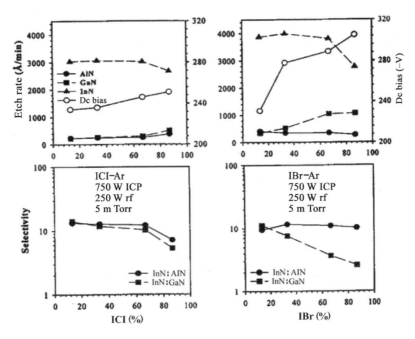

Figure 2.20. Nitride etch rates (top) and etch selectivities for InN–AlN and InN–GaN (bottom) in ICl–Ar or IBr–Ar discharges (750 W source power, 250 W rf chuck power, 5 mTorr) as a function of interhalogen content

2.4.3 Methane–Hydrogen–Argon Plasmas

ECR-generated CH_4–H_2–Ar plasma etch rates for GaN, InN and AlN were <400 Å/min at ~–250 V dc bias.[30] Vartuli *et al.* [83] reported ICP GaN, InN and AlN etch rates approaching 2500 Å/min in CH_4–H_2Ar and CH_4–H_2–N_2 plasmas. Etch rates increased with increasing dc bias or ion flux and were higher in CH_4–H_2–Ar plasmas. Anisotropy and surface morphology were good over a wide range of conditions. Compared with Cl-based plasmas, etch rates were consistently slower, which may make the CH_4–H_2-

based processes applicable for devices where etch depths are relatively shallow and etch control is extremely important.

Vartuli *et. al.* [68] compared etch selectivities in CH_4–H_2–Ar and Cl_2–Ar plasmas in both RIE- and ECR-generated plasmas. For CH_4–H_2–Ar plasmas, InN:GaN and InGaN:GaN etch selectivities ranged from 1:1 to 6:1, whereas etch selectivities of 1:1 or favoring GaN over the In-containing films were reported for Cl_2–Ar plasmas.

2.5 Etch Profile and Etched Surface Morphology

Etch profile and etched surface morphology can be critical to post-etch processing steps, including the formation of metal contacts, deposition of interlevel dielectric or passivation films, or epitaxial regrowth. Figure 2.21 shows SEM micrographs of GaN, AlN and InN etched in Cl_2-based plasma. The GaN (Figure 2.21(a)) was etched at 5 mTorr chamber pressure, 500 W ICP power, 22.5 sccm Cl_2, 2.5 sccm H_2, 5 sccm Ar, 25°C, and a dc bias of –280±10 V. Under these conditions, the GaN etch rate was ~6880 Å/min with highly anisotropic, smooth sidewalls. The sapphire substrate was exposed during a 15% overetch. Pitting of the sapphire surface was attributed to defects in the substrate or growth process. The AlN (Figure 2.21(b)) and InN (Figure 2.21(c)) features were etched at 2 mTorr chamber pressure, 500 W ICP power, 25 sccm Cl_2, 5 sccm Ar, 25°C, and a cathode rf power of

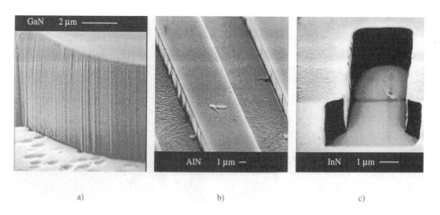

Figure 2.21. SEM micrographs of (a) GaN, (b) AlN, and (c) InN etched in Cl2-based ICP plasmas

250 W. Under these conditions, the AlN etch rate was ~980 Å/min and the InN etch rate was ~1300 Å/min. Anisotropic profiles were obtained over a wide range of plasma chemistries and conditions, with sidewall striations present.

Sidewall morphology is especially critical in the formation of laser mesas for ridge waveguide emitters or for buried planar devices. The vertical striations observed in the GaN sidewall in Figure 2.21(a) were due to striations in the photoresist mask which were transferred into the GaN feature during the etch. The sidewall morphology, and in particular the vertical striations, was improved in a Cl_2–BCl_3 ICP at –150 V dc bias. The etch conditions were at 2 mTorr chamber pressure, 500 W ICP power, 32 sccm Cl_2, 8 sccm BCl_3, 5 sccm Ar, 25°C, and a dc bias of –150±10 V. Ren et al.[84] have demonstrated improved GaN sidewall morphology etched in an ECR using an SiO_2 mask. Vertical striations in the SiO_2 mask were reduced by optimizing the lithography process used to pattern the SiO_2. The SiO_2 was then patterned in an SF_6–Ar plasma in which a low-temperature dielectric overcoat was used to protect the resist sidewall during the etch.

In several studies, AFM has been used to quantify the etched surface morphology as RMS roughness. Rough etch morphology often indicates a nonstoichiometric surface due to preferential removal of either the Group III or Group V species.

A summary of etch rate results for the nitrides with different chemistries and different techniques is shown in Table 2.2.

Table 2.2. Summary of etch rate results for GaN, AlN and InN with different plasma chemistries in different techniques

Gas chemistry	Etching technique	GaN		Ref.	AlN		Ref.	InN		Ref.
			Etch rate (nm/min) at given bias							
$SiCl_4$ [w/Ar, SiF_4]	RIE	55	−400 V	(18)	−		−	−		−
BCl_3	RIE	105	−230 V	(26)	−		−	−		−
HBr [w/Ar, H_2]	RIE	60	−400 V	(27)	−		−	−		−
CHF_3, C_2ClF_5	RIE	45	500 W	(28)	−		−	−		−
SF_6	RIE	17	−400 V	(27)	−		−	−		−
CHF_3, C_2ClF_5	RIE	60	−500 V	(28)	−		−	−		−
BCl_3–Ar	ECR	30	−250 V	(29)	17	−250 V	(29)	17	−300 V	(29)
CCl_2F_2–Ar	ECR	20	−250 V	(29)	18	−300 V	(29)	18	−300 V	(29)
CH_4–H_2–Ar	ECR	40	−250 V	(20)	2.5	−300 V	(29)	10	−300 V	(29)
Cl_2–H_2–Ar	ECR	200	−180 V	(32)	110	−150 V	(32)	150	−180 V	(32)
$SiCl_4$–Ar	ECR	95	−280 V	(35)	−		−	−		−
HI–H_2	ECR	110	−150 V	(15)	120	−150 V	(15)	100	−150 V	(15)
HBr–H_2	ECR	70	−150 V	(15)	65	−150 V	(15)	17	−150 V	(15)
ICl–Ar	ECR	1300	−275 V	(38)	200	−272 V	(38)	1150	−275 V	(38)
IBr–Ar	ECR	300	−170 V	(67)	160	−170 V	(67)	325	−170 V	(67)
BCl_3	MRIE	350	<−100 V	(57)	125	<−100 V	(15)	100	<−100 V	(15)
Cl_2–H_2–Ar	ICP	688	−280 V	(44)	−		−	−		−
Cl_2–Ar	ICP	980	−450 V	(53)	670	−450 V	(53)	150	−100 V	(48)
Cl_2–N_2	ICP	65	−100 V	(48)	39	−100 V	(48)	30	−100 V	(48)
BBr_3	ICP	150	−380	(49)	50	−200	(49)	500	−380	(49)
BI_3	ICP	200	−175	(49)	100	−175	(49)	700	−240	(49)
ICl	ICP	30	−300	(49)	30	−300	(49)	600	−300	(49)
IBr	ICP	20	−300	(49)	30	−300	(49)	600	−300	(49)
Ar ion	Ion milling	110	500 eV	(19)	29	500 eV	(25)	61	500 eV	(25)
Cl_2 [Ar ion]	CAIBE	210	500 eV	(59)	62	500 eV	(26)	−		−
HCl [Ar ion]	CAIBE	190	500 eV	(58)	−		−	−		−
Cl_2	RIBE	150	500 eV	(50)	−		−	−		−
HCl	RIBE	130	500 eV	(50)	−		−	−		−
Cl_2–Ar	RIBE	50	−400	(50)	50	−400	(50)	80	−400	(50)
HCl	Photo-asisted	0.04Å/pulse		(66)	−		−	−		−
H_2, Cl_2	LE4	50–7	1–15 eV	(64, 65)	−		−	−		−

2.6 Plasma-Induced Damage

Plasma-induced damage often degrades the electrical and optical properties of compound semiconductor devices. Since GaN is more chemically inert than GaAs and has higher bonding energies, more aggressive etch condi-

2.6 Plasma-Induced Damage

tions (higher ion energies and plasma flux) may be used with potentially less damage to the material. Limited data have been reported for plasma-induced damage of the Group III nitrides.[85–88] Pearton and *et al.* [85] reported increased plasma-induced damage as a function of ion flux and ion energy for InN, InGaN and InAlN in an ECR. The authors also reported: (a) more damage in InN films compared with InGaN; (b) more damage in lower doped materials; (c) more damage under high ion energy conditions due to formation of deep acceptor states which reduce the carrier mobility and increased resistivity. Post-etch annealing processes removed the damage in the InGaN, whereas the InN damage was not entirely removed.

Ren *et al.* [86] measured electrical characteristics for InAlN and GaN FET structures to study plasma-induced damage for ECR BCl_3, BCl_3–N_2 and CH_4–H_2 plasmas. They reported: (a) doping passivation in the channel layer in the presence of hydrogen; (b) high ion bombardment energies can create deep acceptor states that compensate the material; (c) preferential loss of N can produce rectifying gate characteristics. Ping *et al.* [87] studied Schottky diodes for Ar and $SiCl_4$ RIE plasmas. More damage was observed in pure Ar plasmas and under high dc bias conditions. Plasma-induced damage of GaN was also evaluated in ICP and ECR Ar plasma using photoluminescence (PL) measurements as a function of cathode rf power and source power.[88] The peak photoluminescence (PL) intensity decreased with increasing ion energy independent of etch technique. As a function of source power or plasma density the results were less consistent. The PL intensity showed virtually no change at low ICP source power and then decreased as the plasma density increased. In the ECR plasma, the PL intensity increased by ~115% at low ECR source power and improved at higher ECR source powers but at a lower rate. The effect of post-etch annealing in Ar varied depending on initial film conditions; however, annealing at temperatures above 440°C resulted in a reduction in the PL intensity.

Surface stoichiometry can also be used to evaluate plasma-induced damage. Nonstoichiometric surfaces can be created by preferential loss of one of the lattice constituents. This may be attributed to higher volatility of the respective etch products, leading to enrichment of the less volatile species, or preferential sputtering of the lighter element. AES can be used to measure surface stoichiometry. Figure 2.22 shows characteristic Auger spectra for (a) as-grown GaN samples and samples exposed to an ECR plasma at 850 W applied microwave power and cathode rf powers of (b) 65 and (c) 275 W. For the as-grown sample, the Auger spectrum showed a Ga:N ratio of 1.5 with normal amounts of adventitious carbon and native oxide on the GaN surface. Following plasma exposure, the Ga:N ratio increased as the cathode rf power increased with some residual atomic Cl from the plasma. Under high ion energy conditions, preferential removal of the lighter N atoms was observed, resulting in Ga-rich surfaces.

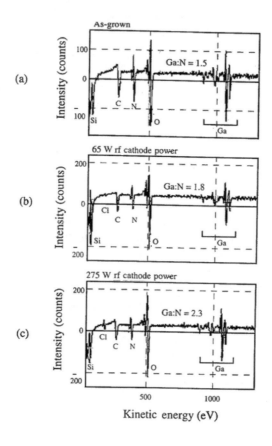

Figure 2.22. AES surface scans of GaN (a) before exposure to the plasma, (b) at 65 W (–120 V bias), and (c) 275 W rf-cathode power (–325 V bias), 1 mTorr, 170°C, and 850 W microwave power in an ECR-generated Cl_2–H_2 discharge

2.6.1 n-Gallium Nitride

The etching requirements for electronic devices are more demanding than those for photonic devices, at least from the electrical quality viewpoint. One of the most sensitive tests of near-surface electrical properties is the quality of rectifying contacts deposited on the etched surface. There has been relatively little work in this area to date. Ren and co-workers [84,86] found that rectifying contacts on ECR plasma etched GaN and InAlN surfaces were very leaky, though some improvement could be obtained by post-etch annealing at 400°C. Ping et al.[87] found that n-GaN surfaces had poor Schottky contact properties, but that plasma chemistries with a chemical component (e.g. Cl_2-based mixtures) produced less degradation than purely physical etching.

The layer structure and contact metals are shown schematically in Figure 2.23. The GaN was grown by rf plasma-assisted MBE on c-plane Al_2O_3 substrates. The Ti–Au ohmic contracts were patterned by lift-off and annealed at 750°C, producing contact resistances in the 10^{-5} Ω cm^{-2} range. Samples were exposed to either pure N_2 or H_2 discharges in a Plasma Therm 790 ICP system at a fixed pressure of 5 mTorr. The gases were injected into the ICP source at a flow rate of 15 sccm. The experimentally varied parameters were source power (300–1000 W) and rf chuck power (40–250 W), which control ion flux and ion energy respectively. In some cases the samples were either annealed in N_2 for 30 s at 300–850°C, or photoelectrochemically etched in 0.2 M KOH solutions at 25°C after plasma exposure. The Pt–Au Schottky metallization was then deposited through a stencil mask by e-beam evaporation. Current–voltage characteristics were recorded on an HP4145A parameter analyzer, and we defined the reverse breakdown voltage V_B as the voltage at which the leakage current was 10^{-3} A. We found in all cases that plasma exposure caused significant increases in forward and reverse current, with ideality factors increasing from typical values of 1.4–1.7 on control samples to >2. For this reason we were unable to extract meaningful values of either ideality factor or barrier height.

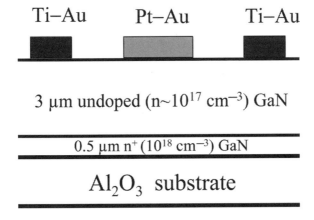

Figure 2.23. Schematic of GaN Schottky diode structure

Figure 2.24 shows a series of I–V characteristics from the GaN diodes fabricated on samples exposed to either H_2 or N_2 discharges at different source powers. It is clear that N_2 plasma exposure creates more degradation of the diode characteristics than does H_2 exposure. This implicates the ion mass ($^{28}N_2^+$, $^2H_2^+$ for the main positive ion species) as being more important in influencing the electrical properties of the GaN surface than a chemical effect, since H_2 would be likely to preferentially remove nitrogen from the GaN as NH_3.

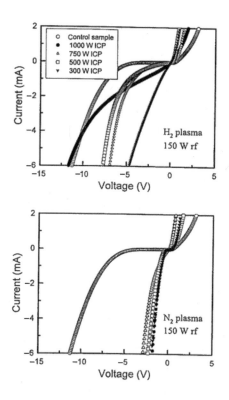

Figure 2.24. I–V characteristics from GaN diodes before and after H_2 (top) or N_2 (bottom) plasma exposure (150 W rf chuck power, 5 mTorr) at different ICP source powers

The variations of V_B of the diodes with the source power during plasma exposure are shown in Figure 2.25. For any exposure to the N_2 discharges V_B is severely reduced. By contrast, there is less degradation with the H_2 plasma to the lower average ion energy at those conditions, as shown at the bottom of Figure 2.25. The average ion energy is approximately equal to the sum of dc self-bias and plasma potential, with the latter being in the range −22 to −28 V as determined by Langmuir probe measurements. Ion-induced damage in GaN displays n-type conductivity and, in addition, the heavy N_2^+ ions are also more effective in preferential sputtering of the N relative to Ga, compared with the H_2^+ ions of similar energy.

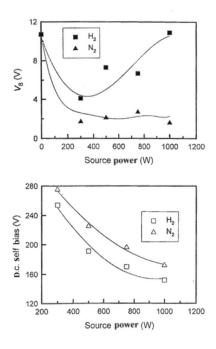

Figure 2.25. Variation of V_B in GaN diodes (top) and dc chuck self-bias (bottom) as a function of ICP source power in H_2 or N_2 plasmas (150 W rf chuck power, 5 mTorr)

Similar conclusions can be drawn from the data on the effect of increasing rf chuck power. There were once again very severe decreases in breakdown voltage and increase in leakage current. The V_B values fall by more than a factor of two even for very low self-biases, and emphasize how sensitive the GaN surface is to degradation by energetic ion bombardment. The degradation saturates beyond ~100 W chuck power, corresponding to ion energies of ~175 eV. We assume that, once the immediate surface becomes sufficiently damaged, the contact properties basically cannot be made any worse and the issue is then whether the damage depth increases with the different plasma parameters. Since ion energy appears to be a critical factor in creating the near-surface damage, we would expect damage depth to increase with ion energy in a nonetching process. In the case of simultaneous etching and damage creation (e.g. in Cl_2–Ar etch processing), higher etch rates would lead to lower amounts of residual damage because the disordered region would be partially removed.

The damage depth was established by photoelectrochemically wet etching different amounts of the plasma-exposed GaN surfaces, and then depositing the Pt–Au metal. Figure 2.26 (top) shows the effect on the I–V characteristics of this removal of different depths of GaN. There is a gradual restoration of the reverse breakdown voltage, as shown at the bottom of the figure. Note that the forward part of the characteristics worsens for removal of 260 Å of GaN, and shows signs of high series resistance. This would be consistent with the presence of a highly resistive region underneath the conducting near-surface layer, created by point-defect diffusion from the surface. A similar model applies to ion-damaged InP, *i.e.* a

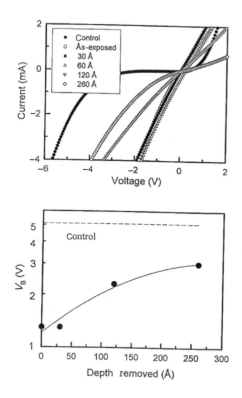

Figure 2.26. I–V characteristics from N_2-plasma-exposed GaN diodes before and after wet etch removal of different amounts of GaN prior to deposition of the Schottky contact (top) and variation of V_B as a function of the amount of material removed (bottom)

nonstoichiometric near-surface region (deficient in P in that case),[89] followed by a transition to a stoichiometric but point-defect-compensated region, and finally to unperturbed InP.

The fact that plasma exposure severely degraded the surface was clear from the AFM data. Exposure to the 500 W source power, 150 W rf chuck power (dc self-bias –221 V), 5 mTorr N_2 discharge increased the RMS surface roughness from 0.8 to 4.2 nm. Subsequent photoelectrochemical etching restored the initial morphology. However, we observed the onset of increasingly rough surfaces for deeper etch depths,[90] reducing a relatively inaccurate measure of how much of the surface had to be removed in order to restore the diode breakdown voltage to its original value. We were able to estimate this depth as ~600 ± 150 Å for the N_2 plasma conditions mentioned above.

Another method for trying to restore the electrical properties of the plasma-exposed surface is annealing. I–V data from annealed samples are shown in Figure 2.27. At the top are characteristics from samples which were plasma exposed (N_2, 500 W source power, 150 W rf chuck power, 5 mTorr), then annealed and the contact deposited. These samples show that increasing annealing temperature up to 750°C brings a substantial improvement in V_B (Figure 2.27, bottom). However, for annealing at 850°C the diode began to degrade, which is consistent with the temperature at which N_2 begins to be lost from the surface. In the case where the samples were exposed to the N_2 plasma, and then the Pt–Au contact was deposited prior to annealing, the I–V characteristics show continued worsening upon annealing (Figure 2.27, center). In this case, the Pt–Au contact is stable to 700°C on unetched samples. The poorer stability in etched samples could be related to the surface damage enhancing interfacial reaction between the Pt and GaN.

The main findings of this study can be summarized as follows:
1. There is a severe degradation in the electrical quality of GaN surfaces after ICP H_2 or N_2 discharge exposure. Under all conditions there is a strong reduction of V_B in diode structures, to the point at which the Schottky contacts show almost ohmic-like behavior. These observations are consistent with the creation of a conducting n-type surface layer resulting from energetic ion bombardment. Heavier ions (N_2^+) create more damage than lighter ions (H_2^+) in this situation, where damage is accumulating without any concurrent etching of the surface.

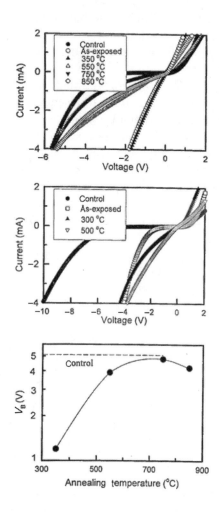

Figure 2.27. I–V characteristics from GaN diodes before and after N_2 plasma exposure (500 W source power, 150 W rf chuck power, 5 mTorr) and subsequent annealing either prior (top) or subsequent (center) to the deposition of the Schottky metallization. The variation of V_B in the samples annealed prior to metal deposition is shown at the bottom of the figure

2. The depth of the damage is approximately 600 Å, as judged by the return of the diode characteristics to their control values.
3. Annealing at 750°C is also effective in helping remove the effects of plasma exposure. Higher temperatures lead to degradation in GaN diode properties for uncapped anneals.

2.6.2 p-Gallium Nitride

The layer structure consisted of 1 μm of undoped GaN ($n \approx 5 \times 10^{16}$ cm^{-3}) grown on a c-plane Al$_2$O$_3$ substrate, followed by 0.3 μm of Mg-doped ($p \approx 10^{17}$ cm^{-3}) GaN. The samples were grown by rf plasma-assisted MBE. Ohmic contacts were formed with Ni–Au deposited by e-beam evaporation, followed by lift-off and annealing at 750°C. The GaN surface was then exposed for 1 min to ICP H$_2$ or Ar plasmas in a Plasma-Therm 790 System. The 2 MHz ICP source power was varied from 300 to 1400 W, while the 13.56 MHz rf chuck power was varied from 20 to 250 W. The former parameter controls the ion flux incident on the sample, and the latter controls the average ion energy. Prior to deposition of 250 μm diameter Ti–Pt–Au contacts through a stencil mask, the plasma-exposed surfaces were either annealed under N$_2$ in a rapid thermal annealing system, or immersed in boiling NaOH solutions to remove part of the surface. As reported previously, it is possible to etch damaged GaN in a self-limiting fashion in hot alkali or acid solutions. The I–V characteristics of the diodes were recorded on an HP 4145A parameter analyzer. The unetched control diodes have reverse breakdown voltages of ~2.5–4 V, depending on the wafer – these values were uniform (±12%) across a particular water.

Figure 2.28 shows the I–V characteristics from samples exposed to either H$_2$ (top) or Ar (bottom) ICP discharges (150 W rf chuck, 2 mTorr) as a function of source power. In both cases there is an increase in both the reverse breakdown voltage and the forward turn-on voltage, with these parameters increasing monotonically with the source power during plasma exposure.

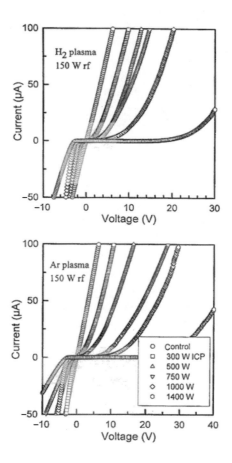

Figure 2.28. I–V characteristics from samples exposed to either H_2 (top) or Ar (bottom) ICP discharges (150 W rf chuck power) as a function of ICP source power prior to deposition of the Ti–Pt–Au contact

Figure 2.29 shows this increase in breakdown voltage as a function of source power, and also the variation of the chuck dc self-bias. As the source power increases, the ion density also increases and the higher plasma conductivity suppresses the dc bias developed. Note that the breakdown voltage of the diodes continues to increase even as this bias (and hence ion energy, which is the sum of this bias and the plasma potential) decreases. These results show that ion flux plays an important role in the change of diode electrical properties. The other key result is that Ar leads to consistently more of an increase in breakdown voltage, indicating that ion mass is important rather than any chemical effect related to removal of N_2 or NH_3 in the H_2 discharges.

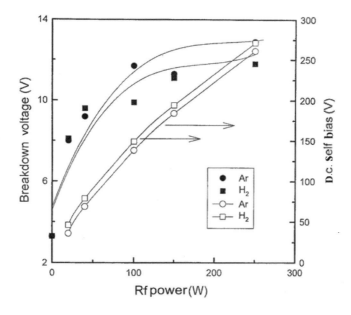

Figure 2.29. Variation of diode breakdown voltage in samples exposed to H_2 or Ar ICP discharges (150 W rf chuck power) at different ICP source powers prior to deposition of the Ti–Pt–Au contact. The dc chuck self-bias during plasma exposure is also shown

The increase in breakdown voltage on the p-GaN is due to a decrease in hole concentration in the near-surface region through the creation of shallow donor states. The key question is whether there is actually conversion to an n-type surface under any of the plasma condition. Figure 2.30 shows the forward turn-on characteristics of the p-GaN diodes exposed to different source power Ar discharge at low source power (300 W); the turn-on remains close to that of the unexposed control sample. However, there is a clear increase in the turn-on voltage at higher source powers, and in fact at ≥750 W the characteristics are those of an n–p junction. Under these conditions the concentration of plasma-induced shallow donors exceeds the hole concentration and there is surface conversion. In other words, the metal–p-GaN diode has become a metal–n-GaN–p-GaN junction. We always find that plasma-exposed GaN surfaces are N_2 deficient relative to their unexposed state; therefore, the obvious conclusion is that nitrogen vacancies create shallow donor levels. This is consistent with thermal annealing experiments, in which N_2 loss from the surface produced increased n-type conduction.

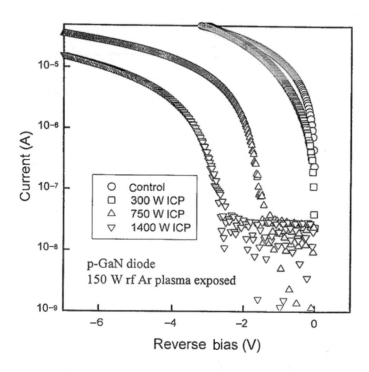

Figure 2.30. Forward turn-on characteristics of diodes exposed to ICP Ar discharges (150 W rf chuck power) at different ICP source powers prior to deposition of the Ti–Pt–Au contact

An important concern is the depth of the plasma-induced damage. We found we were able to etch p-GaN very slowly in boiling NaOH solutions, at rates that depended on the solution molarity (Figure 2.31) even without any plasma exposure of the material. This enabled us to measure directly the damage depth in plasma-exposed samples in two different ways.

The first method involved measuring the etch rate as a function of depth from the surface. Defective GaN resulting from plasma, thermal or implant damage can be wet chemically etched at rates much faster than undamaged material because the acid or base solutions are able to attack the broken or strained bonds present. Figure 2.32 shows the GaN etch rate as a function of depth in samples exposed to a 750 W source power, 150 W rf chuck power Ar discharge. The etch rate is a strong function of the depth from the surface and saturates between ~425 and 550 Å. Within this depth range the etch rate is returned to the "bulk" value characteristic of undamaged p-GaN.

2.6 Plasma-Induced Damage   137

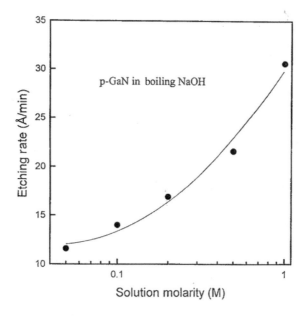

Figure 2.31. Wet etching rate of p-GaN in boiling NaOH solutions as a function of solution molarity

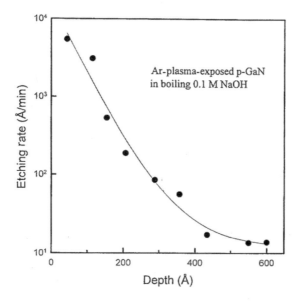

Figure 2.32. Wet etching rate of Ar-plasma-exposed (750 W source power, 150 W rf chuck power) GaN as a function of depth into the sample

The second method to establish damage depth, of course, is simply to measure the I–V characteristics after removing different amounts of material by wet etching prior to deposition of the rectifying contact. Figure 2.33 (top) shows the I–V characteristics from samples exposed to 750 W source power, 150 W rf chuck power (–160 V dc chuck bias) Ar discharges and subsequently wet etched to different depths using 0.1 M NaOH solutions before deposition of the Ti–Pt–Au contact. Figure 2.33 (bottom) shows the effect of the amount of material removed on the diode breakdown voltage. Within the experimental error of ±12%, the initial breakdown voltage is re-established in the range 400–450 Å. This is consistent with the depth obtained from the etch rate experiments described above. These values are also consistent with the damage depths we established in n-GaN diodes exposed to similar plasma conditions.

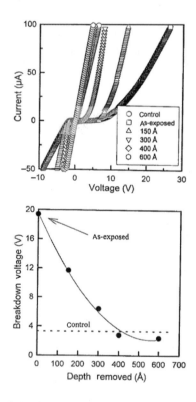

Figure 2.33. I–V characteristics from samples exposed to ICP Ar discharges (750 W source power, 150 W rf chuck power) and subsequently wet etched to different depths prior to deposition of the Ti–Pt–Au contact (top) and breakdown voltage as a function of depth removed (bottom)

2.6 Plasma-Induced Damage

The other method of removing plasma-induced damage is annealing. In these experiments we exposed the samples to the same type of plasma (Ar, 750 W source power, 150 W rf chuck power) and then annealed under N_2 at different temperatures. Figure 2.34 (top) shows the I–V characteristics of these different samples, and Figure 2.34 (bottom) shows the resulting breakdown voltages as a function of annealing temperature. On this wafer, plasma exposure caused an increase in breakdown voltage from ~2.5 to ~18 V. Subsequent annealing at 400°C initially decreased the breakdown voltage, but higher temperature produced a large increase. At temperatures above 700°C, the diode characteristics returned toward their initial values and were back to the control values by 900°C. This behavior is similar to that observed in implant-isolated compound semiconductors, where ion damage compensates the initial doping in the material, producing higher sheet resistance. In many instances the damage site density is larger than that needed to trap all of the free carriers, and trapped electrons or holes may move by hopping conduction. Annealing at higher temperatures removes some of the damage sites, but there are still enough to trap all the conduction electrons/holes. Under these conditions the hopping conduction is reduced and the sample sheet resistance actually increases. At still higher annealing temperatures the trap density falls below the conduction electron or hole concentrations and the latter are returned to their respective bands. Under these conditions the sample sheet resistance returns to its pre-implanted value. The difference in the plasma-exposed samples is that the incident ion energy is a few hundred electron-volts compared to a few hundred kilo-electron-volts in implant-isolated material. In the former case, the main electrically active defects produced are nitrogen vacancies near the surface, whereas in the latter case there will be vacancy and interstitial complexes produced in far greater numbers to far greater depths. In our previous work on plasma damage [30,31] in n-GaN we found that annealing at ~750°C almost returned the electrical properties to their initial values. If the same defects are present in both n- and p-type materials after plasma exposure, then this difference in annealing temperature may be a result of a Fermi level dependence to the annealing mechanism.

The main conclusions of this study may be summarized as follows :
1. The effect of either H_2 or Ar plasma exposure on p-GaN surfaces is to decrease the net acceptor concentration through creation of shallow donor levels, most likely N_V. At high ion fluxes or ion energies there can be type conversion of the initially p-type surface. The change in electrical properties is more pronounced with Ar than with H_2 plasmas under the same conditions.
2. Two different techniques for measuring the damage depth find it to be in the range 400–500 Å under our conditions. After removing this

amount of GaN, both the breakdown voltage and wet chemical etch rates are returned to their initial values.
3. Post-etch annealing in N_2 at 900°C restores the initial breakdown voltage on plasma-exposed p-GaN. Annealing at higher temperatures degraded the electrical properties, again most likely due to N_2 loss from the surface.

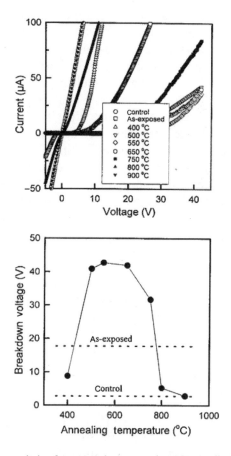

Figure 2.34. I–V characteristics from samples exposed to ICP Ar discharges (750 W source power, 150 W rf chuck power) and subsequently annealed at different temperatures prior to deposition of the Ti–Pt–Au contact (top) and breakdown voltage as a function of annealing temperature (bottom)

2.6.3 Schottky Diodes

Contrary to initial expectations, the surface of GaN is relatively sensitive to energetic ion bombardment or thermal degradation encountered during device processing. In particular, it can preferentially lose N_2, leaving a strong n-type conducting region. While dry etching has been used extensively for patterning of photonic devices (LEDs and LDs) and optoelectronic devices (UV detectors), there has been little work performed on understanding the electrical effects of ion-induced point defects or non-stoichiometric surfaces resulting from the plasma exposure. Several groups have reported increases in sheet resistance of GaN exposed to high-density plasmas, along with decreases in reverse breakdown voltage V_B and reductions in SBH ϕ_B in diodes formed on n-type GaN. In this latter case, low-bias forward currents were increased up to two orders of magnitude after exposure of the diode to pure Ar discharges. Conversely, while the rectifying contact properties were degraded by plasma exposure, the specific resistance of n-type ohmic contacts was improved. Similarly, in p-type GaN, the effect of Ar or H_2 high-density plasma exposure was to decrease the net acceptor concentration to depths of ~500 Å. At high ion fluxes or energies, there was type conversion of the initially p-GaN surface.

Dry etching is needed for a range of GaN electronic devices, including mesa diode rectifiers, thyristors and HBTs for high temperature, high power operation. These applications include control of power flow in utility grids, radar and electronic motor drives. It is critical to understand the depth and thermal stability of dry etch damage in both n- and p-type GaN and its effect on the I–V characteristics of simple diode structures.

In this section we report on a comparison of the effects of Cl_2–Ar and Ar ICP exposure on the electrical properties of n- and p-GaN Schottky diodes. In some cases it was found that Cl_2–Ar discharges could produce even more damage than pure Ar, due to the slightly higher ion energies involved. The damage saturates after a short exposure to either Cl_2–Ar or Ar discharges and is significant even for low ion energies. Annealing between 700 and 800°C restored ≥70% of the reverse breakdown voltage on n-GaN, while the damage depth was again established to be ~500 Å in p-GaN.

The GaN layers were grown by rf plasma-assisted MBE on *c*-plane Al_2O_3 substrates. The Ti–Al (for n-type) and Ni–Au (for p-type) ohmic contacts were patterned by lift-off and annealed at 750°C. The samples were exposed to either 10 Cl_2/5 sccm Ar or 15 sccm Ar ICP discharges in a Plasma-Therm ICP reactor at a fixed pressure of 3 mTorr. We investigated a range of rf chuck powers (25–250 W) and etch times (4–100 s), with a fixed source power of 500 W. In some cases, the samples were either annealed in N_2 for 30 s at 500-800°C or wet etching in 0.1 M NaOH solutions at ~100°C after plasma exposure. The Schottky metallization (Pt–Au in both cases) was then deposited through a stencil mask (ϕ = 70 or 90 μm) by e-beam evaporation. I–V characteristics were recorded on an HP 4145A parameter analyzer, and we defined the reverse breakdown voltage as that at which the leakage current was 10^{-3} A. The forward on-voltage V_F was defined as the voltage at which the forward current was 100 A·cm^{-2}. In all cases the ideality factors increased from 1.3–1.6 on control samples to >2 after plasma exposure, and thus we were unable to extract meaningful values of either barrier height or ideality factor.

Figure 2.35 shows a series of I–V characteristics from n-type GaN diodes fabricated on samples exposed to either Cl_2–Ar (top) or Ar (bottom) discharges at different rf chuck powers. There is a significant reduction in V_B under all conditions, with Ar producing less damage at low chuck powers. This is probably related to two factors: the slightly higher chuck bias with Cl_2–Ar due to the lower positive ion density in the plasma (Cl is more electronegative than Ar) and the heavier mass of the Cl_2^+ ions compared with Ar^+. This is consistent with our past data on the relative effects of N_2 and H_2 plasma exposure, in which ion mass was found to be more important in influencing the electrical properties of the GaN surface than any chemical effects.

The variations of V_B and V_F with the rf chuck power during plasma exposure are shown in Figure 2.36 (top). At powers ≤100 W, the Cl_2–Ar creates more degradation of V_B, as discussed above, while at higher powers the damage saturates. The average ion energy is the sum of dc self-bias (shown at the bottom of the figure) and plasma potential (which is about 22–25 eV under these conditions). Thus, for ion energies <~150 eV, Ar produces less damage than Cl_2/Ar, even though the etch rate with the latter is much higher. This is also reflected in the variation of V_F with rf chuck power.

2.6 Plasma-Induced Damage   143

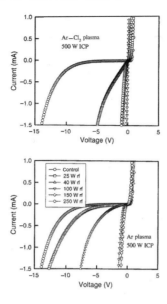

Figure 2.35. I–V characteristics from n-GaN samples exposed to ICP Cl_2–Ar (top) or Ar (bottom) discharges (500 W source power) as a function of rf chuck power prior to deposition of the rectifying contact

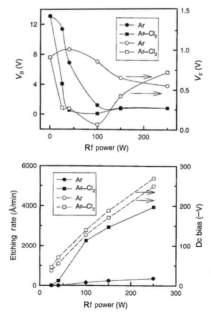

Figure 2.36. Variations of V_B and V_F (top) and of n-GaN etching rate (bottom) as a function of rf chuck power for n-GaN diodes exposed to ICP Cl_2–Ar discharges (500 W source power)

Figure 2.37 shows a series of I–V characteristics from n-type GaN diodes fabricated on samples exposed to the two different plasmas for different times at fixed rf chuck power (150 W) and source power (500 W). It is clear that the damage accumulates rapidly, with the I–V characteristics becoming linear at longer times. It should be remembered that this is damage accumulating ahead of the etch front.

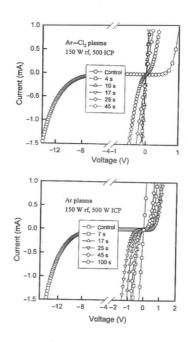

Figure 2.37. I–V characteristics from n-GaN samples exposed to ICP Cl_2–Ar (top) or Ar (bottom) discharges (150 W rf chuck power, 500 W source power) as a function of plasma exposure time prior to deposition of the rectifying contact

To examine the thermal stability of the etch damage, n-type samples were exposed to Ar or Cl_2–Ar discharges at a fixed source power (500 W) and rf chuck power 150 W), and then annealed at different temperatures prior to deposition of the rectifying contact. The annealing produces a significant recovery of the electrical properties for samples exposed to either type of plasma. The V_B values are shown in Figure 2.38, as a function of post-plasma exposure annealing temperature. Annealing temperatures between 700 and 800°C restore >70% of the original V_B value, but clearly annealing alone cannot remove all of the dry-etch-sinduced damage. Annealing temperatures above 800°C were found to lead to preferential loss of N_2 from the surface, with a concurrent degradation in V_B.

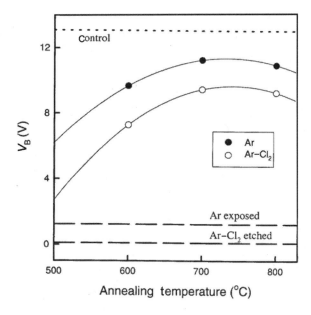

Figure 2.38. Variation of V_B in n-GaN diodes exposed to ICP Cl_2–Ar or Ar discharges (500 W source power, 100 W rf chuck power) with annealing temperature prior to deposition of the rectifying contact

Turning to p-GaN diodes, Figure 2.39 shows I–V characteristics from samples that were wet etched to various depths in NaOH solutions after exposure to either Cl_2–Ar or Ar discharges (500 W source power, 150 W rf chuck power, 1 min). For these plasma conditions we did not observe type conversion of the surface. However, we find that the damaged GaN can be effectively removed by immersion in hot NaOH, without the need for photo- or electro-chemical assistance of the etching. The V_B values increase on p-GaN after plasma exposure due to introduction of shallow donor states that reduce the wet acceptor concentration.

Figure 2.40 shows two methods for determining the depth of the damaged region in p-GaN diodes. At top is a plot of the variation of V_F and V_B with the depth of material removed by NaOH etching. The values of both parameters are returned to their control values by depths of 500–600 Å. What is clear from this data is that the immediate surface is not where the p-doping concentration is most affected, since the maximum values peak at depths of 300–400 Å. This suggests that N_V or other compensating defects created at the surface diffuse rapidly into this region even near room temperature. This is consistent with results in other semiconductors, where damage depths are typically found to be many times deeper than the projected range of incident ions. The bottom part of Figure 2.40 shows the wet etch depth in plasma-damaged p-GaN as a function of etching time. The etch depth saturates at depths of 500–600 Å, consistent with the electrical data. It has previously been shown that the wet etch depth on thermally or ion-damaged GaN was self-limiting. This is most likely a result of the fact that defective or broken bonds in the material are readily attached by the acid or base, whereas in undamaged GaN the etch rate is negligible.

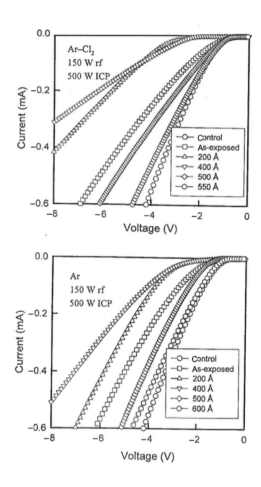

Figure 2.39. I–V characteristics from p-GaN samples exposed to ICP Cl_2–Ar (top) or Ar (bottom) discharges (500 W source power, 150 W rf chuck power) and wet etched in boiling NaOH to different depths prior to deposition of the rectifying contact

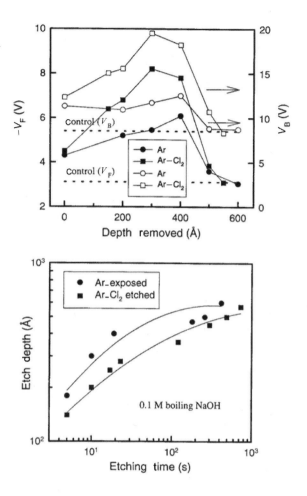

Figure 2.40. Variation of V_B and V_F (top) with depth of p-GaN removed by wet etching prior to deposition of the rectifying contact, and wet etch depth versus etch time in boiling NaOH solutions for plasma-damaged p-GaN (bottom)

The main findings of our study may be summarized as follows:
1. Large changes in V_B and V_F of n- and p-GaN Schottky diodes were observed after exposure to both Cl_2–Ar and Ar ICP discharges. In some cases the electrical properties are more degraded with Cl_2–Ar even though this plasma chemistry has a much higher etch rate.
2. The damage accumulates in the near-surface even for very short exposure times (4 s). The damage depth was established to be 500–600 Å from both the changes in electrical properties and the depth dependence of wet etch rate.

3. Annealing in the range 700–800°C partially restores V_B in n-GaN diodes, but full recovery can only be achieved with an additional wet etch step for removal of the damaged material. The combination of annealing and a wet etch clean-up step looks very promising for GaN device fabrication.

2.6.4 p-n Junctions

Layer structures were grown by MOCVD on c-plane Al_2O_3 substrates at 1040°C. The structure consisted of a low-temperature (530°C) GaN buffer, 1.2 μm of n-GaN (2×10^{17} cm^{-3}, Si-doped) 0.5 μm of nominally undoped ($n\approx10^{16}$ cm^{-3}) GaN and 1.0 μm of p-GaN ($N_A\approx5\times10^{19}$ cm^{-3}, Mg-doped). The p-ohmic metal (Ni–Au) was deposited by e-beam evaporation and lift-off, then alloyed at 750°C. A mesa was then formed by $BCl_3/Cl_2/Ar$ (8/32/5 sccm) ICP etching to a depth of 1.6 μm under different plasma conditions in order to examine the effect of ion energy and ion flux. The ICP reactor was a load-locked Plasma-Therm SLR 770, which used a 2 MHz, three-turn coil ICP source. All samples were mounted using a thermally conductive paste on an anodized Al carrier that was clamped to the cathode and cooled with He gas. The ion energy or dc bias was defined by superimposing an rf bias (13.56 MHz) on the sample. The n-type ohmic metallization (Ti–Al) was then deposited. Reverse I–V measurements were made on 300 μm diameter diodes with an HP 4145B semiconductor parameter analyzer. In this study, the reverse leakage current was measured at a bias of –30 V. Etch rates were calculated from bulk GaN samples patterned with AZ-4330 photoresist. The depth of etched features was measured with an Alpha-step stylus profilometer after the photoresist was removed. Etch profile and surface morphology were analyzed with SEM and AFM.

Figure 2.41 shows the effect of dc chuck bias on the reverse junction leakage current, along with the corresponding GaN etch rates. There is little effect on the current below chuck biases of –250 V. This corresponds to an ion energy of approximately –275 eV, since this energy is the sum of chuck bias and plasma potential (about –25 eV in this tool under these conditions). The reverse current decreases slightly as the dc self-bias is increased from –25 to –50 V. This may result from the sharp increase in etch rate, which leads to faster removal of near-surface damage. The reverse current increases rapidly above –275 V ion energy, which is a clear indication of severe damage accumulating on the sidewall. The damage probably takes the form of point defects such as nitrogen vacancies, which increase the n-type conductivity of the surface. The total reverse current density J_R is the sum of three components, namely diffusion, generation and surface leakage, according to

$$J_R = \left(\frac{eD_h}{l_h N_D} + \frac{eD_e}{l_e N_A}\right) n_i^2 + \frac{eWn_i}{\tau_g} + J_{SL} \qquad (2.1)$$

where e is the electronic charge, $D_{e,h}$ are the diffusion coefficients of electrons or holes, $l_{e,h}$ are the lengths of the n and p regions outside the depletion region in a p–n junction, $N_{D,A}$ are the donor/acceptor concentrations on either side of the junction, n_i is the intrinsic carrier concentration, W the depletion, τ_g the thermal generation lifetime of carriers and J_{SL} is the surface current component, which is bias dependent. The latter component is most affected by the dry etch process, and dominates the reverse leakage in diodes etched in high ion energies.

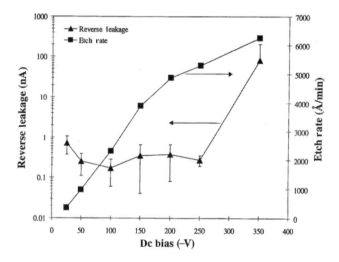

Figure 2.41. Reverse leakage current measured at –30 V for GaN p–i–n junctions etched in ICP 32Cl$_2$/8BCl$_3$/5Ar discharges (500 W source power, 2 mTorr), as a function of dc chuck self-bias

GaN sidewall profiles and etch morphologies have been evaluated from previous results as a function of dc bias. The etch becomes more anisotropic as the dc bias increased from –50 to –150 V due to the perpendicular nature of the ion bombardment energies. However, at –300 V dc bias, a tiered etch profile with vertical striations in the sidewall was observed due to erosion of the mask edge under high ion bombardment energies. The physical degradation (both profile and morphology) of the etched sidewall at –300 V could help explain higher reverse leakage currents above –250 V dc bias. Under high bias conditions, more energetic ions scattering from the surface could strike the sidewalls with significant momentum, thus increasing the likelihood of increased damage and higher reverse leakage currents. Under low bias conditions, the sidewall profile is less anisotropic, implying increased lateral etching of the GaN (undercutting of the mask). Under these conditions the etch process becomes dominated by the chemical component of the etch mechanism, which may account for the slightly higher reverse leakage observed at –25 V dc bias.

Figure 2.42 shows the effect of ICP source power on the junction reverse leakage current. The plasma flux is proportional to source power. In this experiment the ion energy was held constant at –100 V dc bias. There is minimal effect on leakage current for source powers ≤500 W, with severe degradation of the junction characteristics at higher powers even though the GaN etch rate continues to increase. This is an important result, because it shows that the conditions that produce the highest etch rate are not necessarily those that lead to the least damage. Increased sidewall damage under high plasma flux conditions may be due to increased ion scattering as well as more interactions of reactive neutrals with the sidewall of the mesa. SEM micrographs from bulk GaN samples also show a degradation of sidewall profile under high ICP source power conditions. At an ICP source

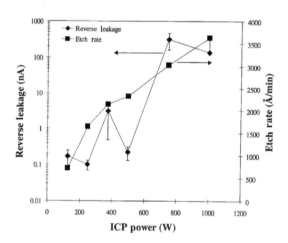

Figure 2.42. Reverse leakage current measured at –30 V for GaN p–i–n junctions etched in ICP 32Cl$_2$/8BCl$_3$/5Ar discharges (–100 V dc chuck self-bias, 2 mTorr), as a function of source power

2.6 Plasma-Induced Damage   151

power of 1000 W, the sidewall has a tiered profile with vertical striations possibly due to erosion of the mask edge. However, sidewall profiles at 250 and 500 W looked reasonably anisotropic and smooth.

Reverse leakage currents were relatively insensitive to chemistry effects in a Cl_2–BCl_3–Ar ICP discharge. The reverse leakage current ranged between ~10 and 40 nA as the Cl_2 changed from 0 to 100%. This is not too surprising given that BCl_3 ions will be the heaviest ions in the discharge under all these conditions, and we expect ion damage to be dominated at this flux. The reverse leakage currents were measured from a different GaN wafer than the other samples used in this study. The surface morphology for the as-grown wafer was significantly higher for this sample and may account for higher reverse leakage currents measured under the standard conditions. Notice the GaN etch rate increased as Cl_2 was added to the BCl_3–Ar plasma up to 80%. In Cl_2–Ar plasma the GaN etch rate decreased due to lower concentrations of reactive Cl neutrals. Etch profiles were relatively anisotropic and smooth except for the Cl_2–Ar plasma where the etch was slightly rough.

Two samples were annealed in this study to determine whether the defects caused by plasma-induced damage to the p–n junction could be removed and low reverse leakage currents recovered. The first sample was initially exposed to the following ICP conditions: 32 sccm Cl_2, 8 sccm BCl_3, 5 sccm Ar, 500 W ICP power, –300 V dc bias and 2 mTorr pressure. The reverse leakage remained essentially constant up to 600°C, where the reverse leakage increased by more than an order of magnitude (see Figure 2.43). (Note, all reverse leakage data were taken at –30 V except for the 600°C data, which was taken at lower voltages due to breakdown at –30 V.) A similar trend was observed for the second sample. Although there was much more scatter in the data, the sample was exposed to the same ICP conditions with the exceptions of 750 W ICP source power and –100 V dc bias. The inability to remove damage from these samples may be due to anneal temperatures which were not high enough. Improved breakdown voltages for dry-etched n- and p-GaN Schottky diodes annealed in the range of 400 to 700°C; have been observed [15]. However, anneal temperatures >800°C were needed to produce near-complete recovery in breakdown voltage.

In summary, there are high-density plasma etching conditions for GaN where there is minimal degradation in the reverse leakage current of p–i–n mesa diodes. Both ion energy and ion flux are important in determining the magnitude of this current, and a high etch rate is not necessarily the best choice for minimizing dry-etch damage.

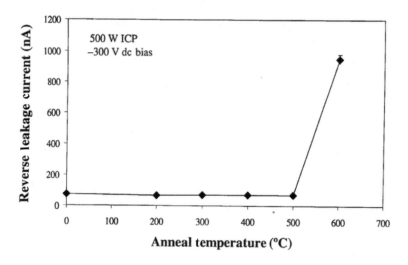

Figure 2.43. Reverse leakage current measured at –30 V for GaN p–i–n junctions etched in ICP 32 sccm Cl_2–8 sccm BCl_3–5 sccm Ar discharges (–300 V dc chuck self-bias, 500 W ICP source power, 2 mTorr), as a function of anneal temperature

2.7 Device Processing

2.7.1 Microdisk Lasers

A novel laser structure is the microdisk geometry, which does not require facet formation. These lasers should, in principle, have low thresholds because of their small active volume. While microcylinder geometries are possible, superior performance is expected when the active disk region sits only on a thin support post or pedestal. To fabricate this latter geometry, it is necessary to have a selective wet etch for the material under the active layer. A microcylinder is initially formed by anisotropic dry etching. We have employed ECR or ICP Cl_2–Ar discharges to produce the initial vertical etch. The undercut is then produced by use of KOH solutions at ~80°C to selectively etch the AlN buffer layer on which the InGaN–GaN quantum well is grown. SEM micrographs of two different lasers are shown in Figure 2.44. In both cases we used an upper cladding layer of AlGaN, which was etched somewhat slower than the pure AlN bottom cladding layer.

2.7 Device Processing   153

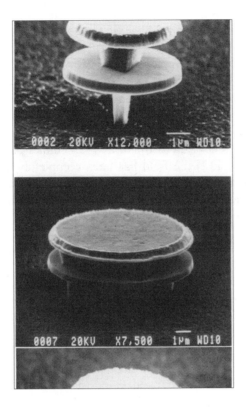

Figure 2.44. SEM micrographs of GaN–InGaN–AlN microdisk laser structures

2.7.2 Ridge Waveguide Lasers

The achievement of CW GaN–InGaN LDs has tremendous technological significance. For commercially acceptable laser lifetimes (typically ≥10,000 h), there is immediate application in the compact disk data storage market. The recording and reading of data on these disks are currently performed with near-infrared (~780 nm) laser diodes. The switch to the much shorter wavelength (~400 nm) GaN-based laser diodes will allow higher recording densities (by $\sim(780/400)^2$ or almost a factor of four). There is also a large potential market in projection displays, where LDs with the three primary colors (red, green and clue) would replace the existing liquid-crystal modulation system. The laser-based system would have advantages in terms of greater design simplicity, lower cost and broader color coverage. The key development is the need to develop reliable green InGaN laser diodes. The high output power of GaN-based lasers and fast off/on times should also have advantages for improved printer technology, with higher resolution than existing systems based on infrared lasers. In underwater military systems, GaN lasers may have application for covert communications because of a transmission passband in water between 450 and 550 nm.

While a number of groups have now reported room-temperature lasers in the InGaN–GaN–AlGaN heterostructure system under pulsed [91–103] and CW operation,[104–112] the field has been completely dominated by Nakamura and co-workers[91–96,98,99,103–111]. The growth is performed by MOCVD, generally at atmospheric pressure. Initial structures were grown on c-plane (0001) sapphire, with a low-temperature (550°C) GaN buffer, a thick n^+-GaN lower contact region, an n^+-InGaN strain-relief layer, an n^+-AlGaN cladding layer, a light-guiding region of GaN, then a multiquantum well region consisting of Si-doped $In_{0.15}Ga_{0.85}N$ wells separated by Si-doped $In_{0.02}Ga_{0.98}N$ barriers. The p side of the device consisted of sequential layers of p-AlGaN, p^+-GaN light-guiding, p-$Al_{0.09}Ga_{0.92}N$ cladding and p^+-GaN contact. A ridge geometry was fabricated by dry etching in most cases (material removed down to the p-$Al_{0.08}Ga_{0.92}N$ layer), followed by dry etching, cleaving or polishing to form a mirror facet. These facets are coated (with TiO_2–SiO_2 in the Nichia case) to reduce laser threshold, while Ni–Au (p type) and Ti–Al (n type) were employed for ohmic metallization.

For this type of structure, threshold current densities are typically ≥4 kA cm^{-2} with operating voltages of ≥5 V at the threshold current. The emission mechanism is still the subject of intense study, but may be related to localization of excitons at compositional fluctuations (leading to potential minima in the band structure) in the InGaN wells.[113–115] These devices display relatively short lifetimes under CW operation, typically tens to hundreds of hours. The failure mechanism is most commonly short-circuiting of the p–n junction, a result of p-contact metallization punch through. It is not that surprising that in this high defect density material that the metal can migrate down threading dislocations or voids under high drive-current conditions. The threshold carrier density of the LDs on sapphire are typically $\sim 10^{20}$ cm^{-3}, well above the theoretical values ($\sim 10^{19}$ cm^{-3}).[115–117]

A major breakthrough in LD lifetime occurred with two changes to the growth. The first was replacement of the AlGaN cladding layers with AlGaN–GaN strained-layer superlattices, combined with modulation doping. These changes had the effect of reducing formation of cracks that often occurred in the AlGaN, and also to reduce the diode operating voltage.[109] The second was the use of epitaxial lateral overgrowth (ELOG).[110,118,119] In this technique, GaN is selectively grown on an SiO$_2$-masked GaN–Al$_2$O$_3$ structure. After ~10 μm of GaN is deposited over the SiO$_2$ stripes, it coalesces to produce a flat surface.[119] For a sufficiently wide stripe width, the dislocation density becomes negligible, compared with $\geq 10^9$cm^{-2} in the window regions. The laser itself is fabricated slightly off-center from the mask regions, due to gaps that occur there due to imperfect coalescence of the GaN. These devices have lower threshold current density ($\leq$4 kA cm^{-2}) and operating voltage (4–6 V) and much longer (10,000 h) room-temperature lifetimes. The reduction in threading dislocation density dramatically changes the lifetime, since the p metal no longer has a direct path for shorting out the junction during operation. The carrier density at threshold is also reduced to ~3×10^{19} cm^{-3}, not far above the expected values. Output power >400 mW, and lifetime >160 h at 30 mW constant output power has been reported.

Subsequent work from Nichia has focused on growth of the LDs on quasi-GaN substrates. The thick (100–200 μm) GaN is grown on ELOG structures by either MOCVD or hydrid (Vapor Phase Epitaxy VPE). The sapphire substrate is then removed by polishing, to leave a free-standing GaN substrate. The mirror facet can then be formed by cleaving. The GaN substrate has better thermal conductivity than sapphire.

One of the most important features of the etching of the ridge waveguide is the smoothness of the sidewall. Figure 2.45 shows SEM micrographs of features etched into pure GaN, using a SiN$_x$ mask and an ICP Cl$_2$–Ar discharge at moderate powers (500 W source power, 150 W rf chuck power). While the sidewalls are reasonably vertical, one can see striations, which result from roughness on the photoresist mask used to pattern the SiN$_x$. Another problem than can occur is illustrated in the SEM micrograph at the top of Figure 2.46. In this, a very high ion energy was employed during the etching, leading to roughening of the feature sidewall. This problem is absent when ion energies below approximately 200 eV are employed, as shown in the micrograph at the bottom of Figure 2.46.

156   Dry Etching of Gallium Nitride and Related Materials

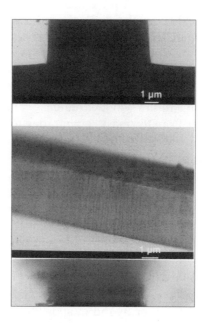

Figure 2.45. SEM micrograph of dry-etched GaN feature

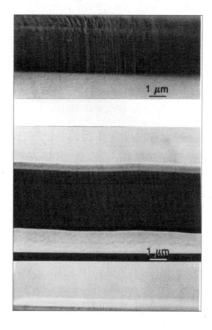

Figure 2.46. SEM micrographs of features etched into GaN at high (top) or moderate (bottom) ion energy

When careful attention is paid to the lithography, the etching of the SiN_x mask, and also the etching of the nitride laser structure, then results like those shown in the SEM micrographs of Figure 2.47 are obtained. The active region of the laser is visible as the horizontal lines along the middle of the sidewall.

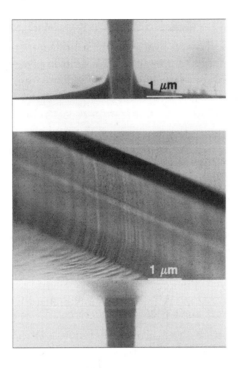

Figure 2.47. SEM micrographs of dry etched GaN–InGaN–GaN ridge waveguide laser structure

2.7.3 Heterojunction Bipolar Transistors

Wide bandgap semiconductor HBTs are attractive candidates for applications in high-frequency switching, communications and radar. While FETs can be used for these same applications [120–122], HBTs have better linearity, higher current densities and excellent threshold voltage uniformity. The GaN–AlGaN system is particularly attractive because of its outstanding transport properties and the experience base that has developed as a result of the success of LEDs [123], LDs [124–126] and UV detectors [127] fabricated from AlGaInN materials. GaN–SiC HBTs have been reported with excellent high-temperature (535°C) performance.[128] Recently, several reports have appeared on operation of GaN–AlGaN HBTs.[129,130] In one case the extrinsic base resistance was reduced through a selective regrowth of GaN(Mg), and devices with 3×20 μm^2 emitters showed dc current gain of ~3 at 25°C.[129] In work from our group, GaN–AlGaN HBTs have been fabricated with a non-self-aligned, low-damage dry-etch process based on that developed for the GaAs–AlGaAs, GaAs–InGaP and InGaAs–AlInAs systems.[131] The performance of GaN–AlGaN devices fabricated by that method also showed low gains at room temperature, typically ≤3. When operated at higher temperatures the gain improved, reaching ~10 at 300°C as more acceptors in the base region became ionized and the base resistance decreased.

In this section we review the fabrication process for GaN–AlGaN HBTs, examine the temperature dependence of the p-ohmic contacts and report measurements of typical background impurity concentrations, determined by SIMS.

Structures grown by two different methods were examined. In the first, rf plasma-assisted MBE at a rate of ~0.5 $\mu m\cdot h^{-1}$ was used to grow the HBT structure on top of a 2 μm thickness undoped GaN buffer that was grown on c-plane (0001) sapphire.[132] An 8000 Å thickness GaN subcollector (Si ~10^{18} cm^{-3}) was followed by a 5000 Å thickness GaN collector (Si ~10^{17} cm^{-3}), a 1500 Å thickness GaN base (Mg acceptor concentration ~10^{18} cm^{-3}), a 1000 Å thickness $Al_{0.15}Ga_{0.85}N$ emitter (Si ~5×10^{17} cm^{-3}), and a 500 Å grade to a 2000 Å thickness GaN contact layer (Si ~8×10^{18} cm^{-3}).

The second structure was grown by MOCVD on c-plane sapphire [133,134], using trimethylgallium, trimethylaluminum and ammonia as the precursors and high purity H_2 as the carrier gas. The growth process has been described in detail previously.[134] The basic layer structure is shown in Figure 2.48.

The process flow for device fabrication is shown schematically in Figure 2.49. First, the emitter metal (Ta–Al–Pt–Au) is patterned by lift-off and used as an etch mask for the fabrication of the emitter mesa. The dry etching was performed in a Plasma Therm 770 ICP system using Cl_2–Ar discharges. The process pressure was 5 mTorr and the source was excited with 300 W of 2 MHz power. This power controlled the ion flux and neutral density, while the incident ion energy was controlled by application of 40 W of 13.56 MHz power to the sample chuck. Base metallization of Ni–Pt–Au was patterned by lift-off, and then the mesa formed by dry etching. The etch rate of GaN under our conditions was ~1100 Å min^{-1}, and was terminated at the subcollector where Ti–Al–Pt–Au metallization was deposited. The contacts were alloyed at 700–800°C.

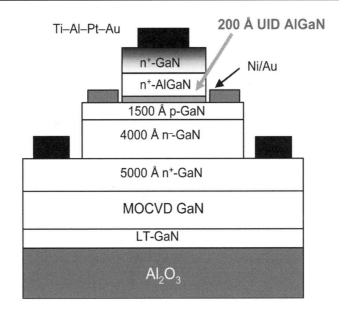

Figure 2.48. Schematic of MOCVD-grown GaN–AlGaN HBT

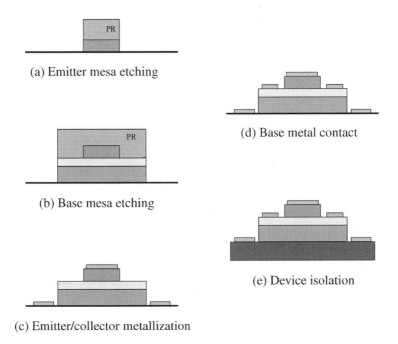

Figure 2.49. Schematic process sequence for GaN–AlGaN HBT

It has been firmly established that high specific contact resistivities are a limiting factor in GaN-based device performance, and in particular the p-ohmic contact. We examined the alloying tezmperature dependence of the I–V characteristics for several different p-metal schemes. The as-deposited contacts are rectifying. Annealing at progressively higher temperatures produced a significant improvement. But even for 800°C anneals the contacts were not purely ohmic when measured at room temperature. This is consistent with past data, showing that p-metallization on GaN is often better described as a leaky Schottky contact.

As the measurement temperature is increased, the hole concentration in the p-GaN increases through higher ionization efficiency of the Mg acceptors. For example, the hole concentration would increase from ~10% of the acceptor density at 25°C to ~60% at 300°C, based on Fermi–Dirac statistics. The p-contact becomes truly ohmic at ≥300°C. From transmission line measurements, we found $\rho_c \approx 2 \times 10^{-2} \times \Omega$ cm^{-2} at this temperature. This indicates that the GaN–AlGaN HBT will perform better at elevated temperatures, where the base contact resistivity is lower. The contact barrier is of order 0.5 eV, whereas the Mg acceptor has an ionization level of 0.18 eV.

The device performance of both the MBE- and MOCVD-grown devices was similar, namely a common-emitter current gain of ≤3 at 25°C, increasing to ~10 at 300°C. In both devices the performance was still limited by the base resistance, and methods to increase the base doping and lower the extrinsic resistance in this region will be critical for future efforts in this area. The common base current gain α was in the range 0.75 (25°C) to 0.9 (300°C), indicating that the base transport factor is close to unity and that I_B is dominated by re-injection to the emitter.

Another important aspect of the realization of GaN–AlGaN HBTs is confinement of the Mg doping to the base. If the p-type spills over into the relatively lightly doped emitter, then the junction is displaced and the advantage of the heterostructure is lost.

In summary, GaN–AlGaN HBTs have been fabricated both on MBE- and MOCVD-grown material, and they display similar performance, *i.e.* a common-emitter current gain of ~10 when operated at ~300°C. However, junction leakage is also higher at elevated temperatures, which is a major drawback in most applications. The fabrication process developed previously for other compound semiconductor systems works well for the GaN–AlGaN system, with the main difference being higher annealing temperatures required for the ohmic contacts. The device performance is still limited by the base doping for both the MBE and MOCVD structures.

2.7.4 Field Effect Transistors

There has been much recent attention on development of AlGaN–GaN heterostructure FETs (HFETs) for high frequency and high power application.[135–140] Both enhancement and depletion mode devices have been demonstrated, with gate lengths down to 0.2 μm. Excellent dc performance has been reported up to 360°C [140], and the best devices have a maximum frequency of oscillation f_{max} of 77 GHz at room temperature.[137] Even better speed performance could be expected from InAlN channel structures, both because of the superior transport properties and the ability to use highly doped $In_xAl_{1-x}N$ ($x = 0 \rightarrow 1$) graded contact layers which should produce low specific contact layers and, consequently, contact resistivities. We have previously demonstrated that nonalloyed Ti–Pt–Au metal on degenerately doped InN ($n = 5\times10^{20}$ cm^{-3}) has $\rho_c \approx 1.8 \; 10^{17} \; \Omega \; cm^2$.[141] While MOCVD has generally been employed for growth of nitride-based photonic devices and for most of the prototype electronic devices [142–144], the ability of the molecular beam techniques to control layer thickness and incorporate higher In concentration in the ternary alloys is well suited to growth of HFET structures.[145,146]

The exceptional chemical stability of the nitrides has meant that dry etching must be employed for patterning. To date, most of the work in this area has concentrated on achievement of higher etch rates with minimal mask erosion, in particular because a key application is formation of dry-etched layer facets. In that case, etch rate, etch anisotropy and sidewall smoothness are the most important parameters, and little attention has been paid to the effect of dry etching on the stoichiometry and electrical properties of the nitride surface.

In these experiments, we used an InAlN and GaN FET structure as a test vehicle for measuring the effect of ECR BCl_3-based dry etching on the surface properties of InAlN and GaN. Preferential loss of N leads to roughened morphologies and creation of a thin n^+ surface layer which degrades the rectifying properties of subsequently deposited metal contacts.

The InAlN samples were grown by MOMBE on 2 inch diameter GaAs substrates using a WAVEMAT ECR N_2 plasma and metalorganic Group III precursors (trimethylamine alane, triethylindium). A low-temperature (~400°C) AlN nucleation layer was followed by a 500 Å thickness AlN buffer layer grown at 700°C. The $In_{0.3}Al_{0.7}N$ channel layer (~5×10^{17} cm^{-2}) was 500 Å thick, and then an ohmic contact layer was produced by grading to pure InN over a distance of ~500 Å.

The GaN layer structure was grown on double-sided polished c-Al_2O_3 substrates prepared initially by HCl–HNO_3–H_2O cleaning and an *in situ* H_2 bake at 1070°C. A GaN buffer <300 Å thick was grown at 500°C and crystallized by ramping the temperature to 1040°C, where trimethylgallium and ammonia were again used to grow ~1.5 μm of undoped GaN ($n < 3\times10^{16}$ cm^{-3}), a 2000 Å channel ($n = 2\times10^{17}$ cm^{-3}) and a 1000 Å contact layer ($n = 1\times10^{18}$ cm^{-3}).

FET structures were fabricated by depositing TiPtAu source/drain ohmic contacts, which were protected by photoresists. The gate mesa was formed by dry etching down to the InAlN or n-GaN channel using an ECR BCl_3 or BCl_3–N_2 plasma chemistry. During this process, we noticed that the total conductivity between the ohmic contacts did not decrease under some conditions. CH_4–H_2 etch chemistry was also studied. To simulate the effects of this process, we exposed the FET substrates to D_2 plasma; we saw strong reductions in sample conductivity. The incorporation of D_2 into the InAlN was measured by SIMS. Changes to the surface stoichiometry were measured by AES. All plasma processes were carried out in a Plasma Therm SLR 770 System with an Astex 7700 low-profile ECR source operating at 500 W. The samples were clamped to an rf-powered, He backside cooled chuck, which was left at floating potential (about −30 V) relative to the body of the plasma.

Upon dry etch removal of the InAlN capping layer, a Pt–Ti–Pt–Au gate contact was deposited on the exposed InAlN to complete the FET processing. If pure BCl_3 was employed as the plasma chemistry, then we observed ohmic and not rectifying behavior for the gate contact. If BCl_3–N_2 was used, then there was some improvement in the gate characteristics. A subsequent attempt at a wet-etch clean-up using either H_2O_2–HCl or H_2O_2–HCl produced a reverse breakdown in excess of 2 V (Figure 2.50). These results suggest that the InAlN surface becomes nonstoichiometric during the dry etch step, and that addition of N_2 retards some of this effect.

Figure 2.51 shows the I_{DS} values obtained as a function of dry etch time in ECR discharges of either BCl_3 or BCl_3–N_2. In the former case the current does not decrease as material is etched away, suggesting that a conducting surface layer is continually being created. By contrast, BCl_3–N_2 plasma chemistry does reduce the drain–source current as expected, even though the breakdown characteristics of gate metal deposited on this surface are much poorer than would be expected.

From the AFM studies of the InAlN gate contact layer surface after removal of the $In_xAl_{1-x}N$ contact layer surface in BCl_3, BCl_3–N_2 or

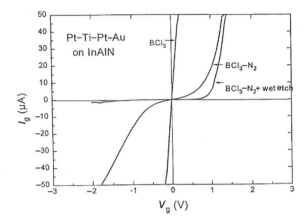

Figure 2.50. I–V characteristics of Pt–TiPt–Au contacts on InAlN exposed to different ECR plasmas

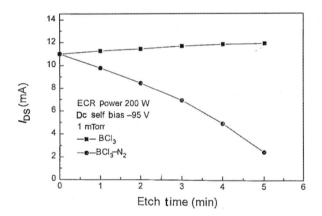

Figure 2.51. I_{DS} values at 5 V bias for InAlN FETs etched for various times in BCl_3 or BCl_3–N_2 ECR plasmas

BCl_3–N_2 plus wet etch, we observed that the mean surface roughness is worse for the former two chemistries, indicating that preferential loss of N is probably occurring during the dry etch step. To try to remove the Group-III-enriched region, the samples were rinsed in a 1:1 HCl/H_2O_2 solution at 50°C for 1 min. This step did increase the reverse breakdown voltage (~2.5 V) compared with dry-etched-only samples (<1 V), but did not produce a completely damage-free surface. This is because the HCl–H_2O_2 does not remove the InAlN immediately below the surface, which is closer to stoichiometry but is still defective.

While BCl_3–N_2 produces less enrichment than pure BCl_3, there is clearly the presence of a defective layer that prevents achievement of acceptable rectifying contacts. From the I–V measurements, we believe this defective layer is probably strong n-type, in analogy with the situation with InP described earlier. At this stage, there is no available wet etch solution for InAlN that could be employed to completely remove the nonstoichiometric layer in the type of clean-up step commonly used in other III–V materials. Other possible solutions to this problem include use of a higher Al concentration in the stop layer, which should be more resistant to nitrogen loss, or employment of a layer structure that avoids the need for gate recess.

Figure 2.52 shows the gate current–voltage characteristics when the gate metal is deposited on the as-etched GaN surface. The Schottky contact is extremely leaky, with poor breakdown voltage. We believe this is caused by the presence of a highly conducting N–deficient surface, similar to the situation encountered on dry etching InP where preferential loss of P produces a metal-rich surface which precludes achievement of rectifying contacts. AES analysis of the etched GaN surface showed an increasing Ga-to-N ratio (from 1.7 to 2.0 in terms of raw counts) upon etching. However, a 5 min anneal at 400°C under N_2 was sufficient to produce excellent rectifying contacts, with a gate breakdown of ~25 V (Figure 2.53). We believe the presence of the conducting surface layer after etching is a strong contributing factor to the excellent ρ_c values reported by Lin *et al.* [145] for contacts on a reactively ion etched n-GaN.

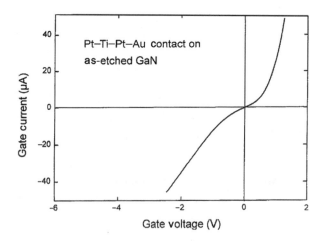

Figure 2.52. I–V characteristic on ECR BCl_3-etched GaN

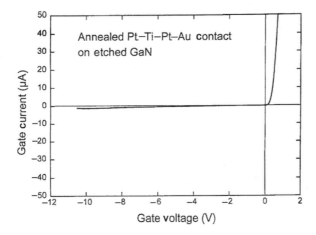

Figure 2.53. I–V characteristic on ECR BCl$_3$-etched GaN annealed at 400°C prior to deposition of the gate metal

The drain I–V characteristics of the 1×50 μm² MESFET are shown in Figure 2.54. The drain–source breakdown was –20 V, with a threshold voltage of –0.3 V. The device displays good pinch-off and no slope to the I–V curves due to gate leakage, indicating that the anneal treatment is sufficient to restore the surface breakdown characteristics. We believe these devices are well suited for high power applications, since GaN is a robust material and the contract metallizations employed are also very stable.

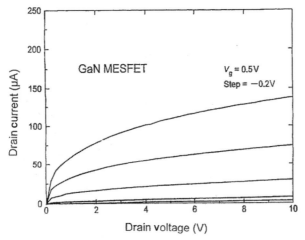

Figure 2.54. Drain I–V characteristics of a 1×50 μm² MESFET

Group III-nitride FET structures are sensitive to several effects during dry etching of the gate mesa. Firstly, if hydrogen is present in the plasma there can be passivation of the doping in the channel layer. Secondly, the ion bombardment from the plasma can create deep acceptor states that compensate the material. Thirdly, even when these problems are avoided through use of H-free plasma chemistries and low ion energies and fluxes, preferential loss of N can produce poor rectifying gate characteristics for metal deposited on the etched surface. Ping *et al.*[146] observed that pure Ar etching produced more damage in Schottky diodes than $SiCl_4$ RIE. The diode characteristics were strongly dependent on plasma self-bias, while annealing at 680°C removed much of the damage.

2.7.5 Ultraviolet Detectors

GaN and its alloys of aluminum gallium nitride (AlGaN) are the most promising semiconductors for development of UV photodetectors for applications such as combustion monitoring, space-based UV spectroscopy and missile plume detection. With a direct bandgap energy of approximately 3.39 eV (366 nm), GaN is an ideal material for the fabrication of photodetectors capable of rejecting near-infrared and visible regions of the solar spectrum while retaining near unity quantum efficiency in the UV. The use of AlGaN materials in photodetector fabrication enables bandgap engineering of the peak responsivity to shorter wavelengths in the deep UV.[147] GaN is also an extremely robust semiconductor suitable for high-temperature (>200°C) applications.

Nitride-based UV photodetectors that have been reported include p–n photodiode devices with 0.05 mm^2 junction area and 0.07 A/W peak responsivity [148], with 0.04 mm^2 junction area devices [149], with 0.25 mm^2 junction area and 0.1 A/W peak responsivity [150], and 0.59 mm^2 junction area and 0.195 A/W peak responsivity.[151] Photodetectors also reported include semi-transparent Schottky junction devices [152] and metal-semiconductor-metal devices.[153,154]

The GaN and AlGaN UV photodiodes were grown on (0001) basal-plane sapphire substrates by MBE using an RF atomic nitrogen plasma source.[155] Details of the growth process for nitride detectors have been reported previously.[156] The p–i–n detector epitaxial layers consisted of a 5×10^{18} cm^{-3} n-GaN layer followed by a 5000 Å intrinsic region with unintentional n-type doping in the 10^{15} cm^{-3} decade. The topmost epitaxial layer consisted of 1000–2000 Å 1×10^{18} cm^{-3} p-GaN. Mesas reaching the n-GaN cathode contact layer were formed by ICP etching with chlorine-based chemistry.[157] Ohmic contacts to the n-type and p-type GaN were made by Ti-based and Ni-based metallizations respectively. All of the GaN p–i–n UV detectors were fabricated with an optical detection area of 0.5 mm^2 and a p–i–n junction area of 0.59 mm^2, which is considerably larger (>12.5 times) than other GaN p–n detectors reported with noise measurements.

In addition to fabricating p–i–n type detectors, shorter UV wavelength MSM photodetectors, operating in a quasi-photoconductive mode, were fabricated from 1.5 μm thickness silicon-doped (~1×10^{17} cm^{-3}) n-AlGaN with a bandgap energy of approximately 320 nm. The MSMs were fabricated by first depositing a Pt-group metallization 1 μm wide with a 5 μm pitch to form the Schottky contacts. Next, a dielectric was deposited to act as an insulator between the AlGaN semiconductor and the bond pads. The dielectric process was not optimized to function as an anti-reflection coating. Reported here are results for AlGaN Metal Semiconductor Metal (MSMs) with active areas of 0.25 mm^2.

Shunt resistance and spectral responsivity data were collected using on-wafer probing. The shunt resistance was determined by the linear trace of the I–V characteristic from −10 mV to +10 mV. The spectral responsivities of the UV photodiodes were measured in photovoltaic mode (zero bias) for p–i–n devices and a photoconductive mode (biased) for MSM devices using a 75 W xenon arc lamp chopped at 700 Hz and filtered by a 1/8 m monochromator set to a 5 nm bandpass. The power of the monochromatic light was measured with a calibrated, NIST traceable, silicon photodiode and then focused onto the GaN wafers resting on a micropositioner stage.

The GaN p–i–n UV photodetector responsivity measurements reported were obtained with the devices operating in the unbiased, photovoltaic mode. Figure 2.55 shows a 25°C spectral responsivity curve for a baseline UV photodetector with 0.194 A/W peak responsivity and four orders of magnitude visible rejection that has been reported previously.[152] The maximum theoretical peak responsivity at the 360 nm bandgap is 0.28 A/W with no reflection and 0.23 A/W including reflection at the GaN surface. Also included on the plot is a trace for a GaN p–i–n UV photodetector with an improved p-type epitaxial process, which yields a greater visible rejection and more constant deep UV responsivity. The improved GaN p–i–n device was fabricated with a 1000 Å p-type cathode layer. The shunt resistances for these improved 0.59 mm^2 devices ranged from 200 MΩ to 50 GΩ, depending on the process. The device exhibits a low dark current and excellent forward-biased diode I–V characteristics with a built-in potential of approximately 3.1 V.

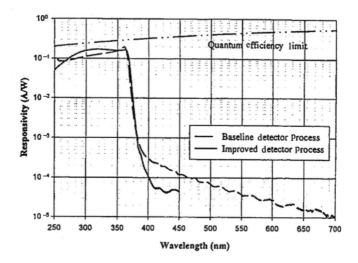

Figure 2.55. Spectral responsivity for GaN p–i–n UV photodetectors plotted against the maximum theoretical value with no reflection

The AlGaN MSM exhibited a substantial photoconductive gain (>700 ×) at 6 V bias, which yielded a responsivity of 7 A/W at 250 nm. The AlGaN device band-edge response does not decay as rapidly as that seen in devices fabricated from GaN material. The MSM devices, fabricated from AlGaN, which exhibited a luminescent peak at 320 nm, exhibited slightly more than three orders of magnitude (1278 ×) rejection of 360 nm light over 320 nm light. The rejection ratio for 250 nm light over 360 nm light was more than 5.5 orders of magnitude (5263×). The increased gain at shorter wavelengths is believed to be the result of greater electron–hole generation near the high electric field regions at the surface of the device. As expected, the MSM devices exhibited excellent shunt resistance (>100 GΩ) and dark current characteristics at low bias, as shown in Figure 2.55.

GaN p–i–n UV photodetectors with an optically active surface area of 0.5 mm^2 and a junction area of 0.59 mm^2 have been fabricated on 3″ diameter GaN p–i–n epitaxial wafers and characterized. Wafer maps of photodetector peak responsivity (maximum of 0.194 A/W at 359 nm) indicated that more than 60% ($\pm 1\sigma$) of all the GaN UV photodetectors performed within a $\pm 12\%$ deviation from the average peak responsivity. Further, the vast majority of GaN UV photodetectors were characterized with shunt resistances that were within one decade of each other.

High-temperature testing of the GaN p–i–n photodetectors up to 300°C indicated no significant increase in visible spectral responsivity or short-term degradation. The room–temperature spectral responsivity of the GaN photodetectors was fully recovered after 300°C testing. The 300°C GaN photodetector $1/f$ noise power densities were measured to be 6.6×10^{-19} and 2.1×10^{-21} A^2/Hz at 100 Hz and 1 kHz, respectively. The room-temperature 100 Hz and 1 kHz noise power density of the GaN p–i–n photodetectors was extrapolated to be on the order of 10^{-30} A^2/Hz ($\sim 10^{-15}$ A/Hz$^{1/2}$ noise current density).

The AlGaN MSM photodetectors, which were fabricated from AlGaN with a near bandgap luminescent peak of 320 nm, exhibited substantial photoconductive gain resulting in 7 A/W responsivity at 250 nm and 1.7 A/W responsivity at 320 nm. The AlGaN MSMs were characterized with 5263 and 1278 responsivity rejection ratios for 250 nm and 320 nm light respectively, versus 360 nm light.

References

1. (1997) GaN and Related Materials, Pearton SJ (ed) Gordon and Breach, New York. (1988) GaN, Pankove JI, Moustakas TD (eds) Academic, San Diego and reference therein.
2. Crawford MG (1997) presented at IUVSTA Workshop on GaN, Hawaii, August
3. Nakamura S, Senoh M, Nagahama S, Iwasa N, Yamada N, Matsushita T, Sugimoto Y, Kikoyu H (1997) Jpn J Appl Phys Part 2 36: L1059
4. Nakamura S (1997) IEEE J Sel Areas Commun 3: 435
5. Nakamura S (1997) presented at 3rd International GaN Conference, Tokushima, Japan October
6. Nakamura S (1998) IEEE J Sel Areas Commun 4: 483
7. Morkoc N (1998) Wide Band Gap Nitrides and Devices, Springer, Berlin
8. Nakamura S, Fosol G (1998) The Blue Laser Diode, Springer, Berlin
9. Mohammad SN, Morkoc H (1996) Prog Quantum Electron 20: 361
10. Mohammad SN, Salvador A, Morkoc H (1995) Proc IEEE 83: 1306
11. Morkoc H, Strite S, Gao GB, Lin ME, Sverdlov B, Burns M (1994) J Appl Phys 76: 1368
12. Ponce FA (1998) In: Encyclopedia of Applied Physics, Trigg GL (ed), VCH, Weinheim
13. Ponce FA, Bour DP (1997) Nature 386: 351
14. Akasaki, Amano H (1994) J Electrochem Soc 141: 2266
15. Pearton SJ, Shul RJ (1998) In: Gallium Nitride I, Pankove JI, Moustakas TD (eds) Academic, San Diego

16. Gillis P, Choutov DA, Steiner PA, Piper JD, Crouch JH, Dove PM, Martin KP (1995) Appl Phys Lett 66: 2475
17. Shul RJ (1999) In: Processing of Wide Bandgap Semiconductors, Pearton SJ (ed) Noyes, Park Ridge, NJ
18. Adesida I, Mahajan A, Andideh E, Asif Khan M, Olsen DT, Kuznia JN (1993) Appl Phys Lett 63: 2777
19. Adesida I, Ping AT, Youtsey C, Sow T, Asif Khan M, Olsen DT, Kuznia JN (1994) Appl Phys Lett 65: 889
20. Aktas O, Fan Z, Mohammad SN, Botchkarev A, Morkoc H (1996) Appl Phys Lett 69: 25
21. Khan MA, Chen Q, Shur MS, McDermott BT, Higgins JA, Burm J, Schaff W, Eastman LF (1996) Electron Lett 32: 357
22. Wu YF, Keller S, Kozodoy P, Keller BP, Parikh P, Kapolnek D, DenBaars SP, Mishra VK (1997) IEEE Electron Device Lett 18: 290
23. Khan MA, Kuznia JN, Shur MS, Eppens C, Burm J, Schaff W (1993) Appl Phys Lett 66: 1083
24. Harrison WA (1980) Electronic Structure and Properties of Solids, Freeman, San Francisco
25. Pearton SJ, Abernathy CR, Ren F, Lothian JR (1994) J Appl Phys 76: 1210
26. Lin ME, Zan ZF, Ma Z, Allen LH, Morkoc H (1994) Appl Phys Lett 64: 887
27. Ping AT, Adesida I, Asif Khan M, Kuznia JN (1994) Electron Lett 30: 1895
28. Lee H, Oberman DB Harris JS Jr (1995) Appl. Phys. Lett. 67, 1754
29. Pearton SJ, Abernathy CR, Ren F, Lothian JR, Wisk PW, Katz A, Constantine C (1993) Semicond Sci Technol 8: 310
30. Pearton SJ, Abernathy CR, Ren F (1994) Appl Phys Lett 64: 2294
31. Pearton SJ, Abernathy CR, Ren F (1994) Appl Phys Lett 64: 3643
32. Shul RJ, Kilcoyne SP, Hagerott Crawford M, Parmeter JE, Vartuli CB, Abernathy CR, Pearton SJ (1995) Appl Phys Lett 66: 1761
33. Vartuli CB, MacKenzie JD, Lee JW, Abernathy CR, Pearton SJ, Shul RJ (1996) J Appl Phys 80: 3705
34. Zhang L, Ramer J, Zheng K, Lester LF, Hersee SD (1996) Mater Res Soc Symp Proc 395: 763 (MRS, Pittsburgh PA 1996)
35. Zhang L, Ramer J, Brown J, Zheng K, Lester LF, Hersee SD (1996) Appl Phys Lett 68: 367
36. Humphreys B, Govett M (1996) MRS Internet J Nitride Semicond Res 1: 6
37. Lee JW, Hong J, MacKenzie JD, Abernathy CR, Pearton SJ, Ren F, Sciortino PF (1997) J Electron Mater 26: 290

38. Vartuli CB, Pearton SJ, Lee JW, Hong J, MacKenzie JD, Abernathy CR, Shul RJ (1996) Appl Phys Lett 69: 1426
39. Shul RJ, Howard AJ, Pearton SJ, Abernathy CR, Vartuli CB, Barnes PA, Bozack JM (1995) J Vac Sci Technol B 13: 2016
40. Vartuli CB, Pearton SJ, Abernathy CR, Shul RJ, Howard AJ, Kilcoyne SP, Parmeter JE, Hagerott Crawford M (1996) J Vac Sci Technol A 14: 1011
41. Shul RJ, Ashby CIH, Rieger DJ, Howard AJ, Pearton SJ, Abernathy CR, Vartuli CB, Barnes PA (1996) Mater Res Soc Symp Proc 395: 751 (MRS, Pittsburgh, PA 1996)
42. Shul RJ (1997) In: GaN and Related Materials, Pearton SJ (ed) Gordon and Breach, The Netherlands
43. Shul RJ, McClellan GB, Pearton SJ, Abernathy CR, Constantine C, Barratt C (1996) Electron Lett 32: 1408
44. Shul RJ, McClellan GB, Casalnuovo SA, Rieger DJ, Pearton SJ, Constantine C, Barratt C, Karlicek RF Jr, Tran C, Schurmann M (1996) Appl Phys Lett 69: 1119
45. Lee YH, Kim HS, Yeom GY, Lee JW, Yoo MC, Kim TI (1998) J Vac Sci Technol A 16: 1478
46. Kim HS, Lee YH, Yeom GY, Lee JW, Kim TI (1997) Mater Sci Eng B 50: 82
47. Hahn YB, Hays DC, Donovan SM, Abernathy CR, Han J, Shul RJ, Cho H, Jung KB, Pearton SJ (1999) J Vac Sci Technol B 17: 1237
48. Cho H, Vartuli CB, Donovan SM, MacKenzie JD, Abernathy CR, Pearton SJ, Shul RJ, Constantine C (1998) J Electron Mater 27: 166
49. Cho H, Vartuli CB, Donovan SM, Abernathy CR, Pearton SJ, Shul RJ, Constantine C (1998) J Vac Sci Technol A 6: 1631
50. Lee JW, Cho H, Hays DC, Abernathy CR, Pearton SJ, Shul RJ, Vawter GA, Han J (1998) IEEE J Sel Top Quantum Electron 4: 557
51. Shul RJ, Willison CG, Bridges MM, Han J, Lee JW, Pearton SJ, Abernathy CR, MacKenzie JD, Donovan SM (1999) Solid-State Electron 42: 677
52. Lee JM, Chang KM, Lee IH, Park SJ (2000) J Vac Sci Technol B 18: 1409
53. Smith SA, Wolden CA, Bremser MD, Hanser AD, Davis RF (1997) Appl Phys Lett 71: 3631; Smith SA, Lampert WV, Rajagopal R, Banks AD, Thompson D, Davis RF (2000) J Vac Sci Technol A 18: 879
54. Shul RJ (1998) In: GaN and Related Materials II, Pearton SJ (ed) Gordon and Breach, New York
55. McLane GF, Meyyappan M, Cole MW, Wrenn C (1991) J Appl Phys 69: 695

56. Meyyappan M, McLane GF, Lee HS, Eckart E, Namaroff M, Sasserath J (1992) J Vac Sci Technol B 10: 1215
57. McLane GF, Casas L, Pearton SJ, Abernathy CR (1995) Appl Phys Lett 66: 3328
58. Ping AT, Schmitz AC, Asif Khan M (1996) J Electron Mater 25: 825
59. Ping AT, Adesida I, Asif Khan M (1995) Appl Phys Lett 67: 1250
60. Ping AT, Asif Khan M, Adesida I (1997) Semicond Sci Technol 12: 133
61. Shul RJ, Vawter GA, Willison CG, Bridges MM, Lee JW, Pearton JS, Abernathy CR (1998) Solid-State Electron 42: 2259
62. Lee JW, Vartuli C, Abernathy CR, MacKenzie J, Mileham JR, Pearton SJ, Shul RJ, Zolper JC, Crawford M, Zavada J, Wilson R, Schwartz R (1996) J Vac Sci Technol B 14: 3637
63. Gillis HP, Choutov DA, Martin KP, Song L (1996) Appl Phys Lett 68: 2255
64. Gillis HP, Choutov DA, Martin HP (1996) J Mater 50: 41
65. Gillis HP, Choutov DA, Martin KP, Pearton SJ, Abernathy CR (1996) J Electrochem Soc 143: 251
66. Leonard RT, Bedair SM (1996) Appl Phys Lett 68, 794
67. Vartuli CB, Pearton SJ, Lee JW, MacKenzie JD, Abernathy CR, Shul RJ (1997) J Vac Sci Technol A 15: 638
68. Vartuli CB, Pearton SJ, Abernathy CR, Shul RJ, Ren F (1998) Proc Electrochem Soc 97-34: 39
69. Lee JW, Hong J, Pearton SJ (1996) Appl Phys Lett 68: 847
70. Shul RJ, Willison CG, Bridges MM, Han J, Lee JW, Pearton JW, Abernathy CR, MacKenzie JD, Donovan SM (1998) Mater Res Soc Symp Proc 482: 802 (MRS, Pittsburgh, PA 1998)
71. Shul RJ, Willison CG, Bridges MM, Han J, Lee JW, Pearton SJ, Abernathy CR, MacKenzie JD, Donovan SM, Zhang L, Lester LF (1998) J Vac Sci Technol A 16: 1621
72. Constantine C, Barratt C, Pearton SJ, Ren F, Lothian JR (1992) Electron Lett 28: 1749
73. Constantine C, Barratt C, Pearton SJ, Ren F, Lothian JR (1992) Appl Phys Lett 61: 2899
74. Lishan DG, Hu EL (1990) Appl Phys Lett 56: 1667
75. Hayes TR (1992) In: Indium Phosphide and Related Materials: Processing, Technology and Devices, Katz A (ed) Artech House, Boston, Chapter 8, pp 277–306
76. Shul RJ, Briggs RD, Han J, Pearton SJ, Lee JW, Vartuli CB, Killeen KP, Ludowise MJ (1997) Mater Res Soc Symp Proc 68: 355 (MRS, Pittsburgh, PA 1997)
77. Ren F, Lothian JR, Kuo JM, Hobson WS, Lopata J, Caballero JA, Pearton SJ, Cole MW (1995) J Vac Sci Technol B 14: 1

78. Ren F, Hobson WS, Lothian JR, Lopata J, Caballero JA, Pearton SJ, Cole MW (1995) Appl Phys Lett 67: 2497
79. Shul RJ, McClellan GB, Briggs RD, Rieger DJ, Pearton SJ, Abernathy CR, Lee JW, Constantine C, Barratt C (1997) J Vac Sci Technol A 15: 633
80. Vartuli CB, Pearton SJ, MacKenzie JD, Abernathy CR, Shul RJ (1996) J Electrochem Soc 143: L246
81. Vartuli CB, Pearton SJ, Lee JW, MacKenzie JD, Abernathy CR, Shul RJ (1997) J Vac Sci Technol B 15: 98
82. Cho H, Hong J, Maeda T, Donovan SM, MacKenzie JD, Abernathy CR, Pearton SJ, Shul RJ, Han J (1998) MRS Internet J Nitride Semicond Res 3: 5
83. Vartuli CB, Pearton SJ, Lee JW, MacKenzie JD, Abernathy CR, Shul RJ, Constantine C, Barratt C (1997) J Electrochem Soc 144: 2844
84. Ren F, Pearton SJ, Shul RJ, Han J (1998) J Electron Mater 27:75
85. Pearton SJ, Lee JW, MacKenzie JD, Abernathy CR, Shul RJ (1995) Appl Phys Lett 67: 2329
86. Ren F, Lothian JR, Pearton SJ, Abernathy CR, Vartuli CB, MacKenzie JD, Wilson RG, Karlicek RF (1997) J Electron Mater 26: 1287
87. Ping AT, Schmitz AC, Adesida I, Khan MA, Chen O, Yang YW (1997) J Electron Mater 26: 266
88. Eddy CR Jr, Molnar B (1999) J Electron Mater 28: 314; Eddy CR Jr, Molnar B (1996) Mat Res Soc Symp Proc 395: 745 (MRS, Pittsburgh, PA 1996)
89. Pearton SJ, Chakrabarti UK, Baiocchi FA (1989) Appl Phys Lett 55: 1633
90. Cho H, Auh KA, Shul RJ, Donovan SM, Abernathy CR, Lambers ES, Ren F, Pearton SJ (1999) J Electron Mater 28: 288
91. Nakamura S, Senoh M, Nagahama S, Iwasa N, Yamada T, Matsushita T, Kiyoku H, Sugimoto Y (1996) Jpn J Appl Phys Part 2 35: L74
92. Nakamura S, Senoh M, Nagahama S, Iwasa N, Yamada T, Matsushita T, Kiyoku H, Sugimoto Y (1996) Jpn J Appl Phys Part 2 35: L217
93. Nakamura S, Senoh M, Nagahama S, Iwasa N, Yamada T, Matsushita T, Kiyoku H, Sugimoto Y (1996) Appl Phys Lett 68: 2105
94. Nakamura S, Senoh M, Nagahama S, Iwasa N, Yamada T, Matsushita T, Kiyoku H, Sugimoto Y (1996) Appl Phys Lett 68: 3269
95. Nakamura S, Senoh M, Nagahama S, Iwasa N, Yamada T, Matsushita T, Kiyoku H, Sugimoto Y (1996) Appl Phys Lett 69: 1477
96. Nakamura S, Senoh M, Nagahama S, Iwasa N, Yamada T, Matsushita T, Kiyoku H, Sugimoto Y, (1996) Appl Phys Lett 69: 1568

97. Itaya K, Onomura M, Nishino J, Sugiura L, Saito S, Suzuki M, Rennie J, Nunoue S, Yamamoto M, Fujimoto H, Kokubun Y, Ohba Y, Hatakoshi G, Ishikawa M (1996) Jpn J Appl Phys Part 2 35: L1315
98. Nakamura S, Senoh M, Nagahama S, Iwasa N, Yamada T, Matsushita T, Sugimoto Y, Kiyoku H (1996) Appl Phys Lett 69: 3034
99. Nakamura S, Senoh M, Nagahama S, Iwasa N, Yamada T, Matsushita T, Sugimoto Y, Kiyoku H (1997) Appl Phys Lett 70: 616
100. Bulman GE, Doverspike K, Sheppard ST, Weeks TW, Kong HS, Dieringer HM, Edmond JA, Brown JD, Swindell JT, Schetzina JF (1997) Electron Lett 33: 1556
101. Mack MP, Abare A, Aizcorbe M, Kozodoy P, Keller S, Mishra UK, Coldren L, DenBaars S (1997) MRS Internet J Nitride Semicond Res 2: 41 (Available from http://nsr.mij.mrs.org/2/5/).
102. Kuramata, Domen K, Soejima R, Horino K, Kubota S, Tanahashi T (1997) Jpn J Appl Phys Part 2 36: L1130; Kobayashi T, Nakamura F, Naganuma K, Toyjo T, Nakajima H, Asatsuma T, Kawai H, Ikeda M (1998) Electron Lett 34: 1494
103. Nakamura S, Senoh M, Nagahama S, Iwasa N, Yamada T, Matsushita T, Sugimoto Y, Kiyoku H (1996) Appl Phys Lett 69: 4056
104. Nakamura S, Senoh M, Nagahama S, Iwasa N, Yamada T, Matsushita T, Sugimoto Y, Kiyoku H (1997) Appl Phys Lett 70: 1417
105. Nakamura S, Senoh M, Nagahama S, Iwasa N, Yamada T, Matsushita T, Sugimoto Y, Kiyoku H (1997) Appl Phys Lett 70: 2753
106. Nakamura S (1997) IEEE J Sel Top Quantum Electron 3: 435
107. Nakamura S, Senoh M, Nagahama S, Iwasa N, Yamada T, Matsushita T, Sugimoto Y, Kiyoku H (1997) Jpn J Appl Phys Part 2 36: L1059
108. Nakamura S (1997) 24th International Symposium on Compound Semiconductors, San Diego, CA, Plen-1, 8–11 September (unpublished)
109. Nakamura S, Senoh M, Nagahama S, Iwasa N, Kozaki T, Umemoto H, Sano M Chocho K (1997) Jpn J Appl Phys Part 2 36: L1568
110. Nakamura S, Senoh M, Nagahama S, Iwasa N, Yamada T, Matsushita T, Kiyoku H, Sugimoto Y, Kozacki T, Umemoto H, Sano M, Chocho K (1998) Appl Phys Lett 72: 2014; (1998) Jpn J Appl Phys Part 2 37: L309
111. Nakamura S, Senoh M, Nagahama S, Iwasa N, Yamada T, Matsushita T, Kiyoku H, Sugimoto Y, Kozacki T, Umemoto H, Sano M, Chocho K (1998) Appl Phys Lett 73: 832; (1998) Jpn J Appl Phys Part 2 37: L627
112. Chichibu S, Azuhata T, Sota T, Nakamura S (1996) Appl Phys Lett 69: 4188
113. Narukawa Y, Kawakami Y, Fujita S, Nakamura S (1997) Phys Rev B 55: 1938R

114. Narukawa Y, Kawakami Y, Funato M, Fujita S, Nakamura S (1997) Appl Phys Lett 70: 981
115. Suzuki M, Uenoyama T (1996) Jpn J Appl Phys Part 1 35: 1420
116. Suzuki M, Uenoyama T (1996) Appl Phys Lett 69: 3378
117. Chow WW, Wright AF, Nelson JS (1996) Appl Phys Lett 68: 296
118. Nam PH, Bremser MD, Zheleva T, Davis RF (1997) Appl Phys Lett 71: 2638
119. Kato Y, Kitamura S, Hiramatsu K, Sawaki N (1994) J Cryst Growth 144: 133
120. Zolper JC (1998) Solid-State Electron 42: 2153
121. Shur MS, Khan MA (1996) In: High Temperature Electronics, Willander M, Hartnagel HL (eds) Chapman and Hall, London, pp 297–321.
122. Brown ER (1998) Solid-State Electron 42: 2119
123. Nakamura S (1997) IEEE J Sel Top in Quantum Electron 3: 345
124. Nakamura S, Senoh M, Nagahama S, Iwasa N, Yamada T, Matsushita T, Sugimoto Y, Kiyoku H (1997) Appl Phys Lett 70: 616
125. Bulman GE, Doverspike K, Sheppard ST, Weeks TW, Kong HS, Dieringer H, Edmond JA, Brown JD, Swindell JT, Schetzina JF (1997) Electron Lett 33: 1556
126. Mack MP, Abare A, Aizcorbe M, Kozodoy P, Keller S, Mishra UK, Coldren L, DenBaars SP (1997) MRS Internet J Nitride Semicond Res 2: 41
127. Van Hove JM, Hickman R, Klaassen JJ, Chow PP, Ruden PP (1997) Appl Phys Lett 70: 282
128. Pankove JI, Leksono M, Chang SS, Walker C, Van Zeghbroeck B (1996) MRS Internet J Nitride Semicond Res 1: 39
129. McCarthy LS, Kozodoy P, DenBaars SP, Rodwell M, Mishra UK (1998) 25th Intl Symp Compound Semicond, Nara, Japan, October; (1999) IEEE Electron Dev Lett EDL-20: 277
130. Ren F, Abernathy CR, Van Hove JM, Chow PP, Hickman R, Klaassen RR, Kopf RF, Cho H, Jung KB, LaRoche JR, Wilson RG, Han K, Shul RJ, Baca AG, Pearton SJ (1998) MRS Internet J Nitride Semicond Res 3: 41; Han J, Baca AG, Shul RJ, Willison CG, Zhang L, Ren F, Zhang AP, Dang GT, Donovan SM, Cao XA, Cho H, Jung KB, Abernathy CR, Pearton SJ, Wilson RG (1999) Appl Phys Lett 74: 2702
131. Ren F, Lothian JR, Pearton SJ, Abernathy CR, Wisk PW, Fullowan TR, Tseng B, Chu SNG, Chen YK, Yang LW, Fu ST, Brozovich RS, Lin HH, Henning CL, Henry T (1994) J Vac Sci Technol B 12: 2916

132. Hickman R, Van Hove JM, Chow PP, Klaassen JJ, Wowchack AM, Polley CJ (1998) Solid-State Electron 42: 2138
133. Han J, Crawford MH, Shul RJ, Figiel JJ, Banas M, Zhang L, Song YK, Zhou H, Nurmikko AV (1998) Appl Phys Lett 73: 1688
134. Han J, Ng T-B, Biefeld RM, Crawford MH, Follstaedt DM (1997) Appl Phys Lett 71: 3114
135. Khan MA, Kuznia JN, Olson DT, Schaff W, Burm J, Shur MS (1994) Appl Phys Lett 65: 1121
136. Binari SC, Rowland LB, Kruppa W, Kelner G, Doverspike K, Gaskill DK (1994) Electron Lett 30: 1248
137. Khan MA, Shur MS, Kuznia JN, Burm J, Schuff W (1995) Appl Phys Lett 66: 1083
138. Khan MA, Chen Q, Sun CJ, Wang JW, Blasingame M, Shur MS, Park H (1996) Appl Phys Lett 68: 514
139. Kruppa W, Binari SC, Doverspike K (1995) Electron Lett 31: 1951
140. Binari SC, Rowland LB, Kelner G, Kruppa W, Dietrich HB, Doverspike K, Gaskill DK (1995) Inst Phys Conf Ser 141: 459, Bristol, UK: Institute of Physics
141. Ren F, Abernathy CR, Chu SNG, Lothian JR, Pearton SJ (1995) Appl Phys Lett 66: 1503
142. Strike S, Lin ME, Morkoc H (1993) Thin Solid Films 231: 197
143. Morkoc H, Strite S, Bao GB, Lin ME, Sverdlov B, Burns M (1994) J Appl Phys 76: 1363
144. Abernathy CR, MacKenzie JD, Bharatan SR, Jones KS, Pearton SJ (1995) Appl Phys Lett 66: 1632
145. Lin ME, Fan ZE, Ma Z, Allen LH, Morkoc H (1994) Appl Phys Lett 64: 887
146. Ping AT, Schmitz AC, Adesida I, Khan MA, Chen O, Yang YW (1997) J Electron Mater 26: 266
147. Van Hove JM, Chow PP, Hickman R, Wowchack AM, Klaassen JJ, Polley CJ (1996) Mater Res Soc Proc 449: 1227 (MRS, Pittsburgh, PA 1996)
148. Xu G, Salvador A, Botchkarev AE, Kim W, Lu C, Tang H, Morkoc H, Smith G, Estes M, Dang T, Wolf P (1998) Mater Sci Forum Part 2 264–268: 1441
149. Kuksenkov DV, Temkin H, Osinsky A, Gaska R, Khan MA (1997) International Electron Devices Meeting Technical Digest IEEE, Piscataway, NJ, p 759
150. Osinsky A, Gangopadhyay S, Gaska R, Williams B, Khan MA (1997) Appl Phys Lett 71: 2334
151. Hickman R, Klaassen JJ, Van Hove JM, Wowchack AM, Polley C, Rosamond MR, Chow PP (1999) MRS Internet J Nitride Semicond Res 4S1: G7.6

152. Khan MA, Kuznia JN, Olson DT, Blasingame M, Bhattaria AR (1993) Appl Phys Lett 63: 1781
153. Carrano JC, Li T, Brown DL, Grudowski PA, Eiting CJ, Dupuis RD, Campbell JC (1998) Appl Phys Lett 73: 2405
154. Ferguson, Liang S, Tran CR, Karlicek RF, Feng ZC, Lu Y, Joseph C (1998) Mater Sci Forum Part 2 264–268: 1437
155. Van Hove JM, Cosimini GJ, Nelson E, Wowchack AM, Chow PP (1995) J Cryst Growth 150: 908
156. Van Hove JM, Hickman R, Klaassen JJ, Chow PP (1997) Appl Phys Lett 70: 2282
157. Cho H, Vartuli CB, Abernathy CR, Donovan SM, Pearton SJ, Shul RJ, Han J (1998) Solid-State Electron 42: 2277

3 Design and Fabrication of Gallium High-Power Rectifiers

3.1 Abstract

GaN power Schottky diodes have numerous advantages over more conventional Si rectifiers, achieving a maximum electric field breakdown strength over 10 times larger and on-state resistance R_{ON} approximately 400 times lower at a given voltage. These characteristics have made GaN devices attractive for hybrid electric vehicles and power conditioning in large industrial motors. In particular, Schottky rectifiers are attractive because of their fast switching speed, which is important for improving the efficiency of inductive motor controllers and power supplies. Both GaN and SiC power Schottky diodes have demonstrated shorter turn-on delays than comparable Si devices. In this chapter we review the design and fabrication of GaN power rectifiers.

3.2 Introduction

There is strong interest in the development of ultrahigh power inverter modules based on GaN and other wide bandgap semiconductors.[1–5] These would have application in pulsed power for avionics and electric ships, in solid-state drivers for heavy electric motors and in advanced power management and control electronics. Schottky rectifiers are a key element of inverter modules because of their high switching speeds and low switching losses. While excellent reverse blocking voltages V_B have been achieved in lateral GaN rectifiers (V_B up to ~9.7 kV), these devices have limited utility because of their low on-state current.[6,7] Recently, a number of reports have appeared of vertical geometry GaN rectifiers fabricated on free-standing substrates.[8–10] These have shown excellent forward current characteristics, with total currents >1.7A for 7 mm diameter devices and low forward turn-on voltages ($V_F \approx 1.8$ V). The reverse breakdown voltages in these structures are still limited by avalanche breakdown at defects and/or surfaces. The rapid progress in improving both defect density and purity of these free-standing substrates makes them the most promising approach for achieving both high V_B and on-state currents, [11,12] in comparison with methods such as MOCVD of lightly doped stand-off layers.[5, 13-16]

A simple model for avalanche breakdown in GaN resulting from impact ionization produces the relation [3]

$$V_B \approx 1.98 \times 10^{15} N^{-0.7} \tag{3.1}$$

where N is the doping concentration in the GaN. Currently, all GaN rectifiers show performance limited by the presence of defects and by breakdown initiated in the depletion region near the electrode corners. In SiC rectifiers, a wide variety of edge termination methods have been employed to smooth out the electric field distribution around the rectifying contact periphery, including mesas, [17] high-resistivity layers created by ion implantation,[18,19] field plates[20,21] and guard rings.[22] The situation is far less developed for GaN, with just a few reports of combined guard rings/field-plate termination.[8,9,23] While SiC has been the workhorse in the research area of high power devices, GaN has been dominant in the commercialization of LED and limited in the application of high power rectifiers due to the lack of free-standing substrates. The recent success of growing GaN free-standing wafers by HVPE technology has geared up the power devices applications of GaN, especially high power rectifiers. In this chapter, the status of GaN high power rectifiers will be presented.

3.3 Background

In this review section, the theoretical calculations are given for reverse breakdown voltage and on-state resistance. Most parameters for GaN are extracted from epi-layer GaN, due to the lack of data from bulk GaN.

3.3.1 Temperature Dependence of Bandgap

3.3.1.1 Gallium Nitride

$$E_g(eV) = 3.475 - 9.39 \times 10^{-4} \frac{T^2}{T+772} = 3.396 + 9.39 \times 10^{-4} \left(\frac{300^2}{300+772} - \frac{T^2}{T+772} \right)$$

For a linear fit between 200 and 400 K

$$E_g \text{ (eV)} = 3.463 - 5.3 \times 10^{-4} (T-300) \tag{3.2}$$

as shown in Figure 3.1

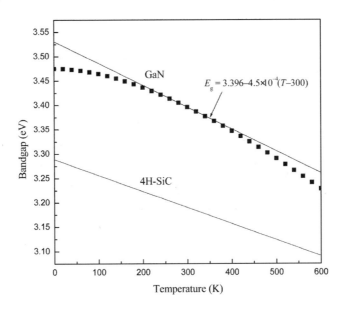

Figure 3.1. Temperature dependence of GaN and 6H-SiC bandgap as a function of temperature. Dotted lines represent the linear fit mainly between 200 and 400 K [after ref. 1] [1]

3.3.1.2 6H-SiC

$$E_g \text{ (eV)} = 3.19 - 3.3 \times 10^{-4} (T - 300) \tag{3.3}$$

3.3.2. Effective Density of States

The effective density of states is derived by the following equation for most semiconductors, and determined by the effective masses (m^*_e for electrons, m^*_h for holes) of carriers and temperature:

$$N_c(T) = 2\left(\frac{2\pi m^*_e kT}{h^2}\right)^{\frac{3}{2}} = 2.50945\times 10^{19}\left(\frac{m^*_e}{m_0}\right)^{\frac{3}{2}}\left(\frac{T}{300}\right)^{\frac{3}{2}}$$

$$N_v(T) = 2\left(\frac{2\pi m^*_h kT}{h^2}\right)^{\frac{3}{2}} = 2.50945\times 10^{19}\left(\frac{m^*_e}{m_0}\right)^{\frac{3}{2}}\left(\frac{T}{300}\right)^{\frac{3}{2}}$$

$$N_C(T) = 2.3\times 10^{18}\left(\frac{T}{300}\right)^{\frac{3}{2}}$$

$$m^*_e = 0.20 m_0 \text{ for wurtzite GaN}$$

$$N_V(T) = 4.6\times 10^{19}\left(\frac{T}{300}\right)^{\frac{3}{2}}$$

$$m^*_h = 1.50 m_0 \text{ for wurtzite GaN [2]} \qquad (3.4)$$

	N_C (cm^{-3})	N_V (cm^{-3})
GaN	2.3×10^{18}	4.6×10^{19}
6H-SiC	1.66×10^{19}	3.29×10^{19}

3.3.3 Intrinsic Carrier Concentration

The intrinsic carrier concentration n_i is described by the following equations, where N_C is the density of states in the conduction band, N_V the density of states in the valence band, and E_g is temperature-dependent bandgap

$$n_i(T) = \sqrt{N_C N_V}\exp\left(-\frac{E_g}{2kT}\right)$$

$$n_i(T) = 1.98\times 10^{16} T^{\frac{3}{2}}\exp\left(-\frac{20488.6}{T}\right) \text{ for GaN}$$

$$n_i(300) = 2.25 \times 10^{-10} \text{ cm}^{-3} \text{ for GaN}$$

$$n_i(300) = 1.6 \times 10^{-6} \text{ cm}^{-3} \text{ for 6H-SiC} \tag{3.5}$$

This small intrinsic carrier concentration at room temperature for wide bandgap semiconductors (Figure 3.2) causes the numerical underflow errors when calculating minority carrier concentration due to

$$np = n_i^2 \tag{3.6}$$

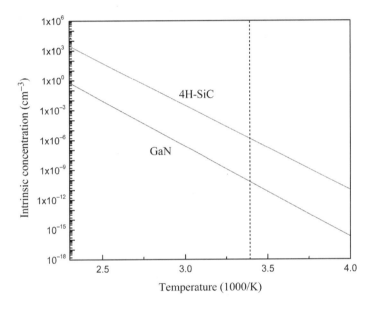

Figure 3.2. Intrinsic carrier concentration in SiC and GaN as a function of temperature [after ref. 31] [3]

3.3.4 Incomplete Ionization of Impurity Atoms

Acceptor dopants for GaN are not fully ionized even at high temperature. This incomplete ionization of impurities can be expressed by the following equation using Fermi–Dirac statistics with impurity ionization levels and degeneracy factors for the conduction and valence bands.

$$N_D^+ \frac{N_D}{1+2\exp\left(\frac{E_{Fn}-E_C+\Delta E_D}{kT}\right)} = \frac{N_D}{1+2\frac{N_D}{N_C}\exp\left(\frac{\Delta E_D}{kT}\right)}$$

$$N_A^+ \frac{NA}{1+4\exp\left(\frac{E_V-F_{Fp}+\Delta E_A}{kT}\right)} = \frac{N_A}{1+4\frac{N_A}{N_V}\exp\left(\frac{\Delta E_A}{kT}\right)}$$

$$E_C - E_{Fn} = kT \ln\left(\frac{N_C}{N_D}\right) \qquad E_{Fn} - E_V = kT \ln\left(\frac{N_V}{N_A}\right)$$

$$\Delta E_D = 0.016 \text{ for Si}$$
$$\Delta E_A = 0.175 \text{ for Mg} \qquad (3.7)$$

In Medici™, the incomplete ionization of impurities is selected by specifying the FERMIDIRAC and INCOMPLE parameters on the MODELS statement. And the band degeneracy is given by the impurity-dependent parameter GB. The donor and acceptor impurity activation energies are incorporated by the parameter EB0 in the IMPURITY statement. Medici can provide for the doping and temperature dependence of the impurity activation energies.

For very high doping concentrations, such as more than 1.0×10^{18} cm^{-3}, the transition from incomplete ionization to complete ionization happens. Medici will take the complete ionization if the parameter HIGH.DOP is specified on the MODELS statement. The complete ionization will be assumed above for impurity concentration greater than HDT.MAX.

3.3.5 Mobility Models

3.3.5.1 Analytical Mobility Model

Temperature- and concentration-dependent mobility model:

$$\mu = \mu_{min} + \frac{\mu_{max}\left(\frac{T}{300}\right)^{\alpha} - \mu_{min}}{1+\left(\frac{T}{300}\right)^{\beta}\left(\frac{N_{tot}}{N_{ref}}\right)^{\gamma}} \qquad (3.8)$$

Carriers	μ_{max} (cm²V⁻¹s⁻¹)	μ_{min} (cm²V⁻¹s⁻¹)	N_{ref} (cm⁻³)	α	β	γ
Electrons	1000	55	2×10¹⁷	−2.0	−3.8	1.0
Holes	170	30	3×10¹⁷	−5.0	−3.7	2.0

3.3.5.2. Field-Dependent Mobility Model

$$\mu_n(E) = \frac{\mu_n}{\left[1 + \left(\frac{\mu_n E}{v_{sat}}\right)^{\beta_i}\right]^{1/\beta_i}}$$

BETAN=2.0   BETAP=1.0 for GaN

$$v_{sat} = \frac{2.7 \times 10^7}{1 + 0.8 \exp\left(\frac{T}{600}\right)} \quad (3.9)$$

Figure 3.3 is the low-field mobility at 300 K as a function of doping concentration.

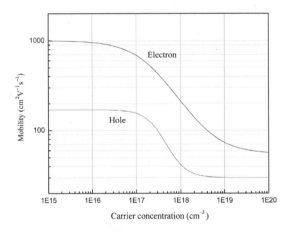

Figure 3.3. Low-field mobility of electrons and holes as a function of doping concentration in GaN at room temperature [after ref. 4] [4]

3.3.6 Generation and Recombination

3.3.6.1 Shockley–Read–Hall Lifetime

The Shockley–Read–Hall (SRH) recombination-generation rate R_{SRH} is given by

$$R_{SRH} = \frac{pn - n_i^2}{\tau_p(n+n_i) + \tau_n(p+n_i)} \quad (3.10)$$

where the lifetimes τ_n and τ_p of electrons and holes dependent upon the doping level, as described by the Scharfetter relation:

$$\tau_n = \frac{\tau_{n0}}{1+(\frac{N_{tot}}{N_n^{SRH}})^{\gamma_{ns}}} \quad (3.11)$$

The lifetimes for GaN are observed in the order of 1–100 ns. The following relation is often used for Si and is also used for GaN:

$$\tau_{n0} = 5\tau_{p0} \quad (3.12)$$

3.3.6.2 Auger Recombination

$$R_{Au} = (C_p p + C_n n)(np - n_i^2) \quad (3.13)$$

Auger recombination constants are $C_n = 1\times 10^{-30}$ and $C_p = 1\times 10^{-31}$ cm^6/s for GaN.

Carriers	$\tau_{n,p}$ (s)	N_n^{SRH}	γ_{ns}	$C_{n,p}$ (cm^6/s)
Electrons	1×10^{-9}	5×10^{16}	1	1×10^{-30}
Holes	1×10^{-9}	5×10^{16}	1	1×10^{-31}

3.3.7 Reverse Breakdown Voltage

The phenomenon of reverse breakdown is explained by avalanche multiplication, which involves impact ionization between host atoms and high-energy carriers. When a high-energy hole or electron under high electric field impacts an electron in the valence band, it will produce a new electron–hole pair (EHP). This newly generated EHP will cause other collisions and multiply carriers very rapidly.[24–35] Avalanche breakdown is defined to occur in theory when

$$\int_0^{W_D} \alpha_p \exp[\int_0^x (\alpha_n - \alpha_p) dx] dx > 1$$

$$\alpha_i = \alpha_0 \exp\left(\frac{-b_0}{E}\right) \quad (3.14)$$

where W_D is the depletion width and α_n and α_p are the ionization rates of electrons and holes.

Oguzman *et al.* [32] and Ruff *et al.* [35] calculated the hole-initiated and electron-initiated ionization rates of electrons and holes, respectively for both wurtzite and zinc blende GaN. Figure 3.4 represents the calculated impact ionization coefficients as a function of inverse electric field for electrons and holes in wurtzite GaN and 6H-SiC.[34–36]

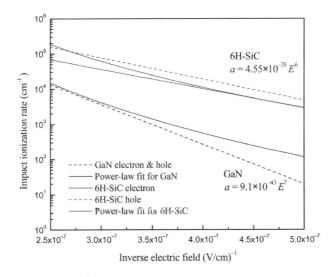

Figure 3.4. Calculated impact ionization coefficients as a function of inverse electric field for electrons and holes in wurtzite GaN and 4H-SiC [32,33]

For GaN:

$$\alpha_{n,p}(\text{cm}^{-1}) = 8.85 \times 10^6 \exp\left(-\frac{2.6 \times 10^7}{E}\right) \text{[2]} \quad (3.15)$$

For 6H-SiC:

$$\alpha_n(\text{cm}^{-1}) = 1.66 \times 10^6 \exp\left(-\frac{1.273 \times 10^7}{E}\right)$$

$$\alpha_p(\text{cm}^{-1}) = 5.18 \times 10^6 \exp\left(-\frac{1.4 \times 10^7}{E}\right) \text{[34]} \quad (3.16)$$

To calculate the integrals of impact ionization without the aid of a computer is time-consuming and almost impossible. Thus, a power-law approximation can make the calculation easier in Fulop's form:

$$\alpha_{\text{eff}} = AE^n = 9.1 \times 10^{-43} E^7 \text{ for GaN [36]} \quad (3.17)$$

$$\alpha_{\text{eff}} = AE^n = 4.55 \times 10^{-35} E^6 \text{ 6H-SiC [37]} \quad (3.18)$$

Simplified breakdown condition is expressed by the following ionization integral:

$$\int_0^{W_C} \alpha_{\text{eff}} \, dx = 1 \quad (3.19)$$

Therefore, the depletion layer width at breakdown for GaN is

$$W_c = \left(\frac{8}{A}\right)^{\frac{1}{8}} \left(\frac{\varepsilon\varepsilon_0}{qN_B}\right)^{\frac{7}{8}} \quad (3.20)$$

The critical electric field (Figure 3.5) can be calculated by the one-dimensional (1D) Poisson equation ($dE/dx = qN_B/\varepsilon\varepsilon_0$) and obtained through numerical substitutions:

$$\frac{d^2V}{dx^2} = -\frac{dE}{dx} = -\frac{qN_B}{\varepsilon\varepsilon_0}$$

$$E_c = \left(\frac{8}{A}\frac{qN_B}{\varepsilon\varepsilon_0}\right)^{\frac{1}{8}} = \left(\frac{8}{AW_c}\right)^{\frac{1}{7}} = 3.4\times10^4 N_B^{\frac{1}{8}} \quad (3.21)$$

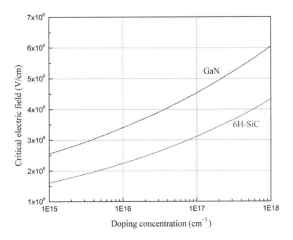

Figure 3.5. The critical electric field for 6H-SiC and wurtzite GaN as a function of doping concentration

If Poisson's equation is solved with the voltage and electric field relationship, the breakdown voltage for the non-punch-through junction case is given by (Figure 3.6)

$$BV_{pp} = 2.87\times10^{15} N_B^{-\frac{3}{4}} \quad (3.22)$$

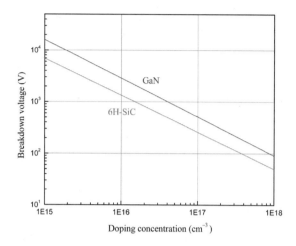

Figure 3.6. The reverse breakdown voltage of non-punch-through junction for 6H-SiC and wurtzite GaN as a function of doping concentration

The depletion layer width, the critical electric field, and the breakdown voltage for 6H-SiC are given by

$$E_c = \left(\frac{7}{A}\frac{qN_B}{\varepsilon\varepsilon_0}\right)^{\frac{1}{7}} = \left(\frac{7}{Aw_c}\right)^{\frac{1}{6}} = 1.16 \times 10^4 N_B^{\frac{1}{7}}$$

$$W_c = \left(\frac{7}{A}\right)^{\frac{1}{7}} \left(\frac{\varepsilon\varepsilon_0}{qN_B}\right)^{\frac{6}{7}} = 6.2 \times 10^4 N_B^{-\frac{6}{7}}$$

$$BV_{pp} = 3.59 \times 10^{14} N_B^{-\frac{5}{7}} \tag{3.23}$$

In the case of a punch-through junction diode, the breakdown voltage is given by

$$BV_{PT} = E_c W_{PT} - \frac{qN_B W_{PT}^2}{2\varepsilon\varepsilon_0} \tag{3.24}$$

Figure 3.7 is a plot of theoretical breakdown voltage of GaN punch-through diode as a function of doping concentration and drift region thickness. It can be seen that a 3 μm epi-layer with doping concentration of 10^{16} cm^{-3} gives more than 900 V of breakdown voltage. The actual experimental value of breakdown voltage is far from these theoretical predictions. The material imperfections, such as the vertically threading dislocations, lead to premature breakdown. Therefore, the edge termination technique should be developed for GaN to prevent the early breakdown, and the crystal quality should be advanced to improve the GaN device performance.

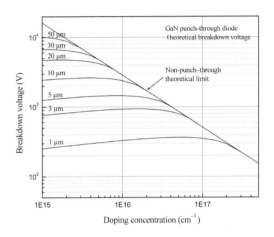

Figure 3.7. The reverse breakdown voltage of a punch-through junction for GaN as a function of doping concentration and drift region thickness

3.3.8 On-State Resistance

The specific on-state resistance of a unipolar diode is the sum of the drift region resistance, the contact resistance and the substrate resistance:

$$R_{diode} = R_{drift} + R_{sub} + R_{contact} \quad (3.25)$$

The specific on-state resistance of drift region is given by

$$R_{ON} = \int \frac{dx}{q\mu N_B} = \frac{W_D}{q\mu N_B} \quad (3.26)$$

where μ is the low-field mobility (μ=1000 cm^2 V^{-1} s^{-1} for GaN), N_B is the doping concentration of the drift region and W_D is the drift region thickness. The on-state resistance of the drift region for GaN can be expressed by the reverse breakdown voltage and is given by

$$R_{ON}(\Omega \cdot cm^2) = 2.4 \times 10^{-12} BV^{2.5} \quad (3.27)$$

as shown in Figures 3.8 and 3.9.

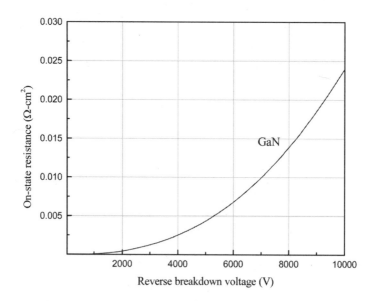

Figure 3.8. The specific on-state resistance for GaN Schottky diode as a function of breakdown voltage

These properties make GaN Schottky diode rectifiers attractive for power distribution in hybrid electric vehicles, such as the Toyota Prius (Figures 3.10–3.12), where a 1000V, 500 kA/cm² at 1.4 V GaN diode and 850V, 500 A/cm² at 1.6 V GaN MISFET are needed for the inverter unit.

For 6H-SiC:

$$R_{ON}(\Omega \cdot cm^2) = 1.45 \times 10^{-11} BV^{2.6} \tag{3.28}$$

For Si:

$$R_{ON}(\Omega \cdot cm^2) = 5.91 \times 10^{-9} \cdot BV^{2.5} \tag{3.29}$$

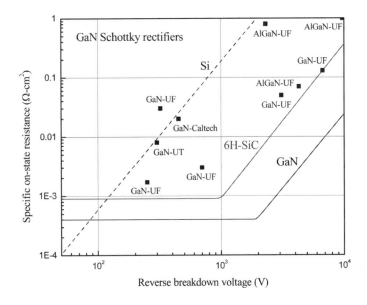

Figure 3.9. The specific on-state resistance for Si, 6H-SiC, and GaN diode as a function of breakdown voltage

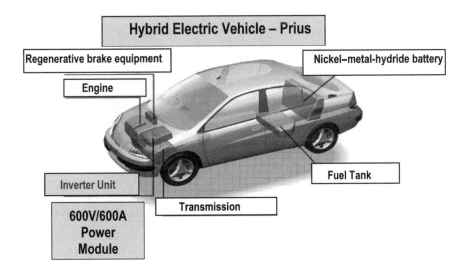

Figure 3.10. Toyota Prius hybrid electric vehicle

194   Design and Fabrication of Gallium High-Power Rectifiers

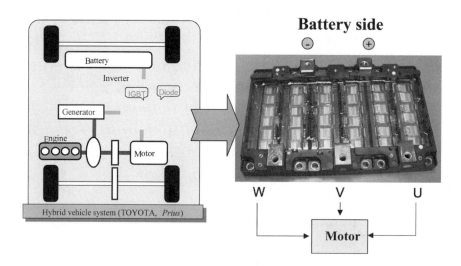

Figure 3.11. Power distribution unit in hybrid electric automobile

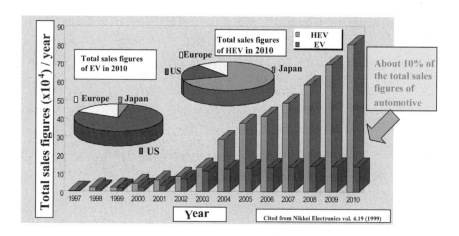

Figure 3.12. Sales figures foe electric and hybrid electric vehicles [after Nikkei Electronics vol. 4.19 (1999)]

3.4 Edge Termination Design

3.4.1 Field Plate Termination

The structure at the top of Figure 3.13 was used as our standard for simulation and is based on the available free-standing GaN bulk substrates, which currently have a doping density of ~10^{16} cm^{-3}. While their thickness is ~200 μm, the depletion depth is limited by the background doping and we used a thickness of 30 μm in the simulations. The parameters we investigated in this study were the dielectric material, its thickness and the extent of metal overlap onto the field plate. The simulations were carried out using the MEDICI^{TM} code. The back n-ohmic contact resistance was assumed to be 10^{-6} Ω cm^2, which is consistent with our past experimental data, and an interface state density of 5×10^{11} eV^{-1} cm^{-2} was assumed for the dielectric–GaN interface. Once again, this is based on our past experimental results. After designing the particular basic structure, a mesh of nodes is created to allow the solutions to the transport equations to be obtained. The program includes Shockley–Read–Hall and Auger recombination, an incomplete ionization model and an average of the available high-field saturation and avalanche models. We assumed a conduction band density of states of 2.6×10^{18} cm^{-3} for the n-type GaN, a surface recombination velocity of 10^3 cm.s^{-1} and Shockley–Read–Hall lifetime of 1 ns.

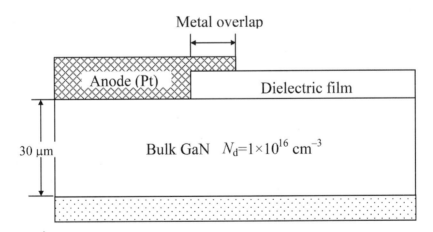

Figure 3.13. Schematic of simulated bulk GaN rectifier structure

The maximum electric field in an unterminated rectifier occurs directly under the corner of the Schottky contact and emphasizes that avalanche breakdown is more likely to initiate at that location. The breakdown voltage was 980 V.

Figure 3.14 shows the calculated breakdown voltages obtained for a 0.7 μm thickness SiO_2 field plate on top of the rectifier, as a function of the extent of the overlap of the Schottky contact onto the SiO_2. Note that the V_B values increase rapidly for metal overlaps up to ~10 μm, with a maximum increase of ~63% in breakdown voltage relative to the unterminated device. Beyond an overlap of 10 μm there is no further improvement in breakdown voltage from this given thickness of SiO_2 field plate. We believe this is due to the fact that the lateral spread of the depletion layer becomes comparable to the depth of this layer, so that extending the field plate into undepleted regions does not affect the breakdown behavior.

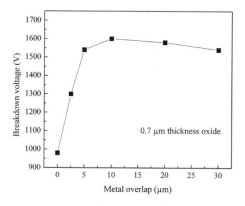

Figure 3.14. Effect of metal overlap distance in V_B for rectifiers with 0.7 μm thickness SiO_2 field plate

The effect of SiO_2 thickness at a given metal overlap distance of 10 μm is shown in Figure 3.15. There is an almost linear increase in V_B with increasing oxide thickness up to 0.7 μm. At thicknesses >1 μm the simulations showed that the electric field inside the oxide began to increase. One must, therefore, choose a thickness such that the field strength inside the oxide does not exceed its breakdown strength. The fact that very thick oxide layers do not lead to an improvement in V_B is an advantage from a practical viewpoint, because such layers would require very long deposition times and introduce problems such as stress.

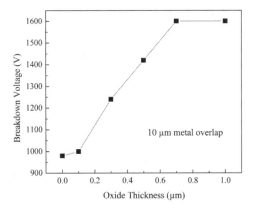

Figure 3.15. Effect of SiO$_2$ thickness on V_B for rectifiers with 10 μm metal overlap

Other dielectrics that have demonstrated reasonably low interface state densities on GaN include AlN, MgO and Sc$_2$O$_3$. SiO$_2$ produces the highest breakdown voltage rectifiers for these conditions because of its large bandgap and low dielectric constant. However, in real devices it should be considered that reliability is of utmost importance, and it is not necessarily the case that SiO$_2$ would be the best choice with this consideration in mind. For example, Sc$_2$O$_3$ appears to produce the most effective passivation of surface states on GaN–AlGaN heterostructure field effect transistors.[28] Obviously, much more work needs to be done to establish experimentally the relative tradeoff between V_B and long-term device stability.

The main findings of this simulation study can be summarized as follows:

1. The use of an optimized SiO$_2$ field-plate edge termination can increase the reverse breakdown volatge of bulk GaN rectifiers by up to a factor of two compared with unterminated devices.
2. The dielectric material, thickness and ramp angle all influence the resulting V_B of the rectifier by determining where the maximum field strength occurs in the device structure. The key aspect in designing the field-plate edge termination is to shift the region of the high-field region away from the periphery of the rectifying contact.

3.4.2 Junction Termination

Figure 3.16 shows a schematic of a rectifier employing planar junction termination with a dielectric field plate. The breakdown point is extended beyond the contact periphery by this approach. Simply using the planar junction without the field plate increases V_B to 760 V, while the addition of the SiO_2 field plate increases it up to 1110 V. Important design parameters influencing V_B are the length of the metal overlap and the SiO_2 thickness. For a given value of the plate length, the SiO_2 thickness must be optimized in order to balance the electric field peaks at the junction and plate edges as reported previously for GaN [38].

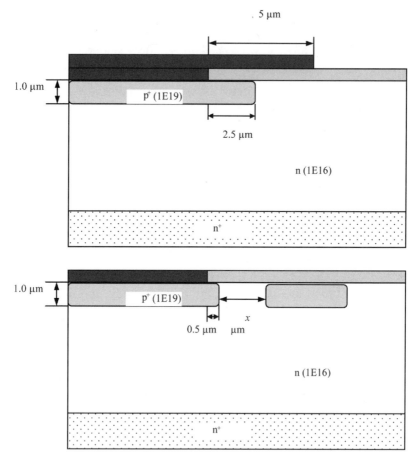

Figure 3.16. Schematic of rectifier with planar junction termination and SiO_2 field plate (top) or with planar junction termination and one p^+ guard ring (bottom)

3.4 Edge Termination Design

Another alternative for edge terminations is the use of equipotential p$^+$ guard rings. As the bias in the main junction increases, the space charge region extends until it reaches the guard ring. The potential of the guard ring is given by

$$V_{FFR} = \left(\frac{2eN_A W_S 2V_A}{\varepsilon} \right)^{\frac{1}{2}} + \frac{eN_A W_S^2}{2\varepsilon} \quad (3.30)$$

where N_A is the acceptor concentration, ε the GaN permittivity, W_S the field ring spacing and V_A the applied Schottky bias on the Schottky contact. As the bias is further increased, this potential increases, tracking the value of the equipotential line from the main junction [22]. At a given bias, the potential of the ring is equal to that of the highest potential around it and lower than the value on the main junction. The net effect is to reduce field crowding near the junction curvature. [22] In our case of one p-guard ring with additional planar junction termination, the calculated V_B is 1180V which is a slight improvement over the planar junction termination with the field plate.

Figure 3.17 shows the effect of guard-ring spacing on the calculated V_B. In our geometry, the maximum V_B is obtained for a spacing of ~3 µm. The benefits of the field spreading are lost at either very small or large spacing. The use of additional guard rings is also beneficial. For example, using two rings, one separated by 2 µm and the other by 3 µm, led to an increase in V_B to 1458 V.

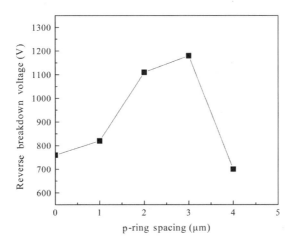

Figure 3.17. Effect of guard ring spacing on V_B

The Junction Termination Extension (JTE) method extends the high-doped side of the main junction by a connected region of lower doping level. The net effect is once again to spread the field lines and avoid premature breakdown. A key design parameter is obviously the doping concentration in the JTE region. In our case, a doping concentration of $4\times10^{17} cm^{-3}$ produced a V_B value of 2720 V. The value of V_B was strongly peaked for a JTE doping around 4×10^{17} cm^{-3}, although V_B values above 2000 V are achieved for a reasonable range of concentrations.

Finally, Figure 3.18 provides a summary of the V_B values calculated for the different edge termination methods. The single JTE approach provides an almost five-fold increase relative to an unterminated rectifier and points out the clear necessity to employ one or more methods in order to maximize the blocking voltage even for GaN rectifiers.

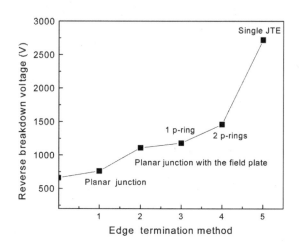

Figure 3.18. Comparison of V_B values for GaN rectifiers employing different edge termination methods

For GaN power rectifiers to mature into a manufacturable technology, attention must be paid to the design and implementation of edge termination methods that maximize the reverse breakdown voltage. Our current study shows that the JTE produces the highest blocking voltages for vertical bulk GaN rectifiers, although the V_B values are highly sensitive to the doping in the JTE region. Guard rings, field plates and planar junctions were also examined in increasing V_B over the value in unterminated rectifiers.

3.5 Comparison of Schottky and p–n Junction Diodes

The minority carrier concentration in GaN p-n junction is extremely small, in the order of 10^{-36} cm^{-3} for $N_D=10^{16}$ cm^{-3} at room temperature, because the intrinsic carrier concentration is about 10^{-10} cm^{-3}:

$$np = n_i^2 \tag{3.31}$$

3.5.1 Reverse Bias

Ideally, the reverse current density J_R can be calculated by the generation current in the space charge region and the diffusion current due to the generation close to the space charge region:

$$J_R = \frac{qn_iW}{\tau_g} + q\sqrt{\left(\frac{\mu_p kT}{q}\right)\frac{1}{\tau_p}} \tag{3.32}$$

where τ_g is the appropriate generation lifetime in the space charge region and τ_p is the minority carrier lifetime of a hole in the n region. The generation current is in the order of 10^{-23} A cm^{-2} for the GaN p–n$^-$–n$^+$ diode at room temperature with a breakdown voltage of 1530 V. The diffusion current is in the order of 10^{-50} A cm^{-2} and diffusion current contribution is negligible. Those large leakage currents arise from the surface leakage, defects in epilayer and junction, and tunneling:

3.5.2 Forward Bias

If a forward bias is applied, then the built-in voltage must be exceeded before a substantial current can flow.

$$V_{bi} = \frac{kT}{q}\ln\left(\frac{N_D^+ N_A^-}{n_i^2}\right) \tag{3.33}$$

$V_{bi} = 3.178$ V for GaN p–n$^-$–n$^+$ diode ($N_D^+ N_A^- = 10^{-34}$ cm^{-3})

$V_{bi} = 2.719$ V for 6H-SiC p–n⁻–n⁺ diode ($N_D^+ N_A^- = 10^{-34}$ cm^{-3})

The forward current density J_F at low forward voltage V_F is determined by a current contribution due to recombination in the space charge region and by a diffusion contribution due to recombination close to the space charge region:

$$J_R = \frac{qn_iW}{\tau_r} \exp\left(\frac{qV_F}{2kT}\right) + q\sqrt{\left(\frac{\mu_p kT}{q}\right)\frac{1}{\tau_p}} \exp\left(\frac{qV_F}{kT}\right) \quad (3.34)$$

Figures 3.19–21 compare the forward I–V characteristics (including temperature dependence) for GaN p–n and Schottky diodes.

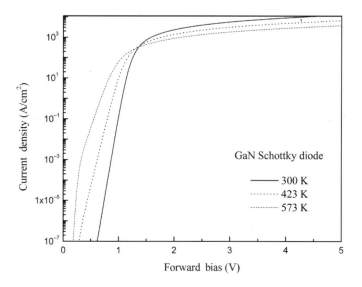

Figure 3.19. The forward I–V characteristics of 5 μm n-region Schottky diode

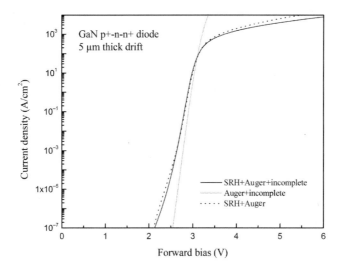

Figure 3.20. The forward I–V characteristics of p^+–n–n^+ diode

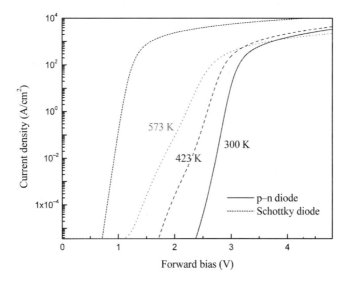

Figure 3.21. The temperature-dependent forward I–V characteristics of p^+–n–n^+ diode

3.6 High Breakdown Lateral Diodes

For laterally depleting devices, the structure consisted of ~3 mm of resistive (10^7 Ω/□) GaN. To form ohmic contacts, Si^+ was implanted at 5×10^{14} cm^{-2}, 50 keV into the contact region and activated by annealing at 150°C for 10 s under N_2. The ohmic and rectifying contact metallization was the same as described above.

Three different edge termination techniques were investigated for the planar diode:
1. Use of a p-guard ring formed by Mg^+ implantation at the edge of the Schottky barrier metal. In these diodes the rectifying contact diameter was held constant at 124 µm, while the distance of the edge of this contact from the edge of the ohmic contact was 30 µm in all cases.
2. Use of p-floating field rings of width 5 mm to extend the depletion boundary along the surface of the SiO_2 dielectric, which reduces the electric field crowding at the edge of this boundary. In these structures, a 10 µm wide p-guard ring was used, and one to three floating field rings were employed.
3. Use of junction-barrier-controlled Schottky (JBS) rectifiers, *i.e.*, a Schottky rectifier structure with a p–n junction grid integrated into its drift region.

In all of the edge-terminated devices the Schottky barrier metal was extended over an oxide layer at the edge to further minimize field crowding, and the guard and field rings were formed by Mg^+ implantation and 1100°C annealing.

Figure 3.22 shows the influence of guard ring width on V_B at 25°C. Without any edge termination, V_B is ~2300 V for these diodes. The forward turn-on voltage was 15–50 V, with a best on-resistance of 0.8 Ω cm^2. The figure-of-merit $(V_B)^2/R_{ON}$ was 6.8 MW/cm^2. As the guard-ring width was increased, we observed a monotonic increase in V_B, reaching a value of ~3100 V for 30 µm width rings. The figure-of-merit was 1535 MW/cm^2 under these conditions. The reverse leakage current of the diodes was still in the nanoamp range at voltages up to 90% V of the breakdown value.

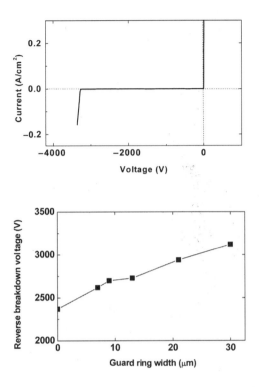

Figure 3.22. Current–voltage characteristics of GaN power rectifiers with p-guard for edge terminations (top), and effect of p-guard ring on the reverse breakdown voltage of GaN power rectifiers (bottom)

Figure 3.23 shows the variation of V_{RB} with Al percentage in the AlGaN active layers of the rectifiers. In this case we are using the V_{RB} values from diodes without any edge termination or surface passivation. The calculated band gaps as a function of Al composition are also shown, and were obtained from the relation:

$$E_g(x) = E_{g,GaN}(1-x) + E_{g,AlN}x - bx(1-x) \qquad (3.35)$$

where x is the AlN mole fraction and b is the bowing parameter with value 0.96 eV. [19]. Note that V_{RB} does not increase in linear fashion with band gap. In a simple theory, V_{RB} should increase as $(E_g)^{1.5}$, but it has been empirically established that factors such as impact ionization coefficients and other transport parameters need to be considered and that consideration of E_g alone is not sufficient to explain measured V_{RB} behavior. The fact that V_{RB} increases less rapidly with E_g at higher AlN mole fractions may indicate increasing concentrations of defects that influence the critical field for breakdown.

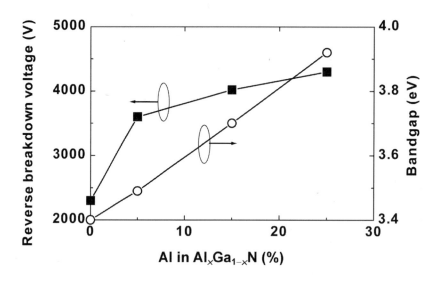

Figure 3.23. Variation of V_{RB} in $Al_xGa_{1-x}N$ rectifiers without edge termination, as a function of Al concentration. The band gaps for the AlGaN alloys are also shown

Figure 3.24 shows the variation of V_{RB} with temperature. The data can be represented by a relation of the form:

$$V_{RB} = V_{RBO}\left[1 + \beta(T - T_0)\right] \quad (3.36)$$

where $\beta = -6.0 \pm 0.4$ V K^{-1} for both types of rectifier. However, in Schottky and p–i–n rectifiers we have fabricated on more conducting GaN; with V_{RB} values in the 400–500 V range, the values were consistently around –0.34 V K^{-1}. Therefore, in present state-of-the-art GaN rectifiers, the temperature coefficient of V_{RB} appears to be a function of the magnitude of V_{RB}. Regardless of the origin of this effect, it is clearly a disadvantage for GaN. While SiC is reported to have a positive temperature coefficient for V_{RB}, there are reports of rectifiers that display negative β values. One may speculate that particular defects present may dominate the sign and magnitude of β, and it will be interesting to fabricate GaN rectifiers on bulk or quasi-bulk substrates with defect densities far lower than in heteroepitaxial material.

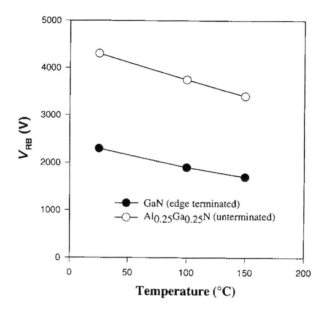

Figure 3.24. Temperature dependence of V_{RB} for GaN and AlGaN rectifiers

3.7 Bulk Diode Arrays

A drawback of Schottky rectifiers fabricated on both conventional homo-epitaxial (Al_2O_3) and free-standing GaN films is a strong dependence of reverse breakdown voltage V_B on device contact diameter.[8] In defect-free material, V_B is related to the maximum electric field at breakdown, E_M and the depletion depth at breakdown W_M through [3]

$$V_B = \frac{E_M W_M}{2} \tag{3.37}$$

However, in the presence of defects such as screw dislocations, nanopipes, or voids, it has been demonstrated that temperature breakdown occurs in both GaN and SiC.[8, 16, 17] For example, in Pt Schottky diodes fabricated on free-standing GaN, we have previously reported a decrease in V_B from 160 V for 75 μm diameter to ~ 6 V for 7 mm diameter devices. [8] In order to overcome this degradation in reverse bias performance when fabricating, we demonstrate a method for interconnecting the output of many (~130) smaller diodes (500×500 μm²) to produce a very high total forward current. Similar parallel connection has been used to achieve high total forward current in SiC devices. [15]

The diode arrays were fabricated in 1 cm², 200 μm thickness, free-standing GaN substrate grown by HVPE on *c*-plane Al₂O₃. The GaN films were subsequently removed from the sapphire by laser heating. [18] Full-area back contacts of Ti–Al–Pt–Au were deposited by e-beam evaporation and annealed at 850 °C for 30 s to minimize contact resistance. Edge termination utilized an SiO₂ (1500 Å) field plate and e-beam-deposited Pt–Au Schottky contacts with 500×500 μm² dimensions. To achieve parallel connection of the diode array, SiN$_x$ (2500 Å) was deposited all over the wafer as isolation and opened with 100 μm diameter current paths for each Schottky metal pad. Metal was further deposited with Ti–Au for adhesion and seed layers and Au was electroplated to a thickness of 3 μm on both the front and back of the rectifiers. A schematic of a completed device is shown in Figure 3.25.

Prior to the interconnection of all the rectifiers, the I–V characteristics at 25 °C were recorded, and fit to the relation for thermionic emission over a barrier:

$$J_F = A^* T^2 \exp\left(-\frac{e\phi_b}{kT}\right) \exp\left(\frac{eV}{nkT}\right) \quad (3.38)$$

where J is the current density, A^* is Richardson's constant for n-GaN, T the absolute temperature, e the electronic charge, ϕ_b the barrier height, k Boltzmann's constant, n the ideality factor and V the applied voltage. From the data, ϕ_b was obtained as 1.08 eV and $n=1.4$. The forward turn-on voltage V_F for a Schottky rectifier is given by

$$V_F = \frac{nkT}{e} \ln\left(\frac{J_F}{A^{**}T^2}\right) + n\phi_B + R_{ON} J_F \quad (3.39)$$

where R_{ON} is the on-state resistance. Defining V_F as the bias at which the forward current density is 100 A·cm⁻², V_F was found to be 2.2 V and R_{ON} was 8 mΩ·cm² for an individual device.

With the thick plated metal on the rectifiers, the forward I–V characteristics in Figure 3.25 for a single 500×500 μm² device showed a slight increase in V_F to ~3 V. This might result from current spreading in the structure. The forward dc at 7 V was 1.2 A for the single device. The highest reported forward current from a GaN rectifier is 1.6 A under pulsed (10% duty cycle) conditions.[20]

3.7 Bulk Diode Arrays 209

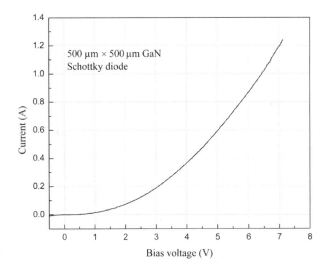

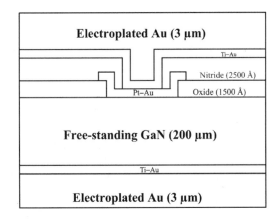

Figure 3.25. Forward I–V characteristic from single 500×500 μm² GaN Schottky rectifier (top) and schematic of free-standing GaN rectifier structure

To connect the output of all of the individual devices, we clamped both sides of the array to Cu disks using a press-packed arrangement and measure the dc I–V characteristics at 25°C using a Tektronix 371B curve tracer. Figure 3.26 shows a typical I–V characteristic, with output currents of 24 A at 3 V and 161 A at 7.12 V. The latter value corresponds to a total on-state resistance of 44 m$\Omega \cdot$cm^2. The on/off ratio was ~8×10^7 at 5 V/ –100 V for this rectifier array at 25°C.

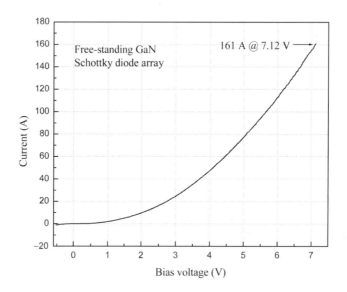

Figure 3.26. Total forward I–V characteristic from rectifier array

The currents demonstrated with the bulk rectifier array show the promise of this technology for even the most demanding applications, such as utility power switching [21] and power distribution in next-generation naval vessels. [22, 23] A simple calculation indicates that V_B values in excess of 16 kV are achievable on free-standing GaN substrates of thickness ≥50 μm provided the carrier concentration is ≤5×10^{15} cm^{-3}, so that both high forward currents and high reverse breakdown voltages should be possible with this approach.

3.8 Conclusions

In summary, the size and geometry dependence of GaN Schottky rectifiers on both sapphire and quasi-bulk substrates have been investigated. The reverse breakdown voltage increases dramatically as contact size is decreased, and is also much larger for vertically depleting devices. The low on-state resistances produce high figures-of-merit for the rectifiers and show their potential for applications involving high power electronic control systems.

References

1. Heydt GT, Skromme BJ (1998) Mat Res Soc Symp Proc Vol 483: 3 (MRS Pittsburgh, PA 1998)
2. Brown ER (1998) Solid. State Electrochem 43: 1918
3. Trivedi M, Shenai K (1999) J Appl Phys 85: 6889
4. Pearton SJ, Ren F, Zhang AP, Lee KP (2000) Mar Sci Eng R 30: 55
5. Bandic ZZ, Bridger DM, Piquette EC, McGill TC, Vaudo RP, Phanse VM, Redwing JM (1999) Appl Phys Lett 74: 1266
6. Zhang AP, Johnson JW, Ren F, Han J, Polyakov AJ, Smirnov NB, Govorkov AV, Redwing JM, Lee KP, Pearton SJ (2001) Appl Phys Lett 78: 823
7. Zhang AP, Dang G, Ren F, Han J, Polyakov AY, Smirnov NB, Govorkov AV, Redwing JM, Cao XA, Pearton SJ (2000) Appl Phys Lett 76: 1767
8. Johnson JW, LaRoche JR, Ren F, Gila BP, Overberg ME, Abernathy CR, Chyi JI, Chuo CC, Nee TE, Lee CM, Lee KP, Park SS, Park JI, Pearton SJ (2001) Solid-State Electrochem 45: 405
9. Johnson JW, Zhang AP, Luo WB, Ren F, Pearton SJ, Park SS, Park YJ, Chyi J-I (2002) IEEE Trans Electron Dev 49: 32
10. Johnson JW, Luo B, Ren F, Palmer D, Pearton SJ, Park SS, Park YJ (2002) Solid-State Electron 46: 911
11. Park SS, Park IW, Choh SH (2000) Jpn J Appl Phys 39: L1141
12. Morkoc H (2001) Mar Sci Eng R 33: 135
13. Zhang AP, Dang GT, Ren F, Cho H, Lee KP, Pearton SJ, Chty J-I, Nee TE, Chuo CC (1999) Solid-State Electron 44: 619
14. Ren F, Zhang AP, Dang G, Cao XA, Cho H, Pearton SJ, Chyi JI, Lee CM, Chuo CC (1999) Solid-State Electron 44: 619
15. Zhu TG, Lambert DJ, Shelton BS, Wong MM, Chowdhurg V, Dupuis RD (2000) Appl Phys Lett 77: 2918
16. Shelton BS, Zhu TG, Lambert DJ, Dupis RD (2001) IEEE Trans Electron Dev 48: 1498
17. Neudeck PG, Larkin DJ, Rowell A, Matus LG (1994) Appl Phys Lett 64: 1386
18. Morisette DR, Cooper JA Jr, Melloch MR, Dolng GM, Shenoy PM, Zakari M, Gladish J (2001) IEEE Trans Electron Dev 48: 349
19. Itoh A, Kimoto T, Matsunami H (1996) IEEE Electron Dev Lett 17: 139

20. Saxena V, Su JN, Steckl AJ (1999) IEEE Trans Electron Dev 46: 456
21. Tarplee MC, Madangarli VP, Zhang Q, Sudarshan TS (2001) IEEE Trans Electron Dev 48: 2659
22. Dyakonova NV, Ivanov PA, Koglov VA, Levinshtein ME, Palmour JW, Pumyantsev SL, Singh R (1999) IEEE Trans Electron Dev 46: 2188
23. Zhang AP, Dang G, Cao XA, Cho H, Ren F, Han J, Chyi J-I, Lee CM, Nee TE, Chuo CC, Chi GC, Chu SNG, Wilson RG, Pearton SJ (2000) MRS Internet J Nitride Semicond Res 551: W11.67
24. Mehandru R, Gila BP, Kim J, Johnson JW, Lee KP, Luo B, Onstine AH, Abernathy CR, Pearton SJ, Ren F (2002) Electrochem Solid-State Lett 5: G51
25. Hong M, Ng HM, Kwo J, Kortran AR, Baillargeon JN, Chu SNG, Mannaerts JP, Cho AY, Ren F, Abernathy CR, Pearton SJ (2000) presented at 197th Electrochemical Society Meeting, May 2000, Toronto, Canada
26. Luo B, Johnson JW, Kim J, Mehandru R, Ren F, Gila B, Onstine AH, Abernathy CR, Pearton SJ, Baca AG, Briggs RD, Shul RJ, Monier C, Han J (2002) Appl Phys Lett 80: 1661
27. Brezeane G, Fernandez J, Millan J, Rebello J, Badila M, Dilimot G (1998) Mater Sci Forum 264–268: 941
28. Teisseyre H S. Smith, J. Gao and D. Lars *et al.* (1994) J Appl Phys 76: 2429
29. Levinshtein ME, Rumyantsev SL, Shur MS (2001) In: *Properties of Advanced Semiconductor Material*, Wiley Interscience
30. Kolessar R, Nee H-P (2001) APEC 2001, Sixteenth Annual IEEE Vol 2 Conf. in Power Electronics
31. Mnatsakanov TT, G. Green, M. Ren, J. Gao and P. Petroff *et al.* (2003) Solid-State Electron 47
32. Oguzman IH, Bellotti E, Brennan KF, Kolnik J, Wang R, Ruden PP (1997) J Appl Phys 81: 12
33. Ruff M, Mitlehner H, Helbig R (1994) IEEE Trans Electron Devices 41: 1040
34. Trivedi M, Shenai K (1999) J Appl Phys 85: 6889
35. Kolnik J, Oguzman H, Brennan KF, Wang R, Ruden PP (1997) J Appl Phys 81: 2
36. Fulop W (1967) Solid-State Electron 10: 39
37. He J, Wang Y, Zhang X, Xi X, Chan M, Huang R, Hu C (2002) IEEE Trans Electron Devices 49: 933
38. Ren F, Zolper JC (2003) *Wide Energy Bandgap Electronic Device*, (World Scientific Publishers, Singapore, 2003,) p. 125

4 Chemical, Gas, Biological, and Pressure Sensing

4.1 Abstract

There is renewed emphasis on development of robust solid-state sensors capable of uncooled operation in harsh environments. The sensors should be capable of detecting chemical, gas, biological or radiation releases, as well as be able to send signals to central monitoring locations. In this chapter we discuss the advances in use of GaN-based solid-state sensors for these applications. AlGaN–GaN high electron-mobility transistors (HEMTs) show a strong dependence of source–drain current on the piezoelectric polarization-induced 2DEG. Furthermore, spontaneous and piezoelectric polarization-induced surface and interface charges can be used to develop very sensitive but robust sensors for the detection of gases, polar liquids and mechanical pressure. It has been demonstrated that AlGaN–GaN HEMT structures exhibit large changes in source–drain current upon exposing the gate region to various block co-polymer solutions. Pt-gated GaN Schottky diodes and Sc_2O_3–AlGaN–GaN metal-oxide semiconductor diodes also show large change in forward currents upon exposure to H_2-containing ambients. Of particular interest are methods for detecting ethylene (C_2H_4), which offers problems because of its strong double bonds and hence the difficulty in dissociating it at modest temperatures. Apart from combustion gas sensing, the AlGaN–GaN heterostructure devices can be used as sensitive detectors of pressure changes. In addition, large changes in source–drain current of the AlGaN–GaN HEMT sensors can be detected upon adsorption of biological species on the semiconductor surface. Finally, the nitrides provide an ideal platform for fabrication of surface acoustic wave (SAW) devices. Given all these attributes, the GaN-based devices appear to be promising for a wide range of chemical, biological, combustion gas, polar liquid, strain and high temperature pressure-sensing applications. In addition, the sensors are compatible with high bit-rate wireless communication systems that facilitate their use in remote arrays.

4.2 Introduction

AlGaN–GaN HEMTs have demonstrated extremely promising results for use in broadband power amplifiers in base station applications due to the high sheet carrier concentration, electron mobility in the 2DEG channel and high saturation velocity.[1–10] The high electron sheet carrier concentration of nitride HEMTs is induced by piezoelectric polarization of the strained AlGaN layer and spontaneous polarization is very large in wurtzite Group III nitrides.[6–9] This suggests that nitride HEMTs are excellent candidates for pressure sensor and piezoelectric-related applications. Payoffs relative to Si FETs include being chemically inert in electrolyte solutions (low linear drift), sensitivity of ~57mV/pH, cf. other potential gate materials such as SiO_2 (32–40 mV/pH), Al_2O_3 (57 mV/pH), Ta_2O_5 (58.5 mV/pH) and the ability to integrate with blue or UV LEDs for directed cell growth or remediation and also integration with chemical/pressure sensors and off-chip communications circuitry (either by photonic or electrical means).

Unlike conventional III–V-based HEMTs, such as those in the AlGaAs–GaAs system, there is no dopant in the typical nitride-based HEMT structure and all the layers are undoped. The carriers in the 2DEG channel are induced by piezoelectric polarization of the strained AlGaN layer and spontaneous polarization, both of these effects being very large in wurtzite Group III nitrides. Carrier concentrations $>10^{13}$ cm^{-3} can be obtained in the 2DEG, which is five times larger than that in the more conventional AlGaAs–GaAs material system. The portion of the carrier concentration induced by the piezoelectric effect is around 45–50%. This makes nitride HEMTs excellent candidates for pressure sensor and piezoelectric-related applications.

There is a strong interest in wide bandgap semiconductor gas sensors for applications including fuel leak detection in spacecraft and release of toxic or corrosive gases. In addition, these detectors would have dual use in automobiles and aircraft, fire detectors, exhaust diagnosis and emissions from industrial processes. Wide bandgap semiconductors, such as GaN and SiC, are capable of operating at much higher temperatures than more conventional semiconductors, such as Si, because of their large bandgap (3.4 eV for GaN, 3.26 eV for the 4H-SiC poly-type vs. 1.1 eV for Si). The ability of these materials to function in high temperature, high power and high flux/energy radiation conditions will enable large performance enhancements in a wide variety of spacecraft, satellite and radar applications. GaN and SiC uncooled electronics and sensors will reduce spacecraft launch weights and increase satellite functional capabilities. Given the high cost per pound of launching payloads into Earth orbit, the weight savings gained by using wide bandgap devices could have large economic and competitive implications in the satellite industry. Existing commercial satellites require thermal radiators to dissipate heat generated by the spacecraft electronics. These radiators could be eliminated with GaN and SiC, and allow greater functionality (more transponders in a commercial satellite) by utilizing the space and weight formerly occupied by the thermal management system. In addition, the radiation hardness of these materials would reduce the weight of shielding normally used to protect spacecraft electronic components from radiation. Simple Schottky diode or FET structures, fabricated in GaN and SiC are sensitive to a number of gases, including hydrogen and hydrocarbons. An additional attractive attribute of GaN and SiC is the fact that gas sensors based on these materials could be integrated with high-temperature electronic devices on the same chip. While there has been extensive development of SiC-based gas sensors [11–25], the work on GaN is at an earlier stage [26,27], but there has been much recent activity based on the relative advantages of GaN, for sensing.[26–40] These advantages include the presence of the polarization-induced charge, the availability of a heterostructure and the more rapid pace of device technology development for GaN, which borrows from the commercialized and LD businesses.

Wide bandgap electronics and sensors based on GaN can be operated at elevated temperatures (600°C, or 1112°F, glowing red hot) where conventional Si-based devices cannot function, these being limited to <350°C. This is due to its low intrinsic carrier concentration at high temperature, as shown in Figure 4.1. The ability of these materials to function in high temperature, high power and high flux/energy radiation conditions will enable large performance enhancements in a wide variety of spacecraft, satellite, homeland defense, mining, automobile, nuclear power and radar applications. In addition, there appear to be applications for high-temperature pressure sensing for coal and fossil energy applications. AlGaN–GaN heterojunction HEMTs grown on SiC substrates have demonstrated extremely promising results for use as power amplifiers in many analog applications due to the high sheet carrier concentration, electron mobility in the 2DEG channel and high saturation velocity, as illustrated in Figure 4.2.[41] The demonstrated power densities from these devices are well in excess of those possible with GaAs, Si or SiC. The ability to integrate these with GaN-based LEDs is a powerful driver for highly integrated sensor systems. Figure 4.3 shows examples of concepts that employ GaN LEDs for directed cell growth or for causing fluorescence for identification of DNA fragments in a microfluidic array. In the latter case, the fluid containing the DNA is placed in a microfluidic channel and the uncharged DNA fragments in the laminar flow region in the center of the channel are not affected by voltages applied to the gold tabs on the surface of the channel. However, the applied field acts to pull charged fragments to the non-laminar region at the edge of the channel, where they can be separated from the main flow. The integrated, embedded LEDs can be used to induce fluorescence from these separated fragments.

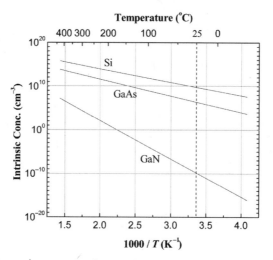

Figure 4.1. Intrinsic carrier concentration as a function of temperature

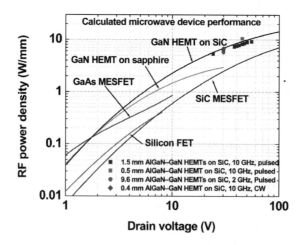

Figure 4.2. Performance of GaN–AlGaN-based power amplifier compared with GaAs-, Si- and SiC-based devices

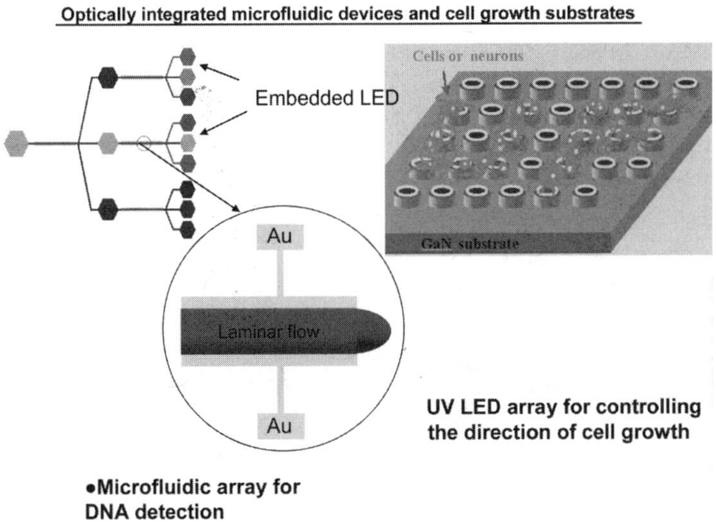

Optically integrated microfluidic devices and cell growth substrates

Figure 4.3. Schematic of optically integrated microfluidic devices and cell-growth substrates based on GaN

The charges in the 2D channel of AlGaN–GaN HEMTs are induced by spontaneous and piezoelectric polarization, which are balanced with positive charges on the surface. Figure 4.4 shows schematics of the direction of the spontaneous and piezoelectric polarization in both Ga and N face wurtzite GaN crystals.[42] The induced sheet carrier concentration of undoped Ga-face AlGaN–GaN can be calculated thus [43]:

$$n_s(x) = \frac{\sigma(x)}{e} - \left(\frac{\varepsilon_0 \varepsilon(x)}{d_d e^2}\right)\left(e\phi_b(x) + E_F(x) - \Delta E_C(x)\right) \qquad (4.1)$$

where ε_0 is the electric permittivity, $\varepsilon(x) = 9.5-0.5x$ is the relative permittivity, x is the Al mole fraction of $Al_xGa_{1-x}N$, d_d is the AlGaN layer thickness, $e\phi_b$ is the Schottky barrier of the gate contact on AlGaN ($e\phi_b(x) = 0.84 + 1.3x(eV)$), E_F is the Fermi level
$(E_F(x) = [9he^2 n_s(x)/16\varepsilon_0\varepsilon(x)\sqrt{(8m^*(x))}]^{2/3} + h^2 n_s(x)/4\pi m^*(x))$,
E_c is the conduction band ($E_g(x) = 6.13x+3.42(1-x)-x(1-x)(eV)$), and ΔE_c is the conduction band discontinuity between AlGN and GaN ($\Delta E_c(x) = 0.7[E_g(x)-E_g(0)]$). Note that $m^*(x) \approx 0.228$. Therefore, the sheet charge density in the 2D channel of an AlGaN–GaN HEMT is extremely sensitive to its ambient. The adsorption of polar molecules on the surface of GaN affects both the surface potential and the resulting device characteristics.

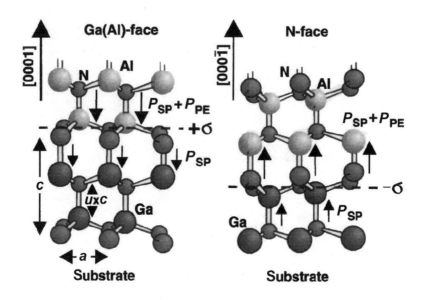

Figure 4.4. Piezoelectric (PE) and spontaneous polarization (SP) effects in Ga(Al) or N-face AlGaN–GaN heterostructures (after Refs [57] and [42])

Without any surface passivation the sheet carrier concentration of the polarization-induced 2DEGs confined at interfaces of AlGaN–GaN HEMTs become sensitive to any manipulation of surface charge. This effect is used to build micro-sensors which are able to detect applied strain and surface polarity change by polar liquids or toxic gases. Besides the basic physics of polarization-induced charges at surfaces and interfaces and the formation of 2DEGs in piezoelectric AlGaN–GaN heterostructures, we will summarize the present status and future concepts of novel AlGaN–GaN-based chemical and biological sensor systems.

In this chapter, the components for building high-temperature chemical, gas and pressure sensors are discussed in addition to the emerging work on using AlGaN–GaN heterostructures as biosensors. Numerous groups have demonstrated the feasibility of AlGaN–GaN heterostructure-based hydrogen detectors with extremely fast time response and capable of operating at high temperature (500–800°C), eliminating bulky and expensive cooling systems.[26–37] In addition, gateless AlGaN–GaN HEMTs show a strong dependence of source–drain current on the polarity and concentration of polymer solutions.[39] There have also been recent reports of the investigation of the effect of external strain on the conductivity of AlGaN–GaN

HEMTs.[40] It is also possible to employ selective substrate removal techniques, such as laser drilling, to etch off the SiC substrate and realize an AlGaN–GaN membrane for a pressure sensor. These can be used for inexpensive, low-cost, low-weight pressure monitoring applications.

4.3 Sensors Based on AlGaN–GaN Heterostructures

4.3.1 Gateless AlGaN–GaN High Electron Mobility Transistor Response to Block Co-Polymers

Gateless AlGaN–GaN HEMT structures exhibit large changes in source–drain current upon exposing the gate region to various block co-polymer solutions.[39] The polar nature of some of these polymer chains leads to a change of surface charges in the gate region on the HEMT, producing a change in surface potential at the semiconductor–liquid interface. The structure was grown by MOCVD deposition at 1040°C on c-plane Al_2O_3 substrates and then fabricated into a gateless HEMT. Silicon nitride was used to encapsulate the source–drain regions, with only the gate region open to allow the polar liquids to reach across the surface.

A schematic of the device is shown in Figure 4.5, with both the plan view layout and cross-sectional view. The source–drain current–voltage characteristics were measured at 25°C using an Agilent 4156C parameter analyzer with the gate region exposed either to air or various concentrations of the block co-polymers and individual polymers. The block copolymers are composed different proportions of the individual polymers, polystyrene (PS) and polystyrene oxide (PEO).

Owing to the parasitic resistance induced by the large gate dimension (20 × 150 μm^2) I, the drain current i did not reach saturation at 40 V. If the device dimension were to be reduced, then larger changes of drain current should be expected. The dipole moments of ethylene and styrene monomers are 1.89 and 0.62 respectively.[33] The dipole moment of the ethylene monomer is three times larger than that of the styrene monomer; however, the effect of PEO solutions on drain current of nitride HEMTs is only half that of the PS. This could be due to PEO being extremely linear, with its dipole oriented along the polymer chain. The 20 × 150 μm^2 gate opening is much larger than the individual monomer in the PEO chain. Therefore, the dipole effect on the device is very locally within the big gate opening and some of the net surface charges induced by the PEO are cancelled out. In the case of PS the polymer chain is very bulky and not very linear, and the net dipole induced charge may be higher than that of PEO;

therefore, larger drain current changes were observed. If the gate dimension reduces to the size of the individual monomer, then the opposite trends may be obtained.

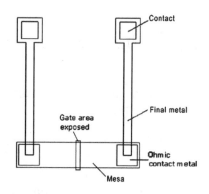

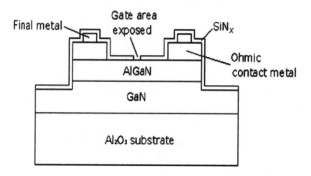

Figure 4.5. Schematic layout of gateless HEMT structures (top) and device cross-section (bottom)

The concentration of the block co-polymer also affected the changes of the drain current, as showed in Figure 4.6 (top). The molecular weight of the co-polymer is 58.8 kg/mol and it contains 71% PS and 29% PEO. The concentration of the co-polymer was varied from 1.0917 mg/ml to 0.08734 mg/ml. As the co-polymer concentration increased, the drain current reduced more. Co-polymer with similar molecular weight, but different compositions also had significant impact on the drain I–V characteristics. As shown in Figure 4.6 (bottom), the copolymer with large percentage of PS reduced more drain current.

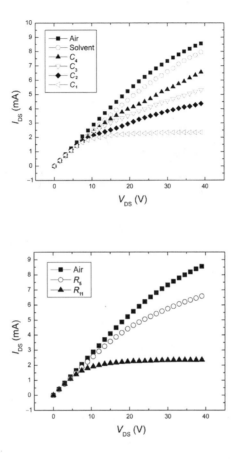

Figure 4.6. Drain I–V characteristics for the different concentrations of the copolymer solutions are as follows: $C_1 = 1.0917$ mg/ml, $C_2 = 0.7612$ mg/ml, $C_3 = 0.5504$ mg/ml, $C_4 = 0.08734$ mg/ml (top). The drain I–V characteristics of copolymers with different compositions (R_5: MW=66.6 kg/mol, 53% of PS and 47% of PEO; R_{11}: MW = 58.8 kg/mol, 71% of PS and 29% of PEO) (bottom)

Eickhoff *et al.* [33] have reported on the time dependence of changes in source–drain current of gateless HEMTs exposed to polar liquids with different dipole moments. As shown in Figure 4.7, the HEMTs show very large changes in current as the different polar liquids are introduced to the surface. However, the magnitude of the changes did not correlate with the dipole moment of the liquids, suggesting that steric hindrance effects or the exact orientation of the molecule when it adsorbs on the surface must play a role.

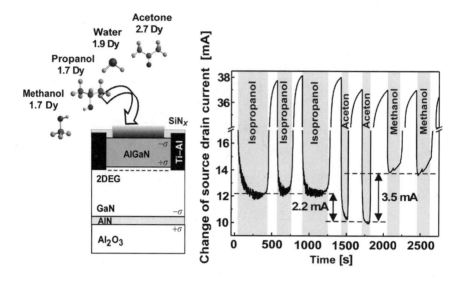

Figure 4.7. Time dependence of changes in source–drain current in gateless HEMTs exposed to ionic liquids with different dipole moments (after Ref. [29])

In summary, gateless AlGaN–GaN HEMTs show a strong dependence of source–drain current on the polarity and concentration of polymer or polar liquid solutions. These can be used as sensitive pH meters. This data suggests the possibility of functionalizing the surface for application as biosensors, especially given the excellent stability of the GaN and AlGaN surfaces, which should minimize degradation of adsorbed cells.

4.3.2 Hydrogen Gas Sensors Based on AlGaN–GaN-Based Metal-Oxide Semiconductor Diodes

Simple GaN Schottky diodes exhibit strong changes in current upon exposure to hydrogen-containing ambients.[27–35] The effect is thought to be due to a lowering of the effective barrier height as molecular hydrogen catalytically cracks on the metal gate (typically made of Pt) and atomic hydrogen diffuses to the interface between the metal and GaN, altering the interfacial charge. Steinhoff et al. [29] found that it was necessary to have a native oxide present between the semiconductor and the gate metal in order to see significant current changes. Thus, it is desirable to specifically incorporate an oxide into GaN-based diodes or HEMTs in order to maximize the hydrogen detection response.

Gas sensors based on metal-oxide semiconductor (MOS) diodes on AlGaN–GaN HEMT layer structures are of interest, because HEMTs are expected to be the first GaN electronic device that is commercialized, as part of next-generation radar and wireless communication systems. These structures have much higher sensitivity than Schottky diodes on a GaN layer, because they are true transistors and, therefore, operate with gain. In addition, the MOS-gate version of the HEMT has significantly better thermal stability than a metal-gate structure [38,44–51] and is well-suited to gas sensing. When exposed to changes in ambient, changes in the surface potential will lead to large changes in channel current.

A typical HEMT gas sensor employs ohmic contacts formed by lift-off of e-beam-deposited Ti(200 Å)–Al(1000 Å)–Pt(400 Å)–Au(800 Å). The contacts are annealed at 850°C for 45 s to minimize the contact resistance. Thin (typically 400 Å thickness) gate dielectrics of MgO or Sc_2O_3 are deposited through a contact window of SiN_x. The dielectrics are deposited by rf plasma-activated MBE at 100°C using elemental Mg or Sc evaporated from a standard effusion all cell at 1130°C and O_2 derived from an Oxford RF plasma source.[49] 200 Å Pt Schottky contacts are deposited on the top of Sc_2O_3. Then, final metal of e-beam–deposited Ti–Au (300 Å–1200 Å) interconnection contacts was employed on the MOS-HEMT diodes. Figure 4.8 shows a schematic (top) and photograph (bottom) of the completed device. The devices are bonded to electrical feed-throughs and exposed to different gas ambients in an environmental chamber. A schematic of the chamber is shown in Figure 4.9.

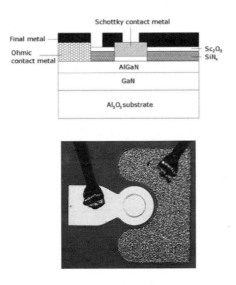

Figure 4.8. Cross-sectional schematic of completed MOS diode on AlGaN–GaN HEMT layer structure (top) and plan-view photograph of device (bottom)

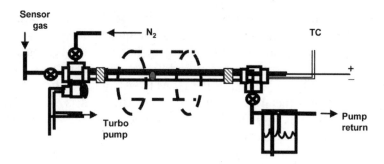

Figure 4.9. Schematic of environmental test chamber (TC) into which different gases can be introduced and the GaN-based sensors tested at different temperatures under different gas ambients

As the detection temperature is increased, the response of the MOS-HEMT diodes increases due to more efficient cracking of the hydrogen on the metal contact. Pt is a good choice as the contact owing to its efficiency at cracking hydrogen at relatively low temperatures; but, for applications requiring more thermally stable contacts, we have also found that W is effective at higher operating temperatures (300°C).

The threshold voltage for a MOSFET is given by [43]

$$V_T = V_{FB} + 2\phi_b + \left(\frac{4eN_D\phi_b\varepsilon_S}{C_i}\right)^{0.5} \qquad (4.2)$$

where V_{FB} is the voltage required for flat band conditions, ϕ_b the barrier height, e the electronic charge and C_i the Sc_2O_3 capacitance per unit area. In analogy with results for MOS gas sensors in other materials systems [20], the effect of the introduction of the atomic hydrogen into the oxide is to create a dipole layer at the oxide–semiconductor interface that will screen some of the piezo-induced charge in the HEMT channel.

To test the time response of the MOS diode sensors, the $10\%H_2/90\%N_2$ ambient was switched into the chamber through a mass flow controller for periods of 10, 20 or 30 s and then switched back to pure N_2. Figure 4.10 (left) shows the time dependence of forward current at a fixed bias of 2 V under these conditions. The response of the sensor is rapid (<1 s), with saturation taking almost the full 30 s. Upon switching out of the hydrogen-containing ambient, the forward current decays exponentially back to its initial value. This time constant is determined by the volume of the test chamber and the flow rate of the input gases and is not limited by the response of the MOS diode itself. Figure 4.10 (top) shows the time response of the forward current at fixed bias to a series of gas injections into the chamber, of duration 10 s each (top) or 30 s each (bottom). The MOS diode shows good repeatability in its changes of current and the ability to cycle this current in response to repeated introductions of hydrogen into the ambient. Once again, the response appears to be limited by the mass transport of gas into and out of the chamber and not to the diffusion of hydrogen through the Pt–Sc_2O_3 stack.

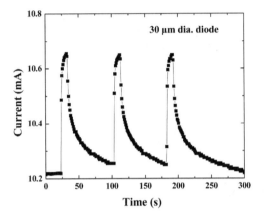

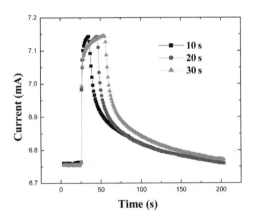

Figure 4.10. Time response at 25°C of MOS-HEMT-based diode forward current at a fixed bias of 2 V when switching the ambient from N_2 to 10%H_2/90%N_2 for periods of 10, 20 or 30 s and then back to pure N_2 (bottom) and three cycles of switching the ambient from N_2 to 10%H_2/90%N_2 for periods of 10 s (top)

AlGaN–GaN MOS-HEMT diodes appear well suited to combustion-gas sensing applications. The changes in forward current are approximately double those of simple GaN Schottky diode gas sensors tested under similar conditions and suggest that integrated chips involving gas sensors and HEMT-based circuitry for off-chip communication are feasible in the AlGaN–GaN system.

4.3.3 Hydrogen-Induced Reversible Changes in Sc_2O_3–AlGaN–GaN High Electron Mobility Transistors

Pt-contacted AlGaN–GaN HEMTs with Sc_2O_3 gate dielectrics show reversible changes in drain–source current upon exposure to H_2-containing ambients, even at room temperature. The changes in current (as high as 3mA for relatively low gate voltage and drain-source voltage) are approximately an order of magnitude larger than for Pt–GaN Schottky diodes and a factor of five larger than Sc_2O_3–AlGaN–GaN MOS diodes exposed under the same conditions. This shows the advantage of using a transistor structure in which the gain produces larger current changes upon exposure to hydrogen-containing ambients. The increase in current is the result of a decrease in effective barrier height of the MOS gate of 30–50 mV at 25°C for $10\%H_2/90\%N_2$ ambients relative to pure N_2 and is due to catalytic dissociation of the H_2 on the Pt contact, followed by diffusion to the Sc_2O_3 – AlGaN interface.

Figure 4.11 (top) shows the MOS-HEMT drain–source current voltage (I_{DS}–V_{DS}) characteristics at 25°C measured in both the pure N_2 or $10\%H_2/90\% N_2$ ambients. The current is measurably larger in the latter case, as would be expected if the hydrogen catalytically dissociates on the Pt contact and diffuses through the Sc_2O_3 to the interface, where it screens some of the piezo-induced channel charge. This is a clear demonstration of the sensitivity of AlGaN–GaN HEMT dc characteristics to the presence of hydrogen in the ambient in which they are being measured. The use of less-efficient catalytic metals as the gate metallization would reduce this sensitivity, but operation at elevated temperatures would increase the effect of the hydrogen because of more efficient dissociation on the metal contact.

Figure 4.11 (bottom) shows some of the recovery characteristics of the MOS-HEMTs upon cycling the ambient from N_2 to $10\%H_2/90\%N_2$. While the change in drain–source current is almost instantaneous (<1 s), the recovery back to the N_2 ambient value is of the order of 20 s. This is controlled by the mass transport characteristics of the gas out of the test chamber, as demonstrated by changing the total flow rate upon switching the gas into the chamber. Given that the current change upon introduction of the hydrogen is rapid, the effective diffusivity of the atomic hydrogen through the Sc_2O_3 must be greater than 4×10^{-12} $cm^2/V.s$ at 25°C. Note the complete reversibility of the drain–source current for repeated cycling of the ambient.

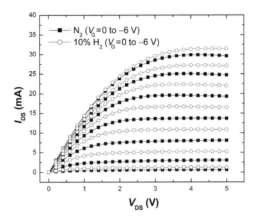

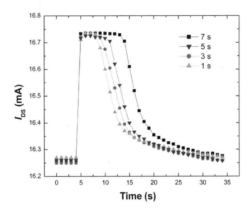

Figure 4.11. I_{DS}–V_{DS} characteristics of MOS-HEMT measured at 25°C under pure N_2 ambient or in 10%H_2/90%N_2 ambient (top) and time dependence of drain–source current when switching from N_2 to 10%H_2/90%N_2 ambient and back again for different injection times of the H_2/N_2 (bottom)

Sc_2O_3–AlGaN–GaN MOS-HEMTs show a marked sensitivity of their drain–source current to the presence of hydrogen in the measurement ambient. This effect is due to the dissociation of the molecular hydrogen on the Pt gate contact, followed by diffusion of the atomic species to the oxide–semiconductor interface, where it changes the piezo-induced channel charge.

The MOS-HEMTs show larger changes in current than their corresponding MOS-diode or Schottky diode counterparts and show promise as sensitive hydrogen detectors.

Of particular interest are methods for detecting ethylene (C_2H_4), which offers problems because of its strong double bonds and hence the difficulty in dissociating it at modest temperatures. The forward diode I–V characteristics at 400°C of the Pt–Sc_2O_3–AlGaN–GaN MOS-HEMT diode, both in pure N_2 and in a 10%C_2H_4/90%N_2 atmosphere, show that, at a given forward bias, the current increases upon introduction of the C_2H_4. In analogy with the detection of hydrogen in comparable SiC and Si Schottky diodes, a possible mechanism for the current increases involves atomic hydrogen either decomposed from C_2H_4 in the gas phase or chemisorbed on the Pt Schottky contacts then catalytically decomposed to release atomic hydrogen. The hydrogen can then diffuse rapidly though the Pt metallization and the underlying oxide to the interface, where it forms a dipole layer [24] and lowers the effective barrier height. We emphasize that other mechanisms could be present; however, the activation energy for the current recovery is ~1 eV, similar to the value for atomic hydrogen diffusion in GaN [52], which suggests that this is at least a plausible mechanism.

Figure 4.12 shows the change in voltage at fixed current as a function of temperature for the MOS diodes when switching from a 100% N_2 ambient to 10%C_2H_4/90%N_2. As the detection temperature is increased, the response of the MOS-HEMT diodes increases due to more efficient cracking of the hydrogen on the metal contact. Note that the changes in both current and voltage are quite large and readily detected. In analogy with results for MOS gas sensors in other materials systems [23,24], the effect of the introduction of the atomic hydrogen into the oxide is to create a dipole layer at the oxide–semiconductor interface that will screen some of the piezo-induced charge in the HEMT channel. The time constant for response of the diodes was determined by the mass transport characteristics of the gas into the volume of the test chamber and was not limited by the response of the MOS diode itself.

Simple Pt or Pd gate GaN Schottky diodes can also provide effective detection of combustion gases. Figure 4.13 shows a plan-view photograph of a two-terminal sensor, along with a device packaged into a header for testing. These devices produce easily measurable changes in forward current upon exposure to hydrogen-containing ambients, as shown in the I–V characteristics of Figure 4.14. This data was taken at 150°C, but the changes are also measurable at room temperature, showing that the sensors do not need an on-chip heater to increase their detection efficiency, although they of course show larger current changes at higher temperatures.

4.3 Sensors Based on AlGaN–GaN Heterostructures   229

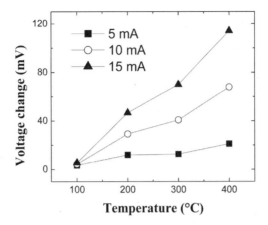

Figure 4.12. Change in MOS diode forward voltage at fixed currents as a function of temperature for measurement in 100%N_2 or 10%C_2H_4/90%N_2

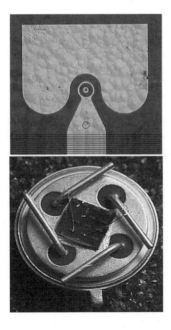

Figure 4.13. Photographs of two-terminal GaN Schottky diode for gas-sensing applications (top) and packaged device (bottom)

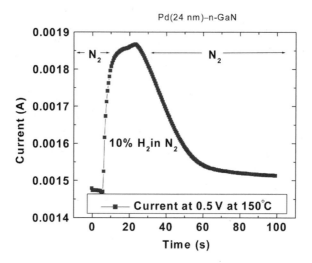

Figure 4.14. Forward current of Pd–GaN Schottky diode at fixed bias of 0.5 V and a temperature of 150°C, upon changing the measurement ambient from pure N_2 to 10% H_2 in N_2

4.3.4 Effect of External Strain on Conductivity of AlGaN–GaN High Electron Mobility Transistors

In order to study the effect of external strain on the conductivity of HEMTs, material systems and transmission line patterns were fabricated with a mesa of 500 Å to ensure the 1Δ current flow during the measurement. Figure 4.15 illustrates the setup for measuring the effect of external strain on the conductivity of the 2DEG channel of the nitride HEMT. Lucite blocks secure the sample and printed circuit board (PCB) for testing. The contact pads were connected to the PCB board, which had BNC connectors on the end for signal outputs, with 1 mm thickness gold wire. A high-precision single-axis traverse was used to bend the sample. The devices were fabricated on half of a 2″ wafer, sawn into 2 mm wide stripes and wire bonded on the test feature, as shown in Figure 4.16. The dc characteristics were obtained from measurements on an Agilent 4156C parameter analyzer.

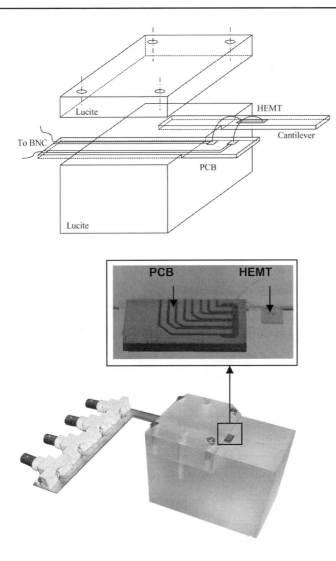

Figure 4.15. Schematic diagram of a pressure sensor package: Experimental setup to detect I–V characteristics connected to the BNC cable according to various mechanical stresses (top). Picture of actual set-up for cantilever mounting package (bottom)

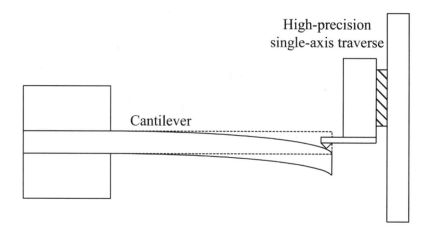

Figure 4.16. Schematic diagram of mechanical stressor with cantilever (top). Picture of mechanical stressor with high-precision single-axis traverse (bottom)

4.3 Sensors Based on AlGaN–GaN Heterostructures

Sheet charges in the AlGaN–GaN HEMTs are induced by spontaneous polarization and piezoelectric polarization.[51–57] Wurtzite GaN and AlGaN are tetrahedral semiconductors with a hexagonal Bravais lattice with four atoms per unit cell. The misfit strain inside a film is measured against its relaxed state. In the misfit strain calculation, the strain is calculated against the relaxed films, a_o, i.e.

$$\varepsilon_{misfit} = \frac{a(x) - a_o(x)}{a_o(x)} \tag{4.3}$$

where $a(x)$ is the lattice constants of $Al_xGa_{1-x}N$ and $a_o(x) = (a_{GaN} - a_{AlN}x)\text{Å} = (3.189 - 0.077x)\text{Å}$.[51] Here, $a(x) - a_o(x)$ is chosen because the AlGaN film is always under tension due to $a(x) \leq a_{GaN}$ and misfit will always be positive. Having defined the elastic strain in the film, a partially relaxed film parameter should be defined. For a perfectly coherent film $a(x) = a_{GaN}$, or

$$\varepsilon_{max.misfit} = \frac{a_{GaN} - a_o(x)}{a_o(x)} = \frac{0.077x}{3.189 - 0.077} \approx \frac{0.077x}{3.189} = 0.024x \tag{4.4}$$

For a partially relaxed film, the ratio of strain comparing to the unrelaxed state defines the degree of strain in the film

$$S(x) = \frac{\varepsilon_{misfit}}{\varepsilon_{max.misfit}} = \frac{a(x) - a_o(x)}{a_{GaN} - a_o(x)} \tag{4.5}$$

Hence, the degree of relaxation is then given by

$$r(x) = 1 - S(x) = \frac{a_{GaN} - a(x)}{a_{GaN} - a_o(x)} \tag{4.6}$$

For around 300 Å of AlGaN layer on the top of GaN, $r(x)$ was measured by Ambacher et al.[57]:

$$r(x) = \begin{cases} 0 & 0 \leq x \leq 0.38 \\ 3.5x - 1.33 & 0.38 \leq x \leq 0.67 \\ 1 & 0.67 \leq x \leq 1 \end{cases} \tag{4.7}$$

The piezoelectric polarization for a partially relaxed strained layer can be expressed by modifying Equation (4.5) by subtracting the portion of relaxation:

$$\sigma(x) = 2(1-r(x))\left(\frac{a_{\text{GaN}} - a_0}{a_0}\right)\left(e_{31} - e_{33}\frac{C_{13}}{C_{33}}\right) \quad (4.8)$$

where e_{31} and e_{33} are the piezoelectric coefficients and C_{13} and C_{33} are the elastic constants.

Figure 4.17 (bottom) plots the relationship between Al concentration of AlGaN and strain induced by AlGaN on GaN for the unrelaxed and partially relaxed conditions. Unlike the linear model for the unrelaxed condition, there is a maximum strain around an Al concentration of 0.35. The piezoelectric polarization-induced sheet carrier concentration of undoped Ga-face AlGaN/GaN can be calculated from Equation (4.3). As illustrated in Figure 4.17 (top), the sheet carrier concentration induced by the piezoelectric polarization is a strong function of Al concentration. In the case of a partially relaxed AlGaN strain layer, there is also a maximum sheet carrier concentration around Al = 0.35. If an external stress can be applied to the AlGaN–GaN material system, then the sheet carrier concentration can be changed significantly and devices fabricated in this fashion could be used in sensor-related applications.

A pure bending of a single beam was used in this work to estimate the strain, since the HEMT structure, around 3 μm, is much thinner than that of a sapphire substrate, 200 μm, and the degree of deflection (maximum deflection is around 2.2 mm) is much shorter than the length of the beam, 27 mm. The strain ε_{xx} of the bending can be estimated from the single beam with thickness of t and unit width. The tensile strain near the top surface of the beam is simply given by [58]

$$\varepsilon_{xx} = td/L^2 \quad (4.9)$$

where t is the sample thickness, d is the deflection and L is the length of the beam.

Figure 4.18 shows the effects of external tensile and compressive strain on the conductivity of an AlGaN–GaN HEMT sample with a mesa. In the case of applying external tensile strain on the HEMT sample, an increase of conductivity was observed. The mesa depth was around 500 Å, which is below the AlGaN–GaN interface (300 Å AlGaN layer on 3 μm GaN layer).

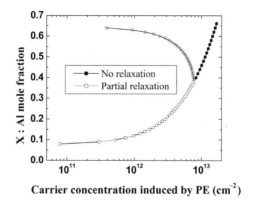

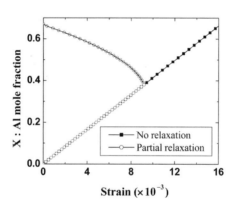

Figure 4.17. Strain induced by AlGaN on GaN for the unrelaxed and partially relaxed AlGaN layer as a function of Al concentration (bottom) and sheet carrier concentration induced by the piezoelectric polarization as a function of Al concentration (top)

The AlGaN layer, therefore, sits above the beam and the external stress applied on the beam should not change the strain of the AlGaN layer. As a result, applying a tensile stress on the beam would pull apart the GaN atoms only and not affect the AlGaN.

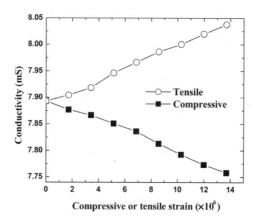

Figure 4.18. The effect of tensile or compressive stress on the conductivity of mesa-isolated AlGaN–GaN HEMTs

4.3.5 Pressure Sensor Fabrication

There are at least two approaches that can be taken: (1) AlGaN–GaN HEMT-based micro-electro-mechanical (MEM) pressure sensors, which can be used in a high temperature ambient; (2) biosensors with a gateless AlGaN–GaN HEMT membrane, in which the excellent stability of the GaN and AlGaN surface should minimize degradation of adsorbed cells.[33] The pressure sensors can be made of a circular membrane of AlGaN–GaN on a SiC substrate by etching a circular hole in the substrate, as illustrated in Figure 4.19 (left). A deflection of the membrane away from the substrate due to differential pressure on the two sides of the membrane produces a tensile strain in the membrane, as shown in Figure 4.19 (right). The differential piezoelectric responses of the AlGaN and GaN layers creates a space charge which induces a 2DEG at the AlGaN–GaN interface. The concentration of the 2DEG is expected to be directly correlated with the tensile strain in the membrane, and hence with the differential pressure. The radial strain is given by [59,60]

$$\varepsilon_r = (S - D)/D = (2\theta R - 2R \sin \theta) = \theta / \sin \theta - 1 \tag{4.10}$$

4.3 Sensors Based on AlGaN–GaN Heterostructures

The total tensile force around the edge of the circular membrane is [61,62]

$$T = \pi D t_{GaN} \sigma_r = \pi D t_{GaN} [E_{GaN}/(1-v)] \varepsilon_r \qquad (4.11)$$

where t_{GaN} is the film of thickness, D is the diameter of the via hole, $\sigma_r = [E_{GaN}/(1-v)]\varepsilon_r$, E_{GaN} is Young's modulus and v is Poisson's ratio of the GaN film. The component of T along the z direction is balanced by the force on the membrane due to a differential pressure $P_i - P_0$, where P_i and P_0 are the inside and outside pressures respectively.

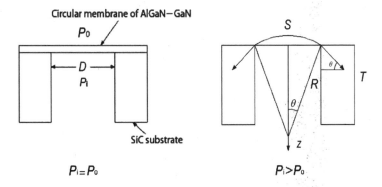

Figure 4.19. Pressure sensors made of a circular membrane of AlGaN–GaN on an SiC substrate by etching a circular hole in the substrate (left). A deflection of the membrane away from the substrate due to differential pressure on the two sides of the membrane produces a tensile strain in the membrane (right)

Hence $T \sin \theta = (P_i - P_0) \pi D^2/4$ and the radial strain ε_r in the nitride film can be expressed as a function of the differential pressure $P_i - P_0$

$$(\theta - \sin \theta) = (P_i - P_0)\left[(1-v)D\right]/(4E_{GaN} t_{GaN}) \qquad (4.12)$$

where E_{GaN} is Young's modulus, D is the diameter of the via hole, t_{GaN} is the film thickness and v is Poisson's ratio of the GaN film. If θ is measured, then the differential pressure $P_i - P_0$ can be estimated with Equation (4.12). We also derive the relationship between conductance σ of AlGaN–GaN HEMT and radial strain ε_r. [63–65]:

$$\sigma = \alpha(r_{AlGaN}) + \beta(f_{GaN}) \times \varepsilon_r \qquad (4.13)$$

where

$$\alpha(r_{\text{AlGaN}}) = \mu_s\{1/[1+(\varepsilon_0\varepsilon(x)/t_{\text{AlGaN}}e^2)h^2/4\pi m*(x)]\}\{-|\Delta_{\text{SP}}|$$
$$-|e_{\text{eff}}|\Delta\varepsilon_{\text{AlGaN}} \quad r_{\text{AlGaN}} - [\varepsilon_0\varepsilon(x)/t_{\text{AlGaN}} e][e\phi_b(x)-\Delta E_c(x)]\} \quad (4.14)$$

and

$$\beta(f_{\text{GaN}}) = \mu_s\{1/[1+(\varepsilon_0\varepsilon(x)/t_{\text{AlGaN}}e^2)h^2/4\pi m*(x)]\}$$
$$\{|e_{\text{eff}}(GaN)|-|e_{\text{eff}}(AlGaN)|\} \quad (4.15)$$

μ_s is the mobility of 2DEG, ε_0 is the electric permittivity, $\varepsilon(x) = 9.5 - 0.5x$ is the relative permittivity, $e\phi_b(x) = 0.84 + 1.3x$ (eV) is the SBH, $e_{\text{eff}} = (e_{31} - e_{33})C_{13}/C_{33}$, h is Planck's constant, e is the electron charge, $m^*(x) \approx 0.228m_e$. By monitoring the conductance of the HEMT on the membrane, the pressure difference, $P_i - P_0$, can be obtained.

Figure 4.20 illustrates four kinds of potential pressure sensors based on GaN. The first three are based on the conductance measurement to monitor the pressure and the fourth based on the nitride film deflection. Since nitride HEMTs have been demonstrated for wireless communication applications, the pressure sensor can be integrated with wireless circuits. The size of the integrated pressure sensor can be as small as 500 × 500 μm^2, and can be placed in pipes or harsh environments for remote sensing.

The first pressure sensor is used to monitor the differential pressure. A via hole can be etched from the back side of the SiC substrate and stop on the front-side GaN layer. The detailed process of via hole formation is described in the next section. Metal electrodes are then deposited on the front side of the membrane and encapsulated with dielectric (SiO$_2$). Previous demonstrations of W- and WSi-based high-temperature metal contacts on GaN and SiC, which can be operated at >800°C, suggest that these are good choices for the metallization.[66,67]

In the second design, the via hole of the second sensor is sealed off with another SiC by wafer bonding. The bonding equipment is commercially available. A brief description of the wafer bonding is given in the next section. By sealing off the via hole, the sensor can be used for measurement of the absolute pressure and vacuum, since the via hole seal is performed under atmosphere. In the third sensor, a reference pressure is incorporated, which is not exposed to the high-pressure side. Since SiC and GaN are good thermal conductors, the reference will experience the same temperature as the front pressure. A correction of temperature effect on the pressure can be made by subtracting the reference signal from the signal from the front-side pressure sensor. The fourth sensor uses a laser beam to reflect from the back of the nitride membrane to monitor the pressure.

With this design, there is no need for metallization on the front side of the sample.

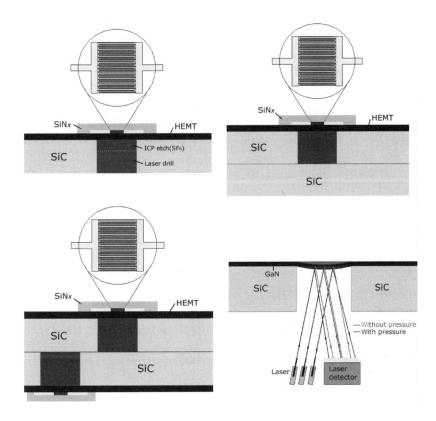

Figure 4.20. Different designs for GaN-based pressure sensors

4.3.6 Selective-Area Substrate Removal

In order to make the AlGaN–GaN Micro Electro-Mechanical Systems (MEMS) based membrane, the SiC substrate must be removed in selected areas. As mentioned previously, due to the extremely high chemical stability of GaN and SiC, there is no convenient wet chemical etchant available. Currently, the etching techniques used for wide bandgap semiconductors are plasma based on dry etching, such as RIE, ECR or ICP etching.[68–79] Typical etch rates for 4H- and 6H-SiC substrates in F_2- or Cl_2-based plasmas range between 0.2 and 1.3 µm/min. Thus, even for a thinned-down substrate of 250 µm, the etch time is generally long and as much as 20–80 h under ion energy conditions where mask erosion is not prohibitive. Another significant drawback with dry etching for creating via holes in SiC is the need for a very robust, typically metal, mask material such as Ni, Al or Cr. The deposition, patterning and subsequent removal of such masks adds considerable complexity to the via fabrication process.

Laser drilling is a convenient method to form through-wafer via holes. This technique has been shown to be capable of very high etch rates and precise control of the via size and sidewall slope. Laser drilling is often used to machine hard materials such as superalloys. A simple model suggests that, above a threshold laser power density, the surface of the material is melted and subsequently ejected by ablation. The development of maskless methods for creating via holes in giant magnetic resistant (GMR) stacks, and in particular for standard thickness substrates, would give added flexibility for creating custom patterns in the substrates through computer control of the laser drilling location and would also eliminate the need for wafer thinning prior to via formation.

We have obtained laser ablation rates for SiC in our initial experiments of 229–870 µm/min, as shown in Figure 4.21 (top). These are significantly faster than the highest plasma etching rates achieved with ICP etching in F_2-based chemistries. Hence, while laser drilling is inherently a serial process when performed in a maskless configuration, the high ablation rates mean that throughput will be comparable to a plasma etch via process in which all the vias are formed simultaneously. A further advantage of the laser drilling technique is that the via hole pattern and the size of the vias can be readily controlled by computer-controlled x–y positioning of the SiC substrate, allowing much more flexibility and reducing costs relative to the need for producing a separate mask in plasma etch processes. We have successfully made laser-drilled vias in materials such as GaN layers grown on SiC substrates, as shown in the Figure 4.21 (bottom). The smallest size of the hole is 20 µm diameter. Since there is no selectivity of laser ablation, we will use laser drilling to remove most of the SiC and leave around 20–30 µm SiC. F_2-based selective plasma etching can be employed to etch off the SiC and stop on the GaN buffer layer.

4.3.7 Biosensors Using AlGaN–GaN Heterostructures

The human immune system is extraordinarily complex, and antigen–antibody interactions are not fully understood at the present time. Antibodies

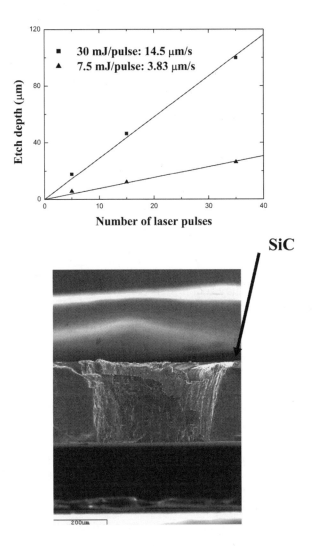

Figure 4.21. Laser ablation depths in SiC substrates as a function of number of laser pulses. The laser was operated at 5 Hz. An effective etch rate of >200 μm/min can be achieved (top); cross-sectional view of a 500 μm diameter through-wafer hole on GaN–SiC sample (bottom)

are protein molecules, which are composed of equal numbers of heavy and light polypeptide amino acid chains held together with disulfide bonds. These highly specialized proteins are able to recognize and bind only certain types of antigen molecule at receptor sites. Sensors based on using antibodies to detect certain specific antigens are called immunosensors. These immunosensors are useful for quantifying how well the human immune system is functioning, and could serve as valuable diagnostic tools. They can also be employed for identifying environmental contaminants and chemical or biological agents.

Silicon-based FETs have been widely used as bio-sensors. For example, the silicon-based ion-sensitive FET (ISFET) is already commercialized to replace conventional electrolyte pH meters. Silicon-based MOSFETs and ISFETs have also been used for immunosensors. However, silicon-based FETs are damaged by exposure to solutions containing ions. Si-FET-based biosensors need an ion-sensitive membrane coated over their surface, and a reference electrode (usually Ag/AgCl) is required to supply the bias voltage. Special precautions must also be taken to apply those biomembranes over the FETs so that they retain their enzymatic activity. On the contrary, GaN-based wide energy bandgap-based semiconductor material systems are extremely chemically stable. (The only known wet chemical etchant for GaN is molten NaOH at 450°C.) The bond between Ga and N is ionic, and proteins can easily attach to the GaN surface. This is one of the key factors for making a sensitive biosensor with useful lifetime, compared with Si-based FETs requiring coating with a membrane to accomplish this goal.

Ground-breaking work on the selectivity of peptide bonding to semiconductor surfaces has shown that preferential bonding does indeed occur [80–82] and is dependent on a number of factors, including crystallographic orientation of the semiconductor surface, ionicity of the semiconductor bonding, *etc*. Previous work has focused primarily on Si, GaAs and InP [80], though studies on metal oxides such as ZnO [81–83] and Cr_2O_3 [84] have also been reported. In most, if not all, of these cases, detailed understanding of the chemistry behind the observed selectivity has not yet been achieved. Similarly, no information has yet been reported on the effects of the peptide bonding on the underlying characteristics of the semiconductor host. Such information is critical to the development of high sensitivity detection schemes that rely upon changes in the position of the near-surface Fermi level for operation.

The development of miniaturized total chemical analysis systems (μTAS) in the past decade is now reflected in the ability of these devices to conduct various chemical and biological analyses in a single chip.[85,86] There is interest in the development of a microfluidic device for fluid handling and reagent delivery. Such a device will enable the attachment of an array of biomolecules on the surface of GaN FET sensors, for the delivery of a variety of reagents during biological assays, and for simultaneous interrogation of a sample with an array of sensors. The advantages of a microfluidics-based delivery system include the following. First, microfluidics enables highly parallel, high-throughput reagent delivery. Miniaturization relieves constraints of real-estate challenge when a large number of sensor arrays are exploited. Second, miniaturization enables integration with other components, such as electronic circuits and sensing components, on the same chip. Therefore, the results can be analyzed and the signal can be interpreted on the device. Third, microfluidics reduces reagent consumption, which is especially beneficial for very expensive antibodies and enzymes. Other advantages from miniaturization include rapid analysis, ease of automation, and potentially low cost.

The chemical sensor array can be integrated with wireless communication circuits for remote sensor applications. This development likely represents assays of the future, which will clearly follow the trend toward increased integration, greater sensitivity, higher specificity, and remote sensing. Given the strong sensitivity of gateless HEMT devices to polar materials, there are currently many efforts to use such structures for the detection of ion currents through cell membranes in general and of action potentials of neuron cells in particular. This suggests the possibility of functionalizing the surface for application as biosensors, especially given the excellent stability of the GaN and AlGaN surfaces, which should minimize degradation of adsorbed cells. As an example, some initial studies of cell adhesion on GaN and AlN are under way. Figure 4.22 shows an optical micrograph of V-79 fibroplast cells cultured on AlN and GaN. As suggested by the coating of fibronectin below the cells, they adhere very well to the Group III nitride surface and survive in a culture medium for a couple of days. So far, very little is known about the interaction of living cell tissue with Group III nitride surfaces, and we are currently performing a systematic investigation of this topic. Eventually it will be necessary to integrate a lipid bilayer membrane onto the surface of the gateless HEMT devices in order to measure single-ion currents associated with high sensitivity bio-detection. A schematic of this approach is shown in Figure 4.23.

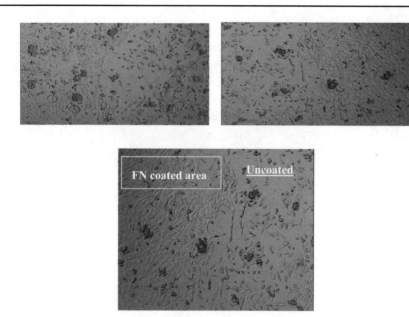

Figure 4.22. Optical microscope images of V-79 fibroplast-cell-coated AlN (left) and GaN (right) surfaces. The image at the bottom shows that providing a nutrient (fibronyacin) to the surface increases the cell density

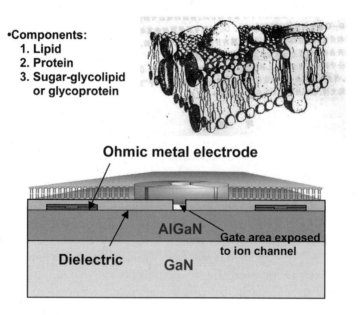

Figure 4.23. Schematic of lipid bilayer membrane (top) and integration onto the gate region of an AlGaN–GaN HEMT

4.3.8 Surface Acoustic Wave-Based Biosensors

SAW-based biosensors for detecting chemical warfare agents, nerve gases, and other toxic agents are of interest for defense industry and homeland security applications. The piezoelectric microbalance or surface acoustic wave transducers can also be used for air sampling, *i.e.* chemical noses. A neural-network-type of pattern recognition approach can be incorporated in the design to provide early warning bio-sensor systems which can be built into aircraft and mobile vehicles to detect noxious agents in the external air. SAW devices have been successfully used as sensors to detect force, acceleration, shock torque, viscosity, pressure and humidity.[87,88] SAW devices have also been widely used to monitor thin film thickness and deposition rate during the film deposition. The detection sensitivity is excellent and can detect 0.8% of a monolayer coverage on the surface. By using various coatings on the surface of the SAW devices, various cells, chemicals, gases and bio-materials can be detected.[89,90] The detection mechanism for the SAW device utilizes the changes to the velocity or phase of an acoustic wave propagating along the substrate due to changes to the characteristics of the propagation path. This is due to the fact that the energy of the SAW is stored only in the top region (less than 1.5 times the wavelength) of the surface of the material in which that wave propagates. Therefore, the surface of a SAW sensor is extremely sensitive to the outer environment. Thus, even slight changes in the environment on the surface can be detected. SAW sensors are not only relatively cheap, very sensitive, and reliable, but they also do not need a dc power supply for certain operations, which makes them perfect for wireless applications.

The sensitivity of a SAW device is proportional to the electromechanical coupling coefficients of the substrate. AlN- and GaN-based semiconductor materials are excellent candidates for SAW devices due to their large electromechanical coupling coefficients (similar in magnitude to quartz and much larger than those of GaAs- or SiC-based semiconductors), as illustrated in Table 4.1. GaN HEMTs can be integrated with SAW devices to create a unique acoustic velocity tuning device with low acoustic loss, high frequency and low loss RF performance. The 2DEG in the GaN HEMT interacts with the lateral electric field, resulting in ohmic loss, which attenuates and slows the SAW. This mechanism can be used to tune the acoustic velocity. Combined with the optical characteristics of the wide and direct band gap (~3.3 eV) semiconductor GaN, the SAW-based chip can also be used for UV optical signal processing. By aligning the device parallel to the c-axis of the GaN film, Rayleigh-type SAWs are excited, while Love-type SAWs are excited. If the devices are aligned perpendicular to the *c*-axis, then the Rayleigh wave mode is suitable for gaseous-type environment sensing. For the Love wave mode operation, there is no vertical wave component, which is ideal for liquid-type environment sensing. The GaN-based SAW chip also offers an acoustic–optical dual mode sensing mechanism. Likewise, the optical properties parallel and perpendicular to the *c*-axis are different, allowing novel optical devices, such as high-contrast modulators. The successful development of the GaN–AlGaN-based HEMT-SAW chip technology will provide industry with state-of-the-art new multifunctional chip technologies. It will not only improve the existing devices, but also develop fundamentally new approaches to multi-mode tunable chemical and biochemical sensors.

Table 4.1. Acoustic velocity and electromechanical coupling coefficient of different SAW materials

Material	Velocity (m/s)	K^2_{eff}
GaN	3693–4001	0.00131
AlN	5790	0.0025
4H-SiC	6832	0.000112
LiNbO$_3$	3488	0.0232
GaAs	2867	0.000593–0.00064
Quartz	3158	0.0011
ZnO–R-Al$_2$O$_3$	4200	0.06

AlN- and GaN-based material systems are almost ideal SAW materials due to their high SAW velocity, high electromechanical coupling coefficients and their compatibility with electronic integration. Nitride-based materials also show excellent resistance to humidity and chemical etching. Nitride-based (AlGaN–GaN) HEMTs have shown a strong piezoelectric effect and have been used to fabricate very sensitive piezoelectric microbalances. Biologically active films can be coated on the microbalance. Microbalances and the influence of adsorption of toxicants and rate of toxicant mass accumulation can affect the acoustic waves propagating in the AlGaN–GaN layer. The high electromechanical coupling coefficients of piezoelectric nitride-based materials, in conjunction with the properties of low acoustic loss and high velocity, offer excellent potential for use in high frequency and low loss RF applications. SAW devices operating in the kilohertz to gigahertz range can be designed and integrated with wireless remote sensing applications.

The GaN-based SAW chip also offers an acoustic–optical dual mode sensing mechanism. Likewise, the optical properties parallel and perpendicular to the *c*-axis are different, allowing novel optical devices, such as high-contrast modulators. It is highly desirable to increase the device operation frequency of SAW devices, as this has been shown to improve the signal-to-noise ratio and the detectivity of such devices.[91] Operation frequency is a function of the device size, determined by lithography, and the velocity of propagation in the SAW media. Conventional piezoelectric materials, such as quartz or lithium niobate, offer limited performance due to their relatively low velocities of ~3000 to ~4000 m/s.[92] Thin piezoelectric films which have high SAW velocity and which can be integrated with a non-piezoelectric substrate such as Si are highly desirable. AlN is an almost ideal SAW material due to its high SAW velocity, ~5600 m/s, as illustrated in Table 4.1, and its compatibility with electronic integration.[93–98] AlN on Si is a particularly attractive combination, as the opposite temperature coefficients of delay of Si and AlN facilitate realization of thermally stable SAW devices. Preliminary work on this material system reported in the literature has shown great promise.[98] This work was carried out using sputtering to deposit the AlN. The resultant microstructure was fine-grained polycrystalline. Significant improvement is expected if higher quality films are employed. This can be achieved by using epitaxial processes, such as MBE.

4.4 Surface Acoustic Wave Device Fabrication

Both inter-digitated transducer (IDT)-based SAW and flexural plate wave (FPW) devices can be employed, for sensing gas or liquid bio-agents respectively, as illustrated in Figure 4.24. For both kinds of SAW device, a pair of IDTs are fabricated on the substrate (AlN or GaN) as the input and output ports of the signals. The fabrication of SAW devices consists of material selection, photolithography, metal deposition, dicing and chip mounting, as depicted in Figure 4.25.

The SAW devices are designed according to the frequency and bandwidth of operation. The acoustics wave can be expressed as a complex constant ($\gamma = \alpha + j\beta$). The attenuation constant, α, and propagation constant, $\beta = 2\pi/\lambda$, are important design parameters of the SAWs, where λ is the acoustic wavelength. Another important design parameter is the electromechanical coupling coefficient K^2, which is a measure of the efficiency in converting an applied microwave signal into mechanical energy. These parameters will determine the magnitude of the observed changes in the SAW phase velocity $\Delta v/v$ and attenuation of the SAW intensity Γ and can be expressed as

$$\Gamma = \frac{K^2}{2}\frac{2\pi}{\lambda}\frac{\sigma/\sigma_m}{1+(\sigma/\sigma_m)^2} \qquad (4.16)$$

and

$$\frac{\Delta v}{v} = \frac{K^2}{2}\frac{1}{1+(\frac{\sigma}{\sigma_m})^2} \qquad (4.17)$$

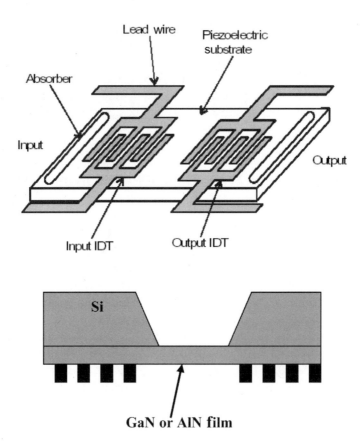

Figure 4.24. Schematics of IDT-based SAW (top) and FPW (bottom) devices

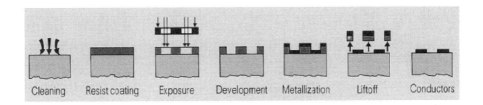

Figure 4.25. Fabrication sequence of the IDT

where σ is the sheet conductivity and σ_m is the critical conductivity at which maximum attenuation occurs.

The physical SAW devices can be designed by choosing the desired operating frequency and bandwidth. As illustrated in Figure 4.26 (top), the operation frequency of the SAW device f_o can be designed by properly choosing the spacing between the inter-digital fingers d such that

$$f_o = \frac{v}{d} \quad (4.18)$$

where v is the wave velocity in the specific substrate. Therefore, the dimensions of the SAW devices depend on the operation frequency, which can vary from a few micrometers for gigahertz operation to millimeter, range for kilohertz operation. SAW devices operating in the kilohertz to gigahertz range can be readily designed and integrated with RF integrated circuits for wireless remote sensing applications. The bandwidth of the acoustic wave is given as

$$B = \frac{v}{2Nd} \quad (4.19)$$

where N is number of inter-digital fingers, as shown in Figure 4.26 (bottom).

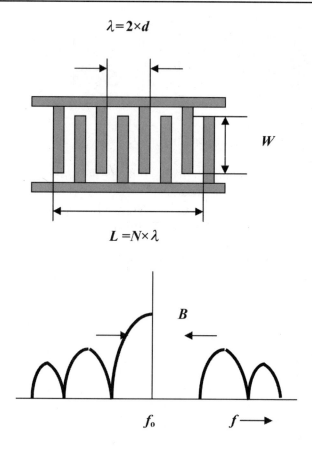

Figure 4.26. The dimension of the SAW device depends on the operation frequency of the SAW (top). The bandwidth of the SAW is a function of the number of the IDT and the frequency (bottom).

4.5 Surface Acoustic Wave Device for Gas Sensing

Gas sensing is achieved by producing a change in the characteristics of the substrate material between the input and output IDT, where the acoustic waves travel. A pair of SAW devices can be used for gas sensing: one as the reference and the other for detection. Depending on the type of chemical and bio-agents needed to be detected, an absorbing layer for the specific chemical or bio-agent is coated on the surface of the SAW between the input and output IDT. Initially, two sets of SAWs are impedance-matched, which means the wave reaching the output IDT will have the same phase and the same velocity. If the bio-agents adhere to the coated materials, the phase of the wave will be affected by the bio-agents, then these two sets will have different phases. By integration with a mixer, one can monitor the change easily. Table 4.2 contains a list of coatings for specific chemical agents. Excellent gas detection should be observed for GaN- and AlN-based SAW devices.

Table 4.2. Coatings for SAW devices

Analyte	Coating	Detection limit
Explosive	Carbowax 1000	1 ppb
Sulfur dioxide	Dinonyl phthalate	5 ppm
Toluene	Polyamidoxime	90 ppm
Phosgene	Methyltrioctyl phosphonium dimethylphosphate	5 µg/l
H_2	Pd	50 ppm
Hg	Au	2×10^{-3} M
H_2S	WO_3	10 ppb
NH_3	Pt	1 µg/ml
NO_2	MnO_2	7 ppm
SO_2	Au	20 ppb
Ozone	Polybutadiene	10 ppb
Candida albicans	Anti-Candida antibody	10^6 cells/cm
Glucose	Hexokinase	1 mM
Human IgG	Goat Antihuman IgG Protein A	12 µg

4.6 Flexural Plate Wave Device for Liquid Sensing

As shown in the Figure 4.24 (bottom), the configuration of the FPW is identical to the gas-sensing-based SAW device, with the exception of an additional via hole fabricated on the substrate to accommodate the liquid sample. GaN membranes can be fabricated by etching off the Si substrate in an ICP system using SF_6-based chemistry. Thick photoresist (>10 µm) is used as the etch mask.

4.7 Surface Acoustic Wave Array

To increase selectivity, it is typical to employ an array of SAW devices, with different coatings used to identify specific chemical or bio-agents.

Since each SAW device is so small, the arrays are portable. In addition, integrated with a power amplifier such as an MMIC, these devices can be transported by an unmanned aircraft into dangerous environments. SAW devices do not need a power supply; so, once they are installed at targeted areas, they can be monitored wirelessly without a battery, which means they can operate semi-permanently. The successful development of the GaN–AlGaN-based HEMT-SAW chip technology will provide industry with state-of-the-art new multifunctional chip technologies. It will not only improve the existing devices, but also develop fundamentally new approaches to multi-mode tunable chemical and biochemical sensors.

4.8 Wireless Sensor Network and Wireless Sensor Array Using Radio Frequency Identification Technology

Owing to their flexibility in field deployment, sensors with wireless data transmission capability are of great interest for applications in security, civil engineering, biometrics, telemedicine, homeland defense, transportation, *etc.* For many applications, multiple wireless sensors of the same function or different functions need to be deployed at multiple locations. A simple wireless sensor network is therefore needed to manage the collection and processing of data from multiple sensors at multiple locations. It is preferable to make the RF transceiver in each wireless sensor as simple as possible to reduce its size and power consumption.

The RF identification (RFID) technology has advanced to a level that an ultra-small RFID tag can be monolithically integrated in a tiny chip. The tag device consumes very little power and, therefore, can operate with a very small battery or even without a battery. In the latter case, the tag device is powered by the external interrogation signal from the base station. Its low data bandwidth, low power operation makes it suitable to integrate with sensors for wireless sensing, since the data bandwidth and operating power of the sensor is also very low. Once the detected sensor data is digitized, it can be merged with the ID code and transmitted together. The RFID part of the integrated wireless sensor functions as a simple RF transceiver which receives a base station signal to turn on the transmitter and sends back the sensor data. This type of wireless sensor is suitable for short-range operation similar to a personal area network (PAN).

It is possible to combine sensor technologies and RFID technologies to build a simple wireless sensor network. The network consists of multiple sensors of the same function or different functions, *e.g.* chemical, biological agent, pressure, temperature, and UV detections. Each sensor is connected to an RFID tag with a unique ID code that identifies its sensing function and location. Two operation modes, active and passive, will be studied. In the active operation mode, it sends out the data to a central monitor station whenever the sensor detection is positive and triggers the RFID tag transmitter. A battery or other power source is needed to operate the wireless sensor. In the passive operation mode, the sensor is activated by the central monitor station's interrogation signal and responds with the data. Extremely low-power wireless sensor design will be explored to investigate the feasibility of powering the device with an RF interrogation signal, which eliminates the need to use a battery. Figure 4.27 shows the block diagram of the integrated wireless sensor with RFID. The sensor can be powered by the external interrogation RF energy from the base station, or an integrated battery. The sensor data is merged with the RFID code, which modulates the antenna load and the reflected signal from the base station.

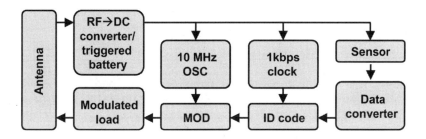

Figure 4.27. Block diagram of wireless sensor with RFID, using 10 MHz oscillator (OSC) and modulator (MUD).

A wireless sensor array using RFID is similar to the above except that multiple wireless sensors are integrated together on one substrate. A typical example is a biomedical or biochemical lab-on-chip. Two types of integration can be employed, *i.e.* hybrid integration and monolithic integration. In hybrid integration, an array of wireless sensors, with each consisting of a printed antenna, an RF tag circuit, and a sensor will be integrated on a PCB substrate. The antenna size is dominant and determines the overall size of the wireless sensor array. To have monolithic integration, the antenna needs to be small and operate at higher frequency.

The advantages of using RFID technology are small size, low cost, and low power. On the other hand, the RFID technology is limited in range, which is in the order of a few meters. For some applications where size and power consumption are not critical, but operating distance is a major concern, the wireless sensor architecture in Figure 4.28 can be used. The device consists of multiple sensors, a memory stores its ID code, a data processor that collects sensor data and merges this with the ID code, and transmitter with power amplifier and antenna array for long-range transmission. The beam-forming antenna array can be pre-programmed to point the antenna beam to the central station.

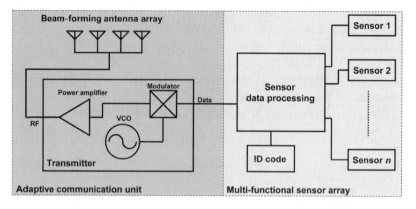

Figure 4.28. Long-range wireless sensor using voltage, controlled oscillator (VCU)

Finally, it is possible to use the signals from a distributed sensor array to track the extent of a gas or chemical release in real time. Figure 4.29 shows a simulation of the spreading of a gas release in an urban environment (the white shapes represent buildings) by combining wind speeds and directions, ambient temperature, nature of the gas, *etc*. A distributed network of GaN-based sensors with integrated wireless data transmission would be capable of real-time tracking of the extent of chemical contamination.

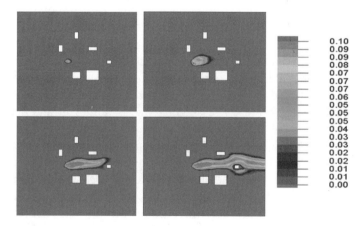

Figure 4.29. Simulation of time dependence of spreading of a gas release between buildings

4.9 Summary

AlGaN–GaN HEMTs show a strong dependence of source–drain current on the piezoelectric polarization-induced 2DEGs to variations in electrostatic boundary conditions of the free surface above the 2DEG (polar liquids, decomposition of hydrogen by catalytic Pt contacts, and applied strain). These particularly attractive features of AlGaN–GaN heterostructures make it possible for these sensors to function with excellent thermal, chemical, and mechanical stability, which suggests that nitride HEMTs are excellent candidates for pressure sensor and piezoelectric-related applications. Much more research is needed to develop AlGaN–GaN heterostructure-based sensors which really can compete in real applications.

References

1. Eastman LF, Mishra UK (2002) IEEE Spectrum 39: 28
2. Tarakji A, Hu X, Koudymov A, Simin G, Yang J, Khan MA, Shur MS, Gaska R (2002) Solid-State Electron 46: 1211
3. Koudymov A, Hu X, Simin K, Simin G, Ali M, Yang J, Khan MA (2002) IEEE Election Dev Lett 23 : 449
4. Simin G, Koudymov A, Fatima H, Zhang JP, Yang J, Khan MA, Hu X, Tarakji A, Gaska R, Shur MS (2002) IEEE Electron Dev Lett 23: 458
5. Eastman LF, Tilak V, Smart J, Green BM, Chumbes EM, Dimitrov R, Kim H, Ambacher OS, Weimann N, Prunty T, Murphy M, Schaff WJ, Shealy JR (2001) IEEE Trans Electron Dev 48: 479
6. Johnson JW, Baca AG, Briggs RD, Shul RJ, Monier C, Ren F, Pearton SJ, Dabiran AM, Wowchack AM, Polley CJ, Chow PP (2001) Solid-State Electron 45: 1979
7. Lu W, Yang J, Khan MA, Adesida I (2001) IEEE Trans Electron Dev ED48: 581
8. Pearton SJ, Ren F, Zhang AP, Lee KP (2000) Mater Sci Eng Rep R 30: 55
9. Binari SC, Ikossi K, Roussos JA, Kruppa W, Park D; Dietrich HB, Koleske DD, Wickenden AE, Henry RL (2001) IEEE Trans Electron Dev 48: 465
10. Simin G, Koudymov A, Tarakji A, Hu X, Yang J, Asif Khan M, Shur MS, Gaska R (2001) Appl Phys Lett 79: 2651
11. Spetz Lloyd, A, Baranzahi A, Tobias P, Lundström I (1997) Phys Status Solidi A 162: 493

12. Vasiliev A, Moritz W, Fillipov V, Bartholomäus L, Terentjev A, Gabusjan T (1998) Sens Actuators B 49: 133
13. Savage AM, Konstantinov A, Saroukan AM, Harris C (2000) Proc International Conference Section Contribute and Related Materials ICSCRM 1999, pp 511–515
14. Spetz Lloyd, A Tobias P, Unéus P, Svenningstorp H, Ekedahl L-G, Lundström I (2000) Sens Actuators B 70: 67
15. Connolly EJ, O'Halloran GM, Pham HTM, Sarro PM, French PJ (2002) Sens Actuators A 99: 25
16. Arbab A, Spetz A, Lundström I (1993) Sens Actuators B 15–16: 19
17. Hunter GW, Neudeck PG, Okojie RS, Beheim GM, Thomas V, Chen L, Lukco D, Liu CC, Ward B, Makel D (2002) Proc ECS Vol 01-02: 212 (Electrochemical Society, Pennington, NJ 2002)
18. Chen LY, Hunter GW, Neudeck PG, Knight DL, Liu CC, Wu QH (2002) in Proceedings of the Third International Symposium on Ceramic Sensors (1996) Anderson U, Liu M, Yamazoe N (eds) Electrochemical Society Inc. Pennington, NJ, pp 92–98
19. Ekedahl L-G, Eriksson M, Lundström I (1998) Acc Chem Res 31: 241
20. Svenningstorp H, Tobias P, Lundström I, Salomonsson P, Mårtensson P, Ekedahl L-G, Spetz AL (1999) Sens Actuators B 57: 159
21. Hunter GW, Liu CC, Makel D (2001) *MEMS Handbook*, Hak MG (ed) CRC Press, Boca Raton
22. Chen L, Hunter GW, Neudeck PG (1997) J Vac Sci Technol A 15: 1228; (1998) J Vac Sci Technol A 16: 2890
23. Tobias P, Baranzahi A, Spetz AL, Kordina O, Janzen E, Lundström I (1997) IEEE Electron Dev Lett 18: 287
24. Baranzahi A, Spetz AL, Lundström I (1995) Appl Phys Lett 67: 3203
25. Casady JB, Agarwal AK, Seshadri S, Siergiej RR, Rowland LB, MacMillan MF, Sheridan DC, Sanger PA, Brandt CD (1998) Solid-State Electron 42: 2165
26. Ambacher O, Eickhoff M, Steinhoff G, Hermann M, Gorgens L, Werss V, *et al.* (2002) Proc ECS 214: 27 (Electrochem. Soc., Pennington, NJ 2002)
27. Neuberger R, Muller G, Ambacher O, Stutzmann M (2001) Phys Status Solidi A 185: 85
28. Schalwig J, Muller G, Ambacher O, Stutzmann M (2001) Phys Status Solidi A 185: 39
29. Steinhoff G, Hermann M, Schaff WJ, Eastmann LF, Stutzmann M, Eickhoff M (2003) Appl Phys Lett 83: 177
30. Eickhoff M, Neuberger R, Steinhoff G, Ambacher O, Muller G, Stutzmann M (2001) Phys Status Solidi B 228: 519

31. Schalwig J, Muller G, Eickhoff M, Ambacher O, Stutzmann M (2002) Sens Actuators B 81: 425
32. Stutzmann M, Steinhoff G, Eickhoff M, Ambacher O, Nobel CE, Schalwig J, *et al.* (2002) Diamond Rel at Mater 11: 886
33. Eickhoff M, Schalwig J, Steinhoff G, Weidmann O, Gorgens L, Neuberger R, Hermann M, Baur B, Muller G, Ambacher O, Stutzmann M (2003) Phys Status Solidi C 6: 1908
34. Kim J, Gila B, Chung GY, Abernathy CR, Pearton SJ, Ren F (2003) Solid-State Electron 47: 1069
35. Kim J, Ren F, Gila B, Abernathy CR, Pearton SJ (2003) Appl Phys Lett 82: 739
36. Kim J, Gila B, Abernathy CR, Chung GY, Ren F, Pearton SJ (2003) Solid-State Electron 47: 1487
37. Ambacher O, Eickhoff M, Link A, Hermann M, Stutzmann M, Bernardini, Tilak V, Eastmann LF (2003) Phys Status Solidi C 6: 1878
38. Khan MA, Hu X, Simin G, Lunev A, Yang J, Gaska R, Shur MS (2001) IEEE Electron Dev Lett 21: 63
39. Kang BS, Louche G, Duran RS, Gannou Y, Pearton SJ, Ren F (2004) Solid-State Electron 48: 851
40. Kang BS, Kim S, Kim J, Ren F, Baik K, Pearton SJ, Gila B, Abernathy CR, Pan C, Chen G, Chyi JI, Chandrsekaran V, Sheplak M, Nishida T, Chu SNG (2003) Appl Phys Lett 83: 4845
41. Zhang AP, Rowland LB, Kaminsky EB, Tilak V, Grande JC, Teetsov J, Vertiatchikh A, Eastman LF (2003) J Electron Mater 32: 388
42. Morkoc H, Cingolani R, Gil B (1999) Solid-State Eletron 43: 1909
43. Shur MS (1990) *Physics of Semiconductor Devices*, Prentice Hall, Eaglewood Cliffs, NJ
44. Pala N, Gaska R, Rumyantsev S, Shur MS, Asif Khan M, Hu X, Simin G, Yang J (2000) Electron Lett 36: 268
45. Simin G, Hu X, Ilinskaya N, Kumar A, Koudymov A, Zhang J, Asif Khan M, Gaska R, Shur MS (2000) Electron Lett 36: 2043
46. Tarakji A, Hu X, Koudymov A, Simin G, Yang J, Khan MA, Shur MS, Gaska R (2002) Solid-State Electron 46: 1211
47. Koudymov A, Hu X, Simin K, Simin G, Ali M, Yang J, Khan MA (2002) IEEE Electron Dev Lett 23 : 449
48. Simin G, Koudymov A, Fatima H, Zhang JP, Yang J, Khan MA, Hu X, Tarakji A, Gaska R, Shur MS (2002) IEEE Electron Dev Lett 23: 458
49. Gila BP, Johnson JW, Mehandru R, Luo B, Onstine AH, Krishnamoorthy V, Bates S, Abernathy CR, Ren F, Pearton SJ (2001) Phys Status Solidi A 188: 239

50. Kim J, Mehandru R, Luo B, Ren F, Gila BP, Onstine AH, Abernathy CR, Pearton SJ, Irokawa Y (2000) Appl Phys Lett 80: 4555
51. Kim J, Mehandru R, Luo B, Ren F, Gila BP, Onstine AH, Abernathy CR, Pearton SJ, Irokawa Y (2000) Appl. Phys Lett 81: 373
52. Pearton SJ, Zolper JC, Shul RJ, Ren F (1999) J Appl Phys 86: 1
53. Green M, Chu KK, Chumbes EM, Smart JA, Shealy JR, Eastmann LF (2000) IEEE Electron Dev Lett 21: 268
54. Tarakji A, Hu X, Koudymov A, Simin G, Yang J, Khan MA, Shur MS, Gaska R (2002) Solid-State Electron 46: 1211
55. Zhang AP, Rowland LB, Kaminsky EB, Tilak V, Grande JC, Teetsov J, Vertiatchikh A, Eastmann, LF (2003) J Electron Mater 32: 388
56. Koudymov A, Hu X, Simin K, Simin G, Ali A, Yang J, Khan MA (2002) IEEE Electron Dev Lett 23 : 449
57. Ambacher O, Foutz B, Smart J, Shealy JR, Weimann NG, Chu K, Murphy M, Schaff WJ, Wittmer L, Eastmann LE, Dimitrov R, Mitchell A, Stutzmann M (2000) J Appl Phys 87: 334
58. Rashmi H, Kranti A, Haldar S, Gupta RS (2002) Microelectron J 33: 205
59. Chen WQ, SK (1995) Hark J Appl Phys., 77. 5747
60. Neuberger R, Muller G, Ambacher O, Stutzmann M (2001) Phys Status Solidi A 185: 85
61. Ambacher O, Smart J, Shealy JR, Weimann NG, Chu K, Murphy M, Schaff WJ, Eastmann LE, Dimitrov R, Wittmer L, Stutzmann M, Rieger W, Hilsenbeck J (1999) J Appl Phys 85: 3222
62. Asbeck PM, Yu ET, Lau SS, Sullivan GJ, Van Hove J, Redwing JM (1997) Electron Lett 33: 1230
63. Stoney GG (1909) Proc R Soc London 82: 172
64. Chu SNG (1998) J Electrochem Soc 145: 3621
65. *CRC Handbook of Chemistry and Physics, 77th Edition* (1997) CRC Press, New York
66. Kim J, Ren F, Baca AG, Briggs RD, Pearton SJ (2003) Solid-State Electron 47: 1345
67. Kim J, Ren F, Baca AG, Pearton SJ (2003) Appl Phys Lett 82: 3263
68. Yih PH, Steckl AJ (1995) J Electrochem. Soc 142: 2853
69. Leerungnawarat P, Hays DC, Cho H, Pearton SJ, Strong RM, Zetterling C-M, Östling M (1999) J Vac Sci Technol B 17: 2050
70. Tanaka S, Rajanna K, Abe T, Esashi M (2001) J Vac Sci Technol B 19: 2173
71. Li B, Cao L, Zhao JH (1998) Appl Phys Lett 73: 653
72. Flemish JR, Xie K, Zhao JH (1994) Appl Phys Lett 64: 2315
73. Khan FA, Adesida I (1999) Appl Phys Lett 75: 2268

74. Cho H, Leerungnawarat P, Hays DC, Pearton SJ, Chu SNG, Strong RM, Zetterling C-M, Östling M, Ren F (2000) Appl Phys Lett 76: 739
75. Chabert P, Proust N, Perrin J, Boswell RW (2000) Appl Phys Lett 76: 2310
76. Wang JJ, Lambers ES, Pearton SJ, Östling M, Zetterling C-M, Grow JM, Ren F, Shul RJ (1998) J Vac Sci Technol A 16: 2204
77. McLane GF, Flemish JR (1996) Appl Phys Lett 68: 3755
78. Khan FA, Zhou L, Kumar V (2002) J Electrochem Soc 149: G420
79. Leerungnawarat P, Lee KP, Pearton SJ, Ren F, Chu SNG (2001) J Electron Mater 30: 202
80. Whaley SR, English DS, Hu EL, Barbara PF, Belcher AM (2000) Nature 405: 665
81. Nguyen C, Dai J, Darikaya M, Schwartz DT, Baneyx F (2003) J Am Chem Soc 381: 441
82. Sarikaya M, Tamerler C, Jen AK-Y, Schulten K, Baneyx F (2003) Nature Mater 2: 577
83. Kjaergaard K, Sorensen JK, Schembri MA, Klemm P (2000) Appl Environ Microbiol 66: 10
84. Scembri M, Kjaergaard K, Klemm P (1999) FEMS Microbia Lett 170: 363
85. Manz A, Graber N, Widmer H M (1990) Sens Actuators B 1: 244
86. Ramsey J M, van den Berg A. (2001) *Proceedings of the µTAS 2001 Symposium*, Kluwer Academic Publishers, Dordrecht, The Netherlands
87. Vellekoop MJ, Lubking GW, Sarro PM, Venema A (1994) Sens Actuators A 43: 175
88. Mecea MV (1994) Sens Actuators A 41–42: 630
89. Snow A, Wohltjen H (1984) Anal Chem 56: 1411
90. Suleiman A, Guilbault GG (1984) Anal Chim Acta 162: 97
91. Ballantine DS, White RM, Martin SJ, Ricco AJ, Zellers ET, Frye GC, Wohltjen H (1997) In *Acoustic Wave Sensors*, Stern R, Levy M (eds) Academic, San Diego, p 36
92. *Microwave Acoustic Handbook, Surface Wave Velocities,* Slobodnik J, Conway ED, Demonico, RT (eds) Air Force Cambridge Research Laboratories (LZM), Bedford, MA, p 1973 (1989)
93. Lundquist PM, Lin WP, Xu ZY, Wong GK, Rippert ED, Helfrich JA, Ketterson JB (1994) Appl Phys Lett 65: 1085
94. Okano H, Tanaka N, Takahashi Y, Tanaka T, Shibata K, Nakano S, (1994) Appl Phys Lett 64: 166
95. Brunner D, Angerer H, Bustarret E, Freudenberg F, Höpler R, Dimitrov R, Ambacher O, Stutzmann M (1997) J Appl Phys 52: 5090
96. Strite S, Morkoc H (1992) J Vac Sci Technol B 10: 1237

97. Takikawa H, Kimura K, Miyano R, Sakakibara T, Bendavid A, Martin PJ, Matzumuro A, Tsutsumi K (2001) Thin Solid Films 386: 276
98. Caliendo C, Imperatori P (2003) Appl Phys Lett 83: 1641

5 Nitride-Based Spintronics

5.1 Abstract

The field of semiconductor spintronics seeks to exploit the spin of charge carriers in new generations of transistors, lasers and integrated magnetic sensors. There is strong potential for new classes of ultra-low power, high-speed memory, logic and photonic devices based on spintronics. The utility of such devices depends on the availability of materials with practical magnetic ordering temperatures, and most theories predict that the Curie temperature will be a strong function of bandgap. In this chapter we review the current state of the art in producing room-temperature ferromagnetism in GaN-based materials, the origins of the magnetism and its potential applications.

5.2 Introduction

There is great current interest in the emerging field of semiconductor spin transfer electronics (spintronics), which seeks to exploit the spin of charge carriers in semiconductors. It is widely expected that new functionalities for electronics and photonics can be derived if the injection, transfer and detection of carrier spin can be controlled above room temperature. Among this new class of devices are magnetic devices with gain, spin transistors operating at very low powers for mobile applications that rely on batteries, optical emitters with encoded information through their polarized light output, fast non-volatile semiconductor memory and integrated magnetic/electronic/photonic devices ("electromagnetism-on-a-chip"). Since the magnetic properties of ferromagnetic semiconductors are a function of carrier concentration in the material in many cases, then it will be possible to have electrically or optically controlled magnetism through field-gating of transistor structures or optical excitation to alter the carrier density. A number of recent reviews have covered the topics of spin injection, coherence length and magnetic properties of materials systems, and the general areas of spin injection from metals into semiconductors and applications of the spintronic phenomena.[1–4] Although there have been recent reports of successful and efficient spin injection from a metal to a semiconductor even at room temperature by ballistic transport (*i.e.* Schottky barriers and tunneling), the realization of functional spintronic devices requires materials with ferromagnetic ordering at operational temperatures compatible with existing semiconductor materials.

5.3 Potential Semiconductor Materials for Spintronics

There are two major criteria for selecting the most promising materials for semiconductor spintronics. First, the ferromagnetism should be retained to practical temperatures (*i.e.* >300 K).[5–8] Second, it would be a major advantage if there were already an existing technology base for the material in other applications. Most of the work in the past has focused on (Ga,Mn)As and (In,Mn)As. There are indeed major markets for their host materials in infrared LED and lasers and high-speed digital electronics (GaAs) and magnetic sensors (InAs). In samples carefully grown with single-phase MBE, the highest Curie temperatures reported are ~110 K for (Ga,Mn)As and ~ 35 K for (In,Mn)As. One of the most effective methods for investigating spin-polarized transport is by monitoring the polarized electroluminescence output from a quantum-well LED into which the spin current is injected. Quantum selection rules relating the initial carrier spin polarization and the subsequent polarized optical output can provide a quantitative measure of the injection efficiency.

There are a number of essential requirements for achieving practical spintronic devices in addition to the efficient electrical injection of spin-polarized carriers. These include the ability to transport the carriers with high transmission efficiency within the host semiconductor or conducting oxide, the ability to detect or collect the spin-polarized carriers, and to be able to control the transport through external means, such as biasing of a gate contact on a transistor structure.

We focus on a particular and emerging aspect of spintronics, namely recent developments in achieving practical magnetic ordering temperatures in technologically useful semiconductors.[5–10] While the progress in synthesizing and controlling the magnetic properties of Group III–arsenide semiconductors has been astounding, the reported Curie temperatures are too low to have significant practical impact.

Other materials for which room-temperature ferromagnetism has been reported include (Cd,Mn)GeP$_2$ [6], (Zn,Mn)GeP$_2$ [7], ZnSnAs$_2$ [8], (Zn,Co)O [9] and (Co,Ti)O$_2$ [10]. Some of these chalcopyrites and wide bandgap oxides have interesting optical properties, but they lack a technology and experience base as large as that of most semiconductors.

The key breakthrough that focused attention on wide bandgap semiconductors as being the most promising for achieving practical ordering temperatures was the theoretical work of Dietl *et al.* [11]. They predicted that cubic GaN doped with ~5 at.% of Mn and containing a high concentration of holes (3.5×10^{20} cm^{-3}) should exhibit a Curie temperature exceeding room temperature. In the period following the appearance of this work, there has been tremendous progress on both the realization of high-quality (Ga,Mn)N epitaxial layers and on the theory of ferromagnetism in these so-called dilute magnetic semiconductors (DMSs). The term DMS refers to the fact that some fraction of the atoms in a non-magnetic semiconductor like GaN are replaced by magnetic ions. A key, unanswered question is whether the resulting material is indeed an alloy of (Ga,Mn)N or whether it remains as GaN with clusters, precipitates or second phases that are responsible for the observed magnetic properties.[12]

5.4 Mechanisms of Ferromagnetism

Two basic approaches to understanding the magnetic properties of DMSs have emerged. The first class of approaches is based on mean-field theory. The theories that fall into this general model implicitly assume that the DMS is a more-or-less random alloy, *e.g.* (Ga,Mn)N, in which Mn substitutes for one of the lattice constituents. The second class of approaches suggests that the magnetic atoms form small (a few atoms) clusters that produce the observed ferromagnetism.[12] A difficulty in experimentally verifying the mechanism responsible for the observed magnetic properties is that, depending on the conditions employed for growing the DMS material, it is likely that one could readily produce samples that span the entire spectrum of possibilities from single-phase random alloys to nanoclusters of the magnetic atoms to precipitates and second phase formation. Therefore, it is necessary to decide on a case-by-case basis which mechanism is applicable. This can only be achieved by a careful correlation of the measured magnetic properties with materials analysis methods that are capable of detecting other phases or precipitates. If, for example, the magnetic behavior of the DMS is characteristic of that of a known ferromagnetic second phase (such as MnGa or Mn$_4$N in (Ga,Mn)N), then clearly the mean-field models are not applicable. To date, most experimental reports concerning room-temperature ferromagnetism in DMS employ x-ray diffraction, selected-area diffraction patterns, TEM, photoemission or x-ray absorption (including extended x-ray absorption fine structure, EXAFS, as discussed later) to determine whether the magnetic atoms are substituting for one of the lattice constituents to form an alloy. Given the level of dilution of the magnetic atoms, it is often very difficult to categorically determine the origin of the ferromagnetism. Indirect means, such as superconducting quantum interference device (SQUID) magnetometer measurements, to exclude any ferromagnetic intermetallic compounds as the source of magnetic signals (and even the presence of what is called the anomalous or extraordinary Hall effect), have been widely used to verify a single-phase system, both may be insufficient to characterize a DMS material.

The mean-field approach basically assumes that the ferromagnetism occurs through interactions between the local moments of the Mn atoms, which are mediated by free holes in the material. The spin–spin coupling is also assumed to be a long-range interaction, allowing use of a mean-field approximation.[13–20] In its basic form, this model employs a virtual-crystal approximation to calculate the effective spin-density due to the Mn ion distribution. The direct Mn–Mn interactions are antiferromagnetic, so that the Curie temperature T_C for a given material with a specific Mn concentration and hole density (derived from Mn acceptors and/or intentional shallow level acceptor doping) is determined by a competition between the ferromagnetic and antiferromagnetic interactions. Numerous refinements of this approach have appeared recently, taking into account the effects of positional disorder [16,17], indirect exchange interactions [18], spatial inhomogeneities and free-carrier spin polarization [19,20]. Figure 5.1 shows a compilation of the predicted T_C values, together with some experimental results. In the subsequent period its after two appearance of the Dietl et al. [11] paper, remarkable progress has been made on the realization of materials with T_C values at or above room temperature. While most of the theoretical work for DMS materials has focused on the use of Mn as the magnetic dopant, there has also been some progress on identifying other transition metal atoms that may be effective. Figure 5.2 summarizes the two main theoretical approaches to understanding the ferromagnetism. In the mean-field theories, the Mn ions are considered to be embedded in a high concentration of free carriers that mediate the coupling between the Mn ions. In the bound magnetic polaron (BMP) models, the carrier concentration is much less than the Mn density and BMPs form consisting of carriers localized around large clusters of Mn. As the temperature is lowered, the diameter of the BMPs increases and eventually overlaps at the Curie temperature.

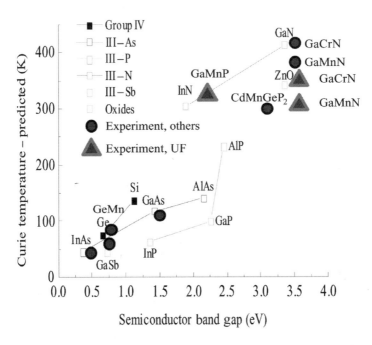

Figure 5.1. Predicted Curie temperatures as a function of bandgap, along with some experimental results (UF refers to University of Florida data)

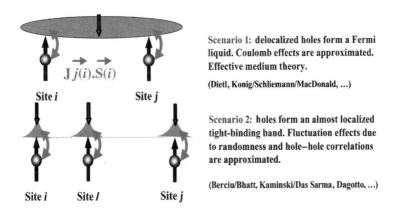

Figure 5.2. Schematic of theoretical approaches for ferromagnetism mechanisms in dilute magnetic semiconductors. The top figure represents the mean-field approach and the bottom the BMP approach

5.5 (Ga,Mn)N

Clear signatures of room-temperature ferromagnetism were observed in epitaxial GaN layers grown on sapphire substrates and then subjected to solid-state diffusion of Mn at temperatures from 250 to 800°C for various periods. [21,22] Anomalous Hall effect was observed at 323 K. The Curie temperature was found to be in the range 220–370 K, depending on the diffusion conditions. The use of ion implantation to introduce the Mn produced lower magnetic ordering temperatures.[23,24]

In (Ga,Mn)N films grown by MBE at temperatures between 580 and 720°C with Mn contents of 6–9 at.%, magnetization M versus magnetic field H curves showed clear hysteresis at 300 K, with coercivities of 52–85 Oe and residual magnetizations of 0.08–0.77 emu/g at this temperature. An estimated T_C of 940 K was reported, using a mean-field approximation to analyze the experimental data.[25] Note that while the electrical properties of the samples were not measured, they were almost certainly n-type [26–28]. As we discussed above, it is difficult to obtain high Curie temperatures in n-type DMS materials according to the mean-field theories, and this is something that needs to be addressed in future refinements of these theories. The 940 K result is now considered to be most likely due to MnGa or Mn_4N inclusions and not to single-phase (Ga,Mn)N.

Room-temperature ferromagnetism in single-phase n-type (Ga,Mn)N grown by MBE has also been reported by Thaler et al. [29], as shown in Figure 5.3. In general, no second phases are found for Mn levels below ~10% for growth temperatures of ~750°C. The (Ga,Mn)N retains n-type conductivity under these conditions. In that case, strenuous efforts were made to exclude any possible contribution from the sample holder in the SQUID magnetometer or other spurious effects. It is also worthwhile pointing out that, for the studies of (Ga,Mn)N showing ferromagnetic ordering by magnetization measurements, a number of materials characterization techniques did not show the presence of any second ferromagnetic phases within detectable limits. In addition, the values of the measured coercivities are relatively small. If indeed there were undetectable amounts of nano-sized clusters, due to geometrical effects, the expected fields at which these clusters would switch magnetically would be expected to be much larger than what has been observed. In accordance with most of the theoretical predictions, magnetotransport data showed the anomalous Hall effect, negative magnetoresistance and magnetic resistance at temperatures that were dependent on the Mn concentration. For example, in films with very low (<1%) or very high (~9%) Mn concentrations, the Curie temperatures were between 10 and 25K. An example is shown in Figure 5.3 for an n-type (Ga,Mn)N sample with ~7% Mn. The sheet resistance shows negative magnetoresistance below 150 K, with the anomalous Hall coefficient disappearing below 25 K. When the Mn concentration was decreased to 3 at.%, the (Ga,Mn)N showed the highest degree of ordering per Mn atom. Figure 5.3(a) shows hysteresis present at 300 K, while the magnetization as a function of temperature is shown in Figure 5.3(b). Data from samples with different Mn concentrations is shown in Figure 5.3(c) and indicates ferromagnetic coupling, leading to a lower moment per Mn. Data from field-cooled and zero-field-cooled conditions was further suggestive of room-temperature magnetization. The significance of these results is that there are many advantages from a device viewpoint to having n-type ferromagnetic semiconductors. EXAFS measurements have been performed on (Ga,Mn)N samples grown by MBE on sapphire at temperatures of 400–650°C with Mn concentrations of $~7\times 10^{20}$ cm^{-3} (*i.e.* slightly over 2 at.%).[30] Most of the Mn substituted for Ga on substitutional lattice positions. In the samples grown at 650°C, <1 at.% of the total amount of Mn was found to be present as Mn clusters. However, at lower growth temperatures (400°C), the amount of Mn that could be present as clusters increased up to ~36 at.% of the total Mn incorporated. The ionic state of the substitutional Mn was found to be primarily Mn(2), so that these impurities act as acceptors when substituting for the Ga with valence three. However, when the electrical properties of these samples were measured, they were found to be resistive.[29] This result emphasizes how much more needs to be understood concerning the effects of compensation and unintentional doping of (Ga,Mn)N, since the EXAFS data indicated the samples should have shown very high p-type conductivity due to incorporation of Mn acceptors.

However, there are still strong variations in the reported magnetic behavior, with some films exhibiting only paramagnetism and even those

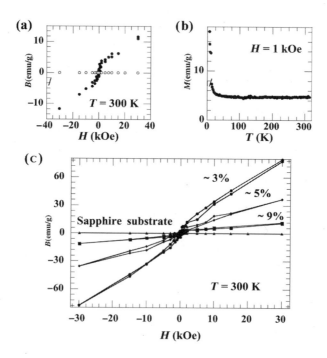

Figure 5.3. (a) *B–H* from MBE-grown (Ga,Mn)N with 9.4 at.% Mn (closed circles) and from sapphire substrate (open circles), (b) *M–T* of (Ga,Mn)N, and (c) *B–H* from (Ga,Mn)N as a function of Mn concentration

with ferromagnetism showing a wide range of apparent Curie temperatures T_C.[31–40] In particular, the origin of this ferromagnetism is not clear. Current hypotheses include a mean-field model (based on the Ruderman–Kittel–Kasuya–Yoshida, RKKY, interaction) in which the ferromagnetism results from carrier mediation by delocalized or weakly localized holes in p-type material or various types of small clusters of the Mn such as Mn_xN. Given that the $Mn^{3+/2+}$ acceptor level is deep ($\sim E_V+$ 1.8 eV), it is not expected that free-carrier-mediated magnetism is significant, but tightly bound carriers could play a role. The first class of approaches based on mean-field theory, which originates in the original model of Zener magnetism.[41–43] The theories that fall into this general model implicitly assume that the DMS is a more-or-less random alloy, *e.g.* (Ga,Mn)N, in which Mn substitutes for one of the lattice constituents. The second class of approaches suggests that the magnetic atoms form small (a few atoms) clusters that produce the observed ferromagnetism. A difficulty in experimentally verifying the mechanism responsible for the observed magnetic properties is that, depending on the growth conditions employed for growing the DMS material, it is likely that one could readily produce samples that span the entire spectrum of possibilities from single-phase random alloys to nanoclusters of the magnetic atoms to precipitates and second phase formation. Therefore, it is necessary to decide on a case-by-case basis which mechanism is applicable. This can only be achieved by a careful correlation of the measured magnetic properties with materials analysis methods that are capable of detecting other phases or precipitates. If, for example, the magnetic behavior of the DMS is characteristic of that of a known ferromagnetic second phase (such as MnGa or Mn_4N in (Ga,Mn)N [44]), then clearly the mean-field models are not applicable.

5.6 Role of Second Phases

There is clearly a need to examine the properties of (Ga,Mn)N with and without second phases, at least as detected by common analysis methods such as x-ray diffraction. The magnetic properties of (Ga,Mn)N grown by MBE with a broad range of Mn concentrations (5 or 50 at.%) and which exhibited either one phase or multiple phases were examined. The first sample had an Mn concentration of ~5 at.% as determined by both AES and Rutherford backscattering. The growth conditions (temperature, rate and V/III ratio) were optimized to produce single-phase material. The second sample also had a Mn concentration of ~5 at.%, but was grown under conditions where we observe second phases. The third sample had a Mn concentration of ~50 at.% and was designed to contain large concentrations of second phases.

Figure 5.4 shows the resulting x-ray diffraction powder scans from the three samples. Second-phase peaks are observed for the unoptimized 5 at.% Mn sample and the 50 at.% sample. In the optimized 5 at.% Mn material, only peaks due to hexagonal *c*-axis-aligned GaN and GaMn were observed. In separate experiments we have observed a linear variation in (Ga,Mn)N lattice constant with Mn concentration provided that the material does not develop secondary phases. The only second phases observed in the unoptimized or high Mn concentration samples are Ga_xMn_y. Within this family of compounds, most are reported to be ferromagnetic in bulk form, namely Mn_2Ga (T_C = 690 K), Mn_3Ga (T_C=743 K), Mn_5Ga_8 ($T_C \approx 210$ K) and MnGa (T_C>300 K). We have not observed Mn_4N or other Mn_xN_y phases under our growth conditions. In growth of (Ga,Mn)N by MBE using a single precursor of [Et$_2$Ga(N$_3$)NH$_2$CH$_3$], the dominant second phase was Mn$_3$GaN (T$_C$≈200 K).[26]

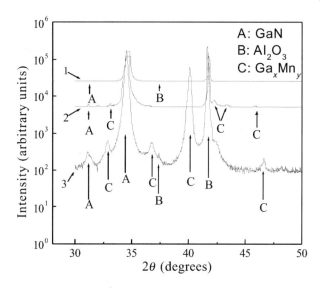

Figure 5.4. X-ray diffraction scans from (Ga,Mn)N with either 5 at.% Mn (optimized growth, sample 1), 5 at.% Mn (unoptimized growth, sample 2) or 50 at.% Mn (sample 3)

All of the samples exhibited hysteresis in 300 K magnetization versus field loops (Figure 5.5), with coercivities in the region of 100 G. A more instructive measurement is that of the temperature dependence of the field-cooled (FC) and zero field-cooled (ZFC) magnetization, performed in a Quantum Design SQUID magnetometer. As shown in Figure 5.6 (top), the single-phase (Ga,Mn)N is ferromagnetic to >300 K, as evidenced by the separation in FC and ZFC curves. By sharp contrast, the x=0.5 sample shows behavior typical of a spin glass at <100 K (Figure 5.6 center), while the multi-phase x=0.05 material shows behavior consistent with the presence of at least two ferromagnetic phases (Figure 5.6 bottom). Note that the 50% sample still exhibits a loop at 300 K, indicating a small difference in FC and ZFC magnetization at this temperature. All of this behavior is consistent with the x-ray diffraction data.

5.6 Role of Second Phases 271

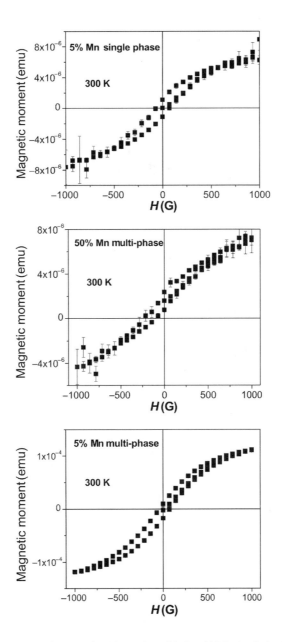

Figure 5.5. 300 K hysteresis loops for (Ga,Mn)N with 5 at.% Mn (optimized growth) at top, 50 at.% Mn (at center) or 5 at.% Mn (unoptimized growth) at bottom

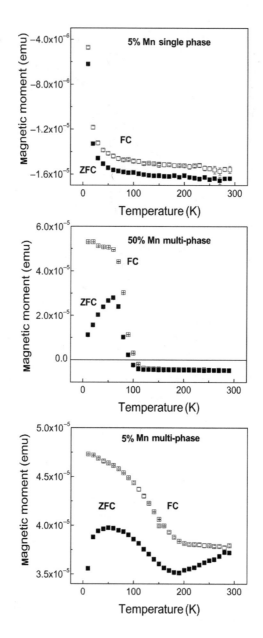

Figure 5.6. Temperature dependence of FC (top curve in each case) and ZFC (bottom curve in each case) magnetic moment for (Ga,Mn)N with 5 at.% Mn (optimized growth) at top, 50 at.% Mn at center or 5 at.% Mn (unoptimized growth) at bottom

What can be established from these experiments is that (Ga,Mn)N with 5 at. % Mn, grown by MBE under optimum conditions, shows no detectable second phases in x-ray diffraction spectra and exhibits ferromagnetism to >300 K. In material known to have second phases, the magnetization versus temperature behavior shows either a spin-glass-type transition or cusps corresponding to the presence of multiple phases.

5.7 Electrical and Optical Properties

While a number of groups have reported room-temperature ferromagnetism in GaN doped with Mn or Cr [31,32,34,35], there has been little investigation of the electrical and optical properties of the material. The carrier-induced ferromagnetism model requires hole-induced interactions between the spins of the substitutional transition metal ions. However, the few reports of the position of Mn in the GaN bandgap find it to be very deep, E_V=1.4 eV, where it would be an ineffective acceptor dopant.[45–47] In addition, most of the data reported for GaMnN indicate it is either insulating or n-type.

The resistivity of GaMnN grown by MBE has been examined using both Schottky diode and TLM measurements and optical absorption spectra from GaMnN films as a function of Mn concentration. Features at E_C=–1.9 eV are found in the absorption spectra, corresponding to transitions from Mn to the conduction band. However, this state does not control the Fermi level position in the GaMnN, and the material remains high-resistivity n-type with a thermal activation energy of ~0.1 eV. The results have interesting indications for both the existing theoretical models for DMSs and for the potential technological uses of GaMnN.

Figure 5.7 (top) shows the low bias I–V characteristics at 25°C from the GaMnN–GaN device structure. The results are consistent with the GaMnN having a high resistivity of ρ=3.7×10^7 Ω cm. Even at higher biases, the I–V characteristic is dominated by the high resistance of the GaMnN (Figure 5.7, bottom).

From TLM ohmic metal patterns placed directly on the GaMnN layer, we were able to measure the temperature dependence of the sheet

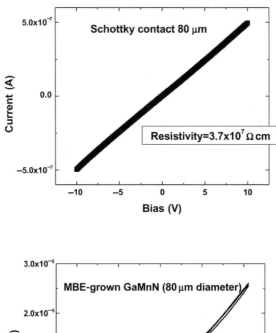

Figure 5.7. Low bias I–V characteristics at 25°C from Schottky diode GaMnN/GaN structure (top) and forward I–V plots from several diodes at highest bias (bottom)

resistivity. Figure 5.8 shows an Arrhenius plot of this data, corrected for the temperature dependence of the mobility contribution to the sheet resistivity. Over a broad range of temperature, the sheet resistivity varies as

$$\rho_S = \rho_{S0} \exp(-E_a / kT) \qquad (5.1)$$

where $E_a=0.11\pm0.02$ eV. This represents the activation energy of the defects or impurities controlling the n-type conductivity of the GaMnN.

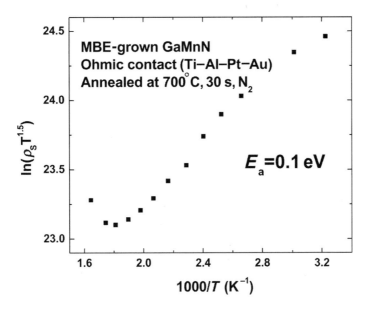

Figure 5.8. Arrhenius plot of sheet resistivity of GaMnN layer, adjusted for the temperature dependence of mobility. The resulting activation energy is ~0.1 eV

These results are consistent with the absorption data. Figure 5.9 shows the spectral dependence of absorption coefficient for GaMnN as a function of Mn concentration. The introduction of Mn produced an absorption band with a threshold near 1.9 eV, which has been previously ascribed to transitions from Mn acceptors to the conduction band. We do not observe the band with a threshold near 1.4 eV corresponding to the transition from the valence band to the Mn acceptor, which is strong evidence for the sample being n-type with the Fermi level in the upper level of the bandgap. Thus, even though there is a very high concentration of Mn acceptors in the GaMnN, these do not control the electrical properties of the materials. One possibility for the 0.1 eV donors are nitrogen vacancies.

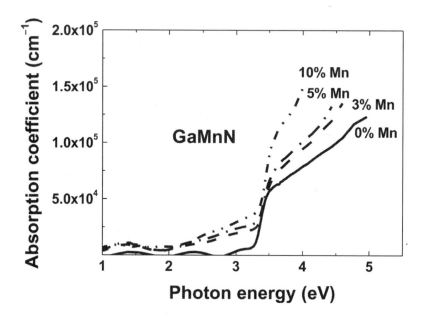

Figure 5.9. Optical absorption spectra from GaMnN layer as a function of Mn concentration

Recent experiments on photoluminescence, optical absorption and photocapacitance spectra from Mn-doped layers grown by MOCVD suggest that Mn forms in GaN a deep acceptor level near E_V=+1.4 eV.[48–51] However, Mn implanted and annealed GaN films with Curie temperatures close to room temperature do not show the semi-insulating behavior expected in such a case. Therefore, some other defect centers must also be formed in high concentration and contribute to conduction mechanisms in implanted films.

The optical transmission spectrum taken for the control sample (Figure 5.10) showed a sharp edge near 3.4 eV, and a dense pattern of interference fringes without any measurable absorption in the below-band-edge region. In contrast, the two Mn-implanted samples showed a strong absorption near 1.8 eV, *i.e.* close to the optical threshold for the transition from the E_V=+1.4 eV Mn acceptor level to the conduction band edge. For the Co-implanted sample the absorption edge was slightly red-shifted to 1.7 eV, indicating the presence of a deep level about 0.1 eV deeper than the Mn acceptor.

5.7 Electrical and Optical Properties

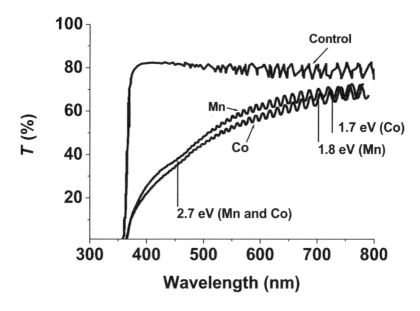

Figure 5.10. Optical transmission spectra at 300 K for the control sample, the 4×10^{16} cm^{-2} Mn-implanted sample and the 4×10^{16} cm^{-2} Co-implanted sample

Capacitance–voltage measurements on the unimplanted GaN gave an electron concentration of 2×10^{16} cm^{-3} and the intercept value in the $1/C^2(V)$ plot of 0.7 V which is standard for Au Schottky diodes prepared on undoped n-GaN films grown by MOCVD. The frequency and temperature dependence of capacitance was very slight. The behavior of the implanted samples was much more complicated. The capacitance at low frequencies was considerably higher than at high frequencies. At temperatures near 400 K the capacitance versus frequency curves showed well-defined low-frequency and high-frequency plateaus. At low frequencies the measured thickness of the space charge region was lower than 0.2 µm (*i.e.* the space charge region boundary was moving within the implanted region). The roll-off frequency in the low-frequency range decreased with decreased measurement temperature in a similar manner for all implanted samples and application of the standard admittance spectroscopy analysis to the shift in frequency with temperature yielded an activation energy of 0.45–0.5 eV. The $1/C^2(V)$ plots were linear for all three implanted samples, the apparent concentration deduced from the slope increasing from $(6–8)\times10^{16}$ cm^{-3} for the Mn-implanted samples to 4.2×10^{17} cm^{-3} for the Co-implanted sample. Thus, at these low frequencies the *C–V* curves are determined by the uncompensated portion of the deep electron trap concentration with level near E_c=–0.5 eV. These are most likely the traps responsible for the strong blue band in the Micro Callodo luminescence (MCL) MCL spectra and for the absorption band near 3 eV in Figure 5.10 (the total concentration of these deep traps could be much higher than the value deduced from C–V measurements since the latter relate to the uncompensated portion of the total density). No such centers were introduced in n-GaN upon implantation of protons, and it seems reasonable to associate the observed centers with complexes of radiation defects with the transition metal acceptors.

Arrhenius plots of sheet resistance of the Mn-implanted (3×10^{16} cm^{-2}) GaN clearly showed more than one defect level present, producing a nonlinear plot. The slopes at the two extremes of the plot corresponded to activation energies of 0.85 eV and 0.11 eV respectively.

In Figure 5.11 we compare the Deep Level Transient Spectroscopy DLTS spectra for the electron traps in the control sample and the 3×10^{16} cm^{-2} Mn-implanted sample. The Mn implantation greatly increases the density of the E_c=–0.25 eV electron traps and E_c=–0.7 eV electron traps known to be related to point defects introduced, for example, during proton implantation into n-GaN. Moreover, the magnitude of the signal taken with reverse bias of –1 V and forward bias of 1 V in the implanted sample was much higher than that taken with reverse bias of –3 V and the forward bias pulse of –1 V. We attribute the difference to the fact that in the former case the space charge boundary during the injection pulse is pushed deeper into the damaged region where the density of deep radiation defects is higher. In Figure 5.12 we compare the DLTS spectra of the Mn-implanted and the Co-implanted samples taken under the same conditions. It can be seen that the concentration of radiation defects in the Co-implanted sample is considerably higher-most likely due to the higher initial density of the radiation defects in the region implanted with Co, Co being heavier than Mn. Our previous studies of the proton-implanted n-GaN samples have shown that the E_c=–0.7 eV radiation defects are very efficient lifetime killers, which explains the extremely low photosensitivity of the Mn- and Co-implanted samples and the very low MCL intensity observed in these samples (the low photosignal and MCL signal were observed even with electron beam excitation with the beam energy of 25 kV, when the excitation region penetrated much deeper than the projected Mn or Co ions range of 0.2 µm, but not deeper than the thicker damaged region).

5.7 Electrical and Optical Properties

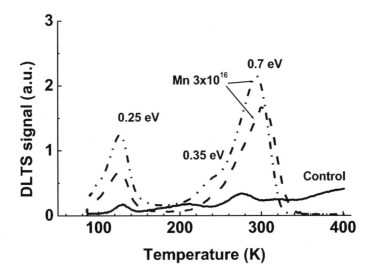

Figure 5.11. DLTS spectra of the control sample at –1 V with +1 V forward pulse (solid curve), of the 3×10^{16} cm^{-2} Mn-implanted sample taken with reverse bias of –1 V and forward bias of +1 V (short dash curve) and sof the same sample measured with reverse bias of –3 V and forward bias of –1 V (dashed curve)

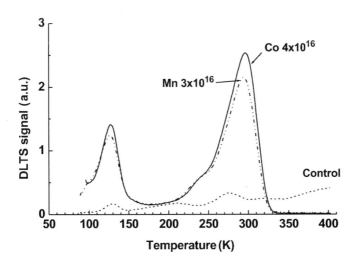

Figure 5.12. DLTS spectra of the Mn-implanted (dashed curve) and the Co-implanted (solid curve) samples

In Figure 5.13 we show the optical DLTS spectra measured on the 3×10^{16} cm^{-2} Mn-implanted sample when the deuterium lamp UV source and the tungsten lamp visible light source were used. It can be seen that with the strongly absorbed UV light of the deuterium lamp the spectrum was dominated by electron traps with energy of 0.28 eV and an overlapping feature produced by the 0.35 eV and 0.5 eV electron traps. With the slightly absorbed visible light of the tungsten lamp the dominant features were the hole traps near E_v=+0.2 eV, E_v=+0.35 eV and E_v=+0.43 eV. The results in the figure refer to the Mn-implanted sample, but the only difference with the Co-implanted sample was the absence of the 0.35 eV electron trap peak in the D-lamp-excited spectrum. The difference might be due to the different penetration depth of the D lamp light and the W lamp light. In the first case electrons and holes are generated mainly in the near-surface implanted region, where a very high density of Mn or Co acceptors exist. The holes are effectively trapped by these acceptors and are disposed of by some very slightly activated process such as hopping via acceptor states with subsequent tunneling at the Schottky diode. Thus, only electron traps are left to be observed in the capacitance transients, and among them the 0.5 eV trap is most likely the one dominating the low-frequency C–V characteristics and producing the blue MCL band.

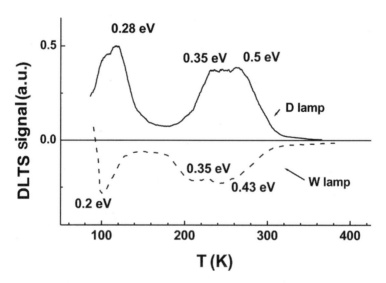

Figure 5.13. Optical DLTS spectra taken on the 3×10^{16} cm^{-2} Mn-implanted sample with the D lamp excitation and the W lamp excitation

We have shown that Mn implantation into GaN gives rise to a strong absorption with the threshold near 1.8 eV, which is close enough to the band observed in MOCVD-grown GaMnN samples in which Mn was shown to produce an acceptor level near E_v=+1.4 eV. The optical threshold for the Co-implanted sample was about 0.1 eV lower-implying that the Co acceptor level could be slightly deeper than the Mn level. This would be in line with general observations on the depth of the transition metals in III–V materials as a function of the d-shell filling; it has been noted that the higher the filling, the deeper the level, so that, in GaAs, the Mn acceptor is at E_v=+0.1 eV and the Co acceptor is at E_v=+0.16 eV, although for the wider bandgap GaP both dopants produce acceptor levels near E_v=+0.4 eV.

The fact that room-temperature ferromagnetism can be achieved in n-type GaMnN is contrary to the main mean-field models, but is attractive from a technological viewpoint since most devices require either n-type material (i.e. FETs) or both n- and p-type material (*i.e.* bipolar transistors and optical emitters). Further optimization of the growth conditions will be needed to reduce the unintentional donor concentration to the point where p-type GaMnN can be produced.

Other reports have also recently appeared on the magnetic properties of GaN doped with other transition metal impurities.[31,32,34,35,48–51] For initially p-type samples directly implanted with either Fe or Ni, ferromagnetism was observed at temperatures of ~200 K and 50 K respectively. (Ga,Fe)N films grown by MBE showed Curie temperatures of <100 K, with EXAFS data showing that the majority of the Fe was substitutional on Ga sites.[29] (Ga,Cr)N layers grown in a similar fashion at 700°C on sapphire substrates showed single-phase behavior, clear hysteresis and saturation of magnetization at 300 K and a Curie temperature exceeding 400 K.[51]

5.8 Transport Properties

A key requirement for successful realization of spin-based devices is an understanding of the transport properties within the DMS, since carriers need to be injected from contacts and cross heterointerfaces in order to be collected. Some potential early demonstration devices include spin LEDs, in which injection and recombination of spin-polarized carriers could lead to polarized light emission, and spin transistors in which electron field gating can be used to control the carrier-induced ferromagnetism. Most GaN is still grown heteroepitaxially on lattice-mismatched substrates, such as sapphire and, therefore, contains high concentrations (usually $\geq 5 \times 10^8$ cm^{-2} as measured by TEM) of threading dislocations and other extended defects.

Numerous reports have demonstrated the deleterious effect of charged dislocations on the transverse carrier mobility in GaN. However, vertical devices are much less degraded by the repulsive band bending around dislocations and the directional dependence of the scattering due to the these dislocations because of the greater average distance between defects in this geometry. This has been confirmed by an investigation of vertical and lateral transport in n-GaN films, which showed vertical electron mobilities of ~950 cm^2/V·s compared with lateral mobilities of 150–200 cm^2/V·s.

Hall measurements showed (Ga, Mn)N electron mobilities in the range of 102 cm^2/V·s at 373 K to 116 cm^2/V·s at 298 K. We believe these values are close to the true mobility in the (Ga, Mn)N, since measurements made on the GaN prior to Mn implantation showed much higher electron mobilities of ~600 cm^2/V·s at 298; K or thus, if most of the current was flowing in the buffer layer, then we would expect to measure an effective mobility closer to this value. This latter value of 600 cm^2/V.s is similar to that reported for high quality n-type GaN.

To obtain the vertical mobilities, temperature-dependent I–V measurements were performed in the Schottky diode structures. The I–V characteristics on (Ga,Mn)N diodes measured from 298 to 373 K are shown in Figure 5.14. The barrier heights ϕ_b extracted from the forward part of the I–V characteristics for both these and the diodes without Mn were obtained, with values ranging from 0.91 eV at 298 K to 0.88 eV at 373 K for the (Ga,Mn)N. By contrast, the measured barrier height on Pt–GaN diodes was 1.08 eV. The saturation current density J_S can be represented as

$$J_S = A^{**}T^2 \exp\left(-\frac{\phi_b}{kT}\right) \qquad (5.2)$$

where A^{**} is the Richardson constant for (Ga, Mn)N, T is the absolute measurement temperature and k is Boltzmann's constant. This can also be written in the form

$$J_S = \left[eN_c\mu_{\text{VERT}}\left(\frac{2eV_{bi}N_d}{\varepsilon}\right)^{1/2}\right]\exp\left(\frac{-e\phi_b}{kT}\right) \qquad (5.3)$$

where N_C is the effective density of states in the conduction band, μ_{VERT} is the electron mobility in the perpendicular direction, e is the electron charge,

V_{bi} is the built-in voltage, ε is the dielectric constant and N_d is the doping density in the (Ga, Mn)N. The vertical values are factors of about three to eight higher at a given temperature than the lateral mobilities obtained from the Hall data. If the active area of the diodes is less than the geometric area then the effective vertical mobilities will be even higher than calculated here.

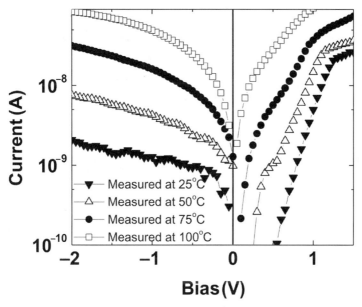

Figure 5.14. I–V characteristics as a function of temperature from n-(Ga,Mn)N Schottky diodes

To give a visual representation of why the lateral mobility is more degraded by threading scattering than is the vertical mobility, Figure 5.15 shows a cross-sectional TEM micrograph of the (Ga,Mn)N–GaN structure. Threading dislocations originating from the GaN–Al$_2$O$_3$ interface can reach the surface and, therefore, electrons traveling laterally through the structure encounter scattering from all of these defects. By contrast, for vertical transport, there is a relatively large fraction of undefective material through which electrons can pass with undegraded mobility. The defects remaining in the (Ga,Mn)N are mostly loops, which have only a second-order effect on the electrical properties, as reported previously for implanted GaAs.

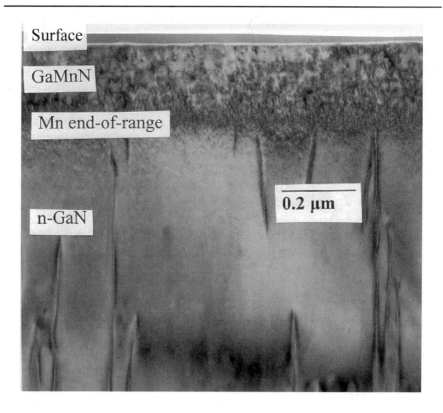

Figure 5.15. TEM cross-section of (Ga,Mn)N layer formed in GaN by high–dose Mn implantation

To place the experimental data in context, Figure 5.16 shows these results along with the calculated electron drift mobility of undoped GaN in both lateral and vertical directions and the individual components from the scattering processes present (acoustic, polar phonon, and piezoelectric). While no quantitative conclusions may be drawn, it is clear that the vertical and lateral mobilities are of comparable magnitude to those in material with minimal scattering. In the existing (Ga,Mn)N the vertical electron mobility is relatively unaffected by charged dislocation scattering and gives an indication of the values that it will be possible to achieve for lateral electron mobilities in material synthesized on low-defect GaN, such as free-standing quasi-substrates.

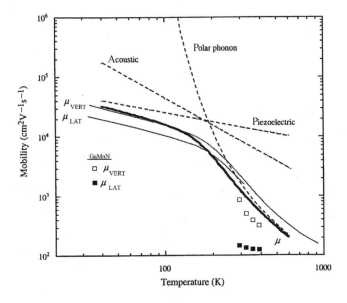

Figure 5.16. Theoretical drift mobilities in pure GaN, along with contribution from the various scattering processes present and the experimentally determined vertical and lateral mobilities for n-(Ga,Mn)N

In summary, the effect of dislocation scattering on electron mobility in (Ga,Mn)N has been examined through a comparison of vertical and lateral transport properties. The vertical electron mobilities are found to be a factor of three to eight higher than the corresponding lateral mobilities at the same temperature. (Ga,Mn)N-based spintronics devices with vertical geometries will be at an advantage relative to lateral devices.

5.9 Contacts to (Ga,Mn)N

One of the possible applications of DMSs is in so-called spin FETs (Spin-FETs) in which electric field gating is used to control the carrier-induced ferromagnetism. This has already been demonstrated in metal–insulator–semiconductor (MIS) structures based in (In,Mn)As at 10 K. Therefore, it is necessary to understand the properties of rectifying contacts on (Ga,Mn)N as a first step towards realizing practical spintronics switches.

From the I–V–T characteristics of Pt Schottky diodes on n-type (Ga,Mn)N, we extracted a value for Richardson's constant for this material of $A^{**} = 91.2$ A·cm^{-2}K^{-2}, but there is a large uncertainty (~±60%) in this due to the narrow range of measurement temperatures from which we had to extrapolate to obtain the estimated Richardson's constant. For comparison, a similar analysis for Pt–n-GaN Schottky diodes reported values of 64.7–73.2 for A^{**}.

From the measured I–V characteristics at each temperature we were able to extract the ϕ_B values for Pt on (Ga,Mn)N. The barrier height at 25°C was 0.82±0.04 eV, compared with a value of 1.08 eV for Pt on n-GaN determined from I-V characteristics. It should also be noted that in the early stage of developing a new materials system it is common to have a wide range of values reported for barrier heights due to the presence of surface defects, interfacial layers and material inhomogeneities. The barrier height was also extracted from the activation energy plots in the measured I–V–T data and the saturation current at each temperature. The plot of $\ln(I_S/AT^2)$ versus inverse temperature yielded a barrier height of 0.91±0.06 eV, which is consistent with the value at 25°C derived from the forward I–V characteristics.

The reverse leakage of the diodes was found to depend on both bias and temperature. From a moderately doped sample of the type studied here, we would expect thermionic emission to be the dominant leakage current mechanism. According to image-force barrier height lowering, this leakage current density J_L can be written as

$$J_L = -J_S \exp\left(\frac{\Delta\phi_b}{kT}\right) \quad (5.4)$$

where $\Delta\phi_B$ is the image-force barrier height lowering, given by $(eE_M/4\pi\varepsilon_S)^{1/2}$, where E_M is the electric field strength at the metal–semiconductor interface and ε_S is the permittivity. The experimental dependence of J_L on bias and temperature is stronger than predicted from this equation. The large bandgap of (Ga,Mn)N makes the intrinsic carrier concentration in a depletion region very small, suggesting that contributions to the reverse leakage from generation in the depletion region are small. Therefore, the additional leakage must originate in contributions from other mechanisms, such as thermionic field emission or surface leakage.

5.10 Aluminum Nitride-Based Ferromagnetic Semiconductors

AlN plays an important role in many areas of solid-state devices [52–67], including thin film phosphors, nitride-based MIS heterostructure transistors, thin-film gas sensors, acoustic wave resonators, UV LED, distributed Bragg reflectors, heat spreaders and heterojunction diodes. AlN may also be promising in the emerging field of spintronics, due to its predicted high Curie temperature T_C when doped with particular transition metals. Room-temperature ferromagnetism has been reported for Cr-doped AlN thin films deposited by reactive sputtering [52] or MBE [53]. Ion implantation provides a versatile and convenient method for introducing transition metals into semiconductors for examination of their effects on the structural and magnetic properties of the resulting material.[68] AlN is an ideal host in this regard, since Kucheyev et al. [69] reported that single-crystal epilayers of AlN grown on sapphire substrates did not become amorphous even at liquid-N_2 temperatures for high doses of kelo-electron-volt heavy ions such as Au. In addition, very high quality AlN on sapphire has recently been reported by several groups [70,71], providing well-characterized material in which to examine the properties of transition metals. The fabrication of ferromagnetic AlN would create a wider range of possible all semiconductor spin-dependent devices. For example, ferromagnetic AlMnN could be used as a magnetic barrier in a tunnel junction, where it would serve as a spin filter. The predicted Curie temperature of AlMnN is greater than 300 K and recently a Curie temperature of more than 340 K has been observed for AlN:Cr.[52,53]

Growth of the films was carried out by gas-source MBE. Solid Al(7N) and Mn(7N) sources were heated in standard effusion cells. Gaseous nitrogen was supplied by an Oxford rf plasma head. All films were grown on (0001)-oriented sapphire substrates, indium mounted to Mo blocks. AlMnN and AlN films were grown at a temperature of 780°C, as indicated by the substrate heater thermocouple. Sapphire substrates were first nitrided for 30 min at a substrate temperature of 1000°C under 1.1 sccm nitrogen (chamber pressure 1.9×10^{-5} Torr). Nucleation at 575°C for 10 min and a 30 min buffer layer at 950°C followed nitriding, both under 1.1 sccm nitrogen. Both AlN and AlMnN films were grown with a substrate temperature of 780°C and an Al effusion cell temperature of 1150°C. The Mn cell temperature was varied from 635 to 658°C. The growth rate of the AlN was 0.2 µm/h and the growth rate of the AlMnN films was 0.16 µm/h.

In situ reflection high-energy electron diffraction (RHEED) was used to monitor films during growth. AlN demonstrated 2D growth and AlMnN films demonstrated 2D/3D growth. AlMnN grown with a Mn cell temperature of 635°C was found to be single phase. The AlMnN with T_{Mn}=658°C formed AlMn as detected by powder x-ray diffraction. A Mn cell temperature of 650°C was found to be the upper limit of single-phase AlMnN under previously mentioned growth conditions. For comparison, a layer of Mn_4N was also grown on sapphire.

The lattice constant was found to decrease as the Mn cell temperature increased for single-phase material. A similar pattern was observed for single-phase GaMnN films grown in the same system under different conditions. GaN implanted with Mn has been reported to exhibit substitutional or near-substitutional incorporation. It is expected that the incorporation of interstitial Mn should either increase or have no effect on the lattice constant. The observation of a decrease in the lattice constant of the AlMnN films suggests that the Mn occupies a substitutional site. This is further confirmed by Hall analysis, which showed pure AlN to be highly resistive as expected and material containing an AlMn second phase to be highly conductive n-type. By contrast, single-phase AlMnN was found to be p-type. If Mn incorporates substitutionally, then one would expect by analogy with its behavior in other III–V materials that it would behave as a deep acceptor. The observation of p-type behavior fits this explanation.

Magnetic remnance and coercivity indicating hysteresis was observed in ternary AlMnN films at 10, 100, and 300 K. Measurements over 300 K were not possible due to limitations in the magnetometer. Saturation magnetization was found to decrease at 300 K compared with 100 K for AlMnN grown at T_{Mn}=650°C. The values of temperature-dependent saturation magnetization of AlMnN are shown in Figure 5.17. This data indicates an approaching T_C for this material. However, hysteresis persisted to 300 K, demonstrating soft ferromagnetism at room temperature. Figure 5.18 depicts the field dependence of magnetization for AlMnN and AlN films. Pure AlN grown under the same conditions as AlMnN demonstrated paramagnetic behavior. This indicates that ferromagnetism arises with the addition of Mn. The diamagnetic background due to the sapphire substrate was subtracted from the raw data and the subsequent corrected data was used for analysis. The magnetization was not normalized to the Mn concentration due to the difficulty in calculation of the precise amount of Mn in the films. Perpendicular measurements were found to have lower values for saturation magnetization in AlMnN films.

The hysteresis observed in the AlMnN is believed to arise from the inclusion of Mn into the AlN lattice. Clusters of second phases, undetectable

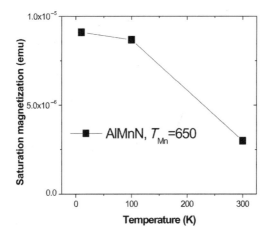

Figure 5.17. Saturation magnetization versus. temperature for single-phase AlMnN grown at 650°C. Saturation magnetization was extracted from SQUID hysteresis loops

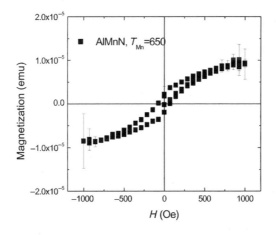

Figure 5.18. Magnetization versus applied field for single-phase AlMnN

by the methods mentioned above, are not thought to be the cause of ferromagnetism observed at 300 K. This is supported by magnetic analysis of material containing the most likely cluster phases, AlMn and Mn_4N. Magnetization as a function of temperature for AlN, single-phase AlMnN, AlMnN with an AlMn phase present, and Mn_4N show substantially different behavior, as shown in Figure 5.19. The reason for the low T paramagnetic behavior seen in AlN and AlMnN films is still unknown. The M vs. T of Mn_4N clearly indicates ferromagnetic behavior, and the formation of clusters has been proposed as the cause of hysteresis in some ferromagnetic III–V materials. However, the formation of Mn_4N clusters does not influence the magnetization above 250 K, since clearly the magnetization drops to zero at that temperature. Also, the magnetization vs. temperature indicates that the formation of AlMn clusters is not the cause of the ferromagnetism observed, evidenced by the order of magnitude difference between the values of magnetization over 150 K. Hence, the incorporation of Mn into the AlN lattice forming the ferromagnetic ternary AlMnN is too most likely the reason for the observed hysteresis.

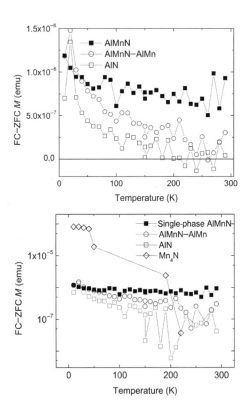

Figure 5.19. Magnetization versus temperature for AlN, AlMnN, Mn_4N and AlMnN–AlMn. The magnetic signal is determined by subtracting the zero-field-cooled trace from the field-cooled curve

5.11 Implanted Aluminum Nitride Films

Ferromagnetism has also been observed in transition-metal-implanted films. Implantation of Cr^+, Co^+ or Mn^+ ions was carried out at an energy of 250 keV (corresponding to a projected range of ~1500 Å in each case) and a fixed dose of 3×10^{16} cm^{-2}. As a rough guide, the peak transition metal concentrations, located at the projected range, are ~3 at.% in the AlN. After implantation, the samples were annealed at 950°C, 2 min under flowing N_2. PL measurements were carried out with a quadrupled Ti:sapphire laser as an excitation source together with a streak camera, providing an excitation power of ~3 mW at 196 nm.[72]

PL spectra taken at 10 K of the AlN implanted with Cr, Mn or Co after annealing at 950°C for 2 min looked basically identical in each case, even without annealing. The unimplanted AlN showed strong band-edge emission at ~6.05 eV and two broad emission bands related to deep level impurities at ~3.0 eV and 4.40 eV, each of which had peak intensity of ~1% of the band-edge emission intensity. The Cr-, Mn- and Co-implanted AlN showed an absence of band-edge emission, which suggests that the point defect recombination centers created during implantation are stable against annealing at 950°C.

Well-defined hysteresis was present in the Co-implanted AlN, with a coercive field of ~160 Oe at 300 K and 230 Oe at 10 K. The diamagnetic contributions from the substrate have been subtracted out of the data. At 300 K, the saturation moment, $M_0 = g\mu_B S$, where g is the degeneracy factor, μ_B the Bohr magnetron and S the total number of spins, was calculated to be ~0.65 μ_B for Cr. This value is lower than the theoretical value of 3 μ_B expected for a half-filled d band of Cr, if all of the Cr ions were participating in the ferromagnetic signal. Disorder effects due to implantation-induced change may contribute to creating a distribution of exchange couplings that favor antiferromagnetism and reduce the effective magnetism.

Similar data is shown in Figure 5.20 for the Co-implanted AlN. Once again there is hysteresis present at 300 K, with a coercive field of ~1750 Oe at 300 K and 240 Oe at 100 K and a calculated saturation moment of 0.52 μ_B for Co. The FC and ZFC magnetization versus temperature are shown at the bottom of the figure. In this case the differences extend to ~100 K.

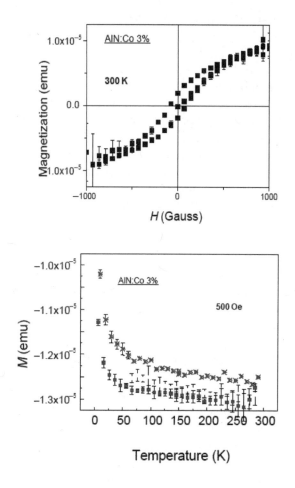

Figure 5.20. 300 K magnetization as a function of field (top) and FC and ZFC magnetization as a function of temperature (bottom) for AlN implanted with 3×10^{16} cm^{-2} Co$^+$ and annealed at 950°C, 2 min

Figure 5.21 (top) shows magnetization versus field at 100 K for Mn-implanted AlN. This was the highest temperature for which clear hysteresis could be obtained. The coercive field was ~220 Oe at both 100 K and 10 K. The FC and ZFC phases are almost coincident at an applied field of 500 Oe, as shown at the bottom of the figure, and consistent with lower overall magnitude of the magnetization. The calculated saturation moment was 0.17 μ_B for Mn at 100 K, compared with the theoretical value of 4 μ_B.

The main θ–2θ XRD peaks of the Cr- or Mn-implanted samples after

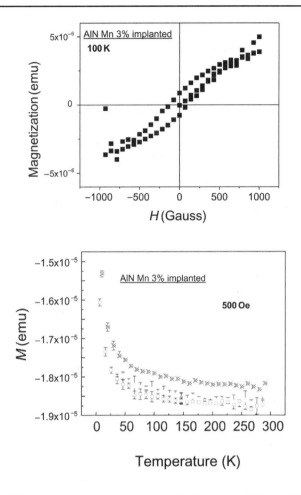

Figure 5.21. 100 K magnetization as a function of field (top) and FC and ZFC magnetization as a function of temperature (bottom) for AlN implanted with 3×10^{16} cm^{-2} Mn$^+$ and annealed at 950°C, 2 min

950°C annealing correspond to the expected AlN (0002) and (0004) lines and Al_2O_3 (0002), (0006) and (0012) substrate peaks and the broad peak at $2\theta = 20°$ is due to short-range disorder from the implantation process and was not observed on the as-grown films. No peaks due to the half-metallic ferromagnetic CrO_2 phase were detected in the Cr-implanted sample, and other potential second phases which could form, such as Cr, CrN [73–75], Cr_2N and Al_xCr_y were not detected and in any case are not ferromagnetic at the temperatures used in these experiments. Similarly, in the case of Co implantation, metallic Co has a Curie temperature of 1382 K and Co_xN phases are all Pauli ferromagnetic. Finally, for Mn implantation, metallic

Mn is antiferromagnetic, whereas Mn_xN is ferromagnetic with a Curie temperature of 745 K. Thus, secondary ferromagnetic phases are not responsible for the observed magnetic properties.

The origin of the observed ferromagnetism is not likely to be carrier mediated, due to the insulating nature of the AlN. Wu et al. [53] suggested that substitutional $Al_xCr_{1-x}N$ random alloys would have Curie temperatures over 600 K, as estimated from a multi-component mean-field theory in which the ferromagnetism occurs in a midgap defect band. Another possible mechanism for the observed magnetic properties is that the Mn is not randomly distributed on Al sites but is present as atomic-scale clusters. Some mean-field theories suggest that Mn clustering can significantly influence T_C as a result of the localization of spin-polarized holes near regions of higher Mn concentration. There is also some support for this assertion from local spin density approximation calculations, which predict it is energetically favorable for the formation of feromagnetic ion dimers and trimers at second nearest-neighbor sites. The percolation network-like model for ferromagnetism in low carrier concentration systems suggested by several groups is another potential mechanism.

High doses (3×10^{16} cm^{-2}) of ion-implanted Co^+, Cr^+ or Mn^+ ions into AlN epi layers on Al_2O_3 substrates severely degrades the band-edge luminescence, which is not recovered by annealing up to 950°C. In each case the implanted AlN shows ferromagnetic ordering, as evidenced by the presence of hysteresis in M vs H loops. The hysteresis persists up to ≥300 K in the case of Cr^+ or Co^+ implantation and 100 K for Mn^+ implantation. Less than ~20% of the implanted ions contribute to the magnetization, but this might be increased by use of much higher annealing temperatures. Simple two-terminal resistivity measurements show that the implanted AlN remains insulating (>10^8 Ω cm), and thus conventional carrier-mediated ferromagnetism is not a likely mechanism for the observed magnetic properties. Implantation provides a versatile method of introducing different transition-metal dopants into AlN for examination of their effect on the structural and magnetic properties.

5.12 Implanted AlGaN Films

There is also interest in the use of transition-metal-doped AlGaN for possible applications in spintronic devices such as polarized light emitters or spin transistors. The latter exploits quantum interference effects provided electrons with a particular spin can be injected into the channel of the device and a gate bias can be applied to cause splitting of spin-up and spin-down states. A key requirement for spin-based semiconductor devices is the achievement of ferromagnetism, preferably above room temperature. The properties of implanted transition metals in AlGaN is of particular relevance for realization of polarized light emitters or spin transistors, since it could serve as the cladding layer in the former and the wide bandgap part of the heterostructure in the latter.

θ–2θ x-ray diffraction scans from the n-AlGaN before and after Mn^+, Co^+ or Cr^+ implantation and annealing at 1000°C did not show any observable differences. The highest intensity peaks in all spectra correspond to the expected AlGaN (0002) and (0004) lines and Al_2O_3 (0002), (0006) and (0 0 0 12) substrate peaks. We did not observe any peaks due to second phases that could exhibit ferromagnetism. For example, in the Mn-implanted material, Mn_xN is ferromagnetic with a Curie temperature of 745 K and GaMn is also ferromagnetic with a Curie temperature near 300 K (metallic Mn is antiferromagnetic). In the Co^+-implanted AlGaN, metallic Co has a Curie temperature of 1382 K and Co_xN phases are all Pauli ferromagnetic. Finally, in the Cr^+-implanted AlGaN, CrO is a half-metallic, while Cr, CrN, Cr_2N, Al_xCr_y and Ga_xCr_y are not ferromagnetic.[36] However, in such thin layers, it could be possible for small quantities of second phases to be present and remain undetectable by x-ray diffraction.

Well-defined hysteresis at 300 K was observed for the Co-implanted $Al_{0.38}Ga_{0.62}N$, as shown at the top of Figure 5.22. The coercive field was ~85 Oe at 300 K and ~75 Oe at 10 K. The saturation magnetization was ~0.4 emu/cm^3 or ~0.76 μ_B calculated saturation moment. This is slightly higher than the value reported for Co^+ implantation into pure AlN under similar conditions (0.52 μ_B), which is consistent with the higher vacancy concentrations expected to be created in AlGaN due to its lower bond strength. The bottom part of Figure 5.22 shows the temperature dependence of FC and ZFC magnetization for the Co^+-implanted AlGaN. The fact that these have different values out to ~230 K is a further indication of the presence of ferromagnetism in the material. In both epitaxial and ion-implanted transition-metal-doped semiconductors, we have found the general result that the hysteresis can be detected to higher temperatures than the difference in FC and ZFC magnetization. As mentioned earlier, the samples exhibited low carrier densities (<3×10^{16} cm^{-3} from Hall measurements) after implantation and annealing and, therefore, carrier-mediated ferromagnetism by free electrons is not expected to be operative. In addition, the Co ionization level is expected to be deep in the AlGaN bandgap, so that there will be no significant contribution to the carrier density from the substitutional fraction of these atoms. More recent percolation network models for ferromagnetism in DMS suggest that localized carriers may mediate the interaction between magnetic ions in low carrier density systems.

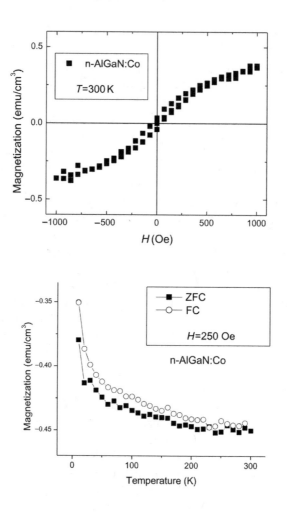

Figure 5.22. 300 K magnetization as a function of field (top) and FC and ZFC magnetization versus temperature for AlGaN implanted with 3×10^{16} Co^+ annealed at 1000°C for 2 min

The Mn-implanted p-type AlGaN also showed a well-defined hysteresis loop at 300 K, with a coercivity of ~60 Oe (Figure 5.23, top). The saturation moment, $M_0 = g\mu_B S$, where g is the degeneracy factor, μ_B the Bohr magneton and S the total number of spins, was calculated to be ~0.57 μ_B.

The theoretical value would be 4 μ_B if all of the implanted Mn was participating towards the ferromagnetism, so the lower experimental value indicates that only a fraction of the Mn is substitutional and magnetically active. The saturation moment for AlGaN is significantly larger than the value of 0.17 μ_B reported for Mn implantation into pure AlN. The temperature dependence of FC and ZFC magnetization is shown at the bottom of Figure 5.23. The ferromagnetism is very weak above ~125 K, but t is detectable through the hysteresis.

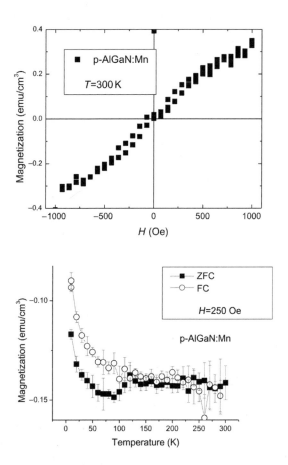

Figure 5.23. 300 K magnetization as a function of field (top) and FC and ZFC magnetization versus temperature for AlGaN implanted with 3×10^{16} Mn$^+$ annealed at 1000°C for 2 min

By sharp contrast to the case of Mn implanted into p-AlGaN, when we performed the same implants into n-AlGaN the resulting differences in FC and ZFC magnetization were very weak and hysteresis loops even at 10 K did not show clear evidence of ferromagnetism. The differences from the p-type material may result from the higher AlN mole fraction in the n-type AlGaN, which makes it harder for the implanted ions to become substitutional upon annealing. An alternative explanation is that holes are more efficient at ferromagnetic coupling between the Mn spins than are electrons. This has been reported previously for both n- and p-type GaAs and GaP doped with Mn. We also did not observe any clear evidence for ferromagnetism in the Cr-implanted n-AlGaN. This is a clear difference from the case of Cr-implanted AlN, where hysteresis was reported at 300 K.

In conventional DMSs, such as (Ga,Mn)As, the magnetization is given by [11]

$$M = \mu_g \mu_B S N_0 x_{eff} B_S \left[\frac{g\mu_B \left(-\partial F_C [M]/\partial M + H \right)}{k_B \left(T + T_{AF} \right)} \right] \quad (5.5)$$

in the mean-field approach, where S is the localized spin, N_0 is the concentration of cation sites, X_{eff} is the effective spin concentration, B_S is the Brillouin function and $F_C(M)$ is the hole contribution to the free-energy functional F (which depends on the magnetization of the localized spin). The validity of this model depends on having a high carrier concentration in the magnetic semiconductor; experimentally, we do not observe this in the AlGaN and, correspondingly, we do not observe a Brillouin-like dependence of magnetization on temperature. In pure AlN, Mn produces an absorption line at ~1.5 eV from the valence band, suggesting the Mn$^{3+/2+}$ acceptor level is deep in the gap and makes the realization of carrier-mediated ferromagnetism unlikely. The mean-field models have also shown that Mn clustering can enhance the Curie temperature through localization of localized carriers at these clustered regions. Some calculations suggest it is energetically favorable to form ferromagnetic transition-metal ion dimers and trimers at second nearest-neighbor sites. Distant pairs would be weakly ferromagnetic or antiferromagnetic. These predictions suggest that the ferromagnetism will be a very strong function of the synthesis conditions used for the magnetic semiconductor. They also suggest that non-equilibrium methods such as ion implantation possess inherent advantages in trying to maximize the Curie temperature, because of their ability to achieve solid solubilities for dopants well above those possible with equilibrium synthesis methods.

Table 5.1 shows a compilation of semiconductors showing room-temperature ferromagnetism.

Table 5.1. Compilation of semiconductors showing room-temperature ferromagnetism PLD is pulsed laser deposition

Material	Bangap of host (eV)	Comments	Ordering temperature (K)	Ref.
$Cd_{1-x}Mn_xGeP_2$	1.72	Solid-phase reaction of evap. Mn	>300	7
(Ga,Mn)N	3.4	Mn incorporated by diff^n	228–370	22
(Ga,Mn)N	3.4	Mn incorporated during MBE; n-type	>300	29
(Ga,Mn)N	3.4	Mn incorporated during MBE	940*	25
(Ga,Cr)N	3.4	Cr incorporated during MBE or bulk growth	>400	34
(ZnO):Co	3.1–3.6	Co incorporated during PLD; ~15% Co	>300	9
(ZnO):Co	3.3	Sintered powders	>425	76
(Al,Cr)N	6.2	Cr incorporated during MBE, sputtering or implantation	>300	52,53,77
(Ga,Mn)P:C	2.2	Mn incorporated by implant or MBE; $p \sim 10^{20}$ cm^{-3}	>330	68
$(Zn_{1-x}Mn_x)GeP_2$	1.83–2.8	Sealed ampoule growth; insulating; 5.6% Mn	312	6
$(ZnMn)GeP_2$	<2.8	Mn incorporated by diffusion	350	7
$ZnSnAs_2$	0.65	Bulk growth	329	8
$ZnSiGeN_2$	3.52	Mn-implanted epi	~300	78
SiC	3.2	Mn or Fe implantation	~300 (by hysteresis)	31,32

*Extrapolated from measurements up to ~750 K.

5.13 Potential Device Applications

Previous articles have discussed some spintronic device concepts, such as spin junction diodes and solar cells, optical isolators and electrically controlled ferromagnets.[78–82] The realization of LED with a degree of polarized output has been used to measure spin injection efficiency in heterostructures. While the expected advantages of spin-based devices include non-volatility, higher integration densities, lower power operation and higher switching speeds, there are many factors still to consider in whether any of these can be realized. These factors include whether the signal sizes due to spin effects are large enough at room temperature to justify the extra development work needed to make spintronic devices and whether the expected added functionality possible will materialize.

Among such devices, the simplest seems to be the concept of an LED with one of the contact layers made ferromagnetic by incorporation of transition-metal impurities, a so called spin-LED.[80–82] Such a device should allow one to modulate the polarization of the light emitted by the spin-LED by application of an external magnetic field. The most straightforward approach to achieve this goal would be to implant Mn into the top contact p-GaN layer of the standard GaN–InGaN LED. The electrical and luminescence properties of such devices show that they do produce electroluminescence, but due to the difficulty in annealing out all radiation defects the series resistance and the turn-on voltage of such spin-LEDs are very much higher than for ordinary LEDs and the electroluminescence intensity (EL) is lower. GaMnN layers produced by MBE have a lower density of defects and may be better suited for spin-LEDs. One of the problems with the latter approach is that the MBE-grown GaMnN films with high Curie temperature have n-type conductivity. Therefore, to incorporate such layers into the GaN-based LEDs one has to reverse the usual order of layers and grow an LED structure with the contact n-layer up. It was shown that problems such as higher series resistance due to lower lateral conductivity of p-GaN compared with n-GaN are inherent to these inverted diodes. Also, it was shown that it is more difficult to attain a high quality of the GaN–InGaN multi-quantum-well active region when growing it on top of a very heavily Mg-doped p-GaN layer. Incorporation of Mn into the top contact layer also produced a relatively high resistivity of the GaMnN and poorer quality of the ohmic contact. In addition, the ferromagnetism in GaMnN is found to be unstable against the type of high temperature (900°C) anneals used to minimize contact resistance.

The reference n-LED structure studied consisted of 2 μm thickness undoped semi-insulating GaN, 2 μm thickness p-GaN(Mg), five undoped quantum wells of InGaN (~40% In, 3 nm), separated by Si-doped n-GaN barriers (10 nm each) and about 170 nm thickness top n-GaN contact layer. All the layers but the top n-GaN layer were grown by MOCVD in a regime similar to the one used to fabricate standard p-LED structures. The n-type layer was grown by MBE at 700°C.

The spin-LED structure differed from the n-LED structure by the structure of the top n-layer, which consisted of 20 nm of n-GaN(Si) and 150 nm of GaMnN with the Mn concentration close to 3%. The GaMnN layer was grown at 700°C. All structures were subjected to rapid thermal anneal in nitrogen to activate the Mg acceptors. However, the spin-LED structure was given only a 750°C anneal, since the degree of magnetic ordering in the GaMnN MBE-grown layer was greatly reduced upon annealing at temperatures exceeding 800°C. A schematic of the LED structure is shown in Figure 5.24.

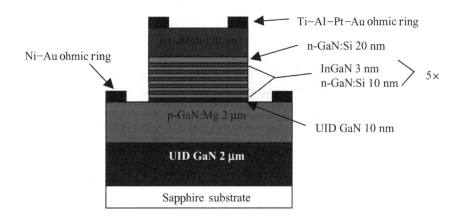

Figure 5.24. Schematic of LED structure. UID is unintentionally doped layer

Measurable room temperature electroluminescence peaked at ~450 nm and was detected at 4 V bias in the reference diode and at ~15 V in the spin-LED structure processed at 750°C (Figure 5.25). However, in spite of the ferromagnetic behavior of the (Ga,Mn)N layer, no polarization of light emission at room temperature was observed in applied magnetic fields ranging from 0 to 5 T during either EL or PL measurements, as shown in Figure 5.26, where the spin-LED structure with 100 nm thickness (Ga,Mn)N layer is taken as an example. This can be attributed either to severe spin losses in the structure during spin injection or to a poor performance of the (Ga,Mn)N layer as a spin aligner at room temperature.

Possible origins of the vanishing spin injection efficiency of the spin LED at room temperature can be clarified by analyzing its properties at low temperatures, which should improve magnetic performance of the DMS layer. Such analysis was performed by utilizing PL measurements, since the freeze-out of free carries at low temperatures prevented us from observation of efficient EL in the structures investigated. The spin injection efficiency was evaluated using an excitation photon energy of about 5 eV, *i.e.* well above the bandgap of the top (Ga,Mn)N layer. Under these conditions the preferential light absorption within the top (Ga,Mn)N layer should ensure that the dominant portion of the carriers and excitons participating in the radiative recombination in the InGaN quantum well is supplied by the magnetic barrier. The results of the measurements performed are summarized in Figure 5.27 (top). The optical (spin) polarization of the quantum well PL was only detected in an applied magnetic field and was generally very weak (<2%). The observed PL polarization reflects combined effects of the spin injection from the magnetic (Ga,Mn)M layer and intrinsic polarization of the InGaN quantum well. The latter has separately been studied by tuning the excitation energy below the bandgap of GaN (also presumably GaMnN), *i.e.* by resonant optical excitation of the InGaN quantum well with a photon energy of 2.9 eV, Figure 5.27 (bottom). It gives up to 5–10% at 2 K with an applied magnetic field of up to 5 T, likely due to population distribution between spin sublevels at a low temperature. This value is consistent with that obtained in the reference samples.

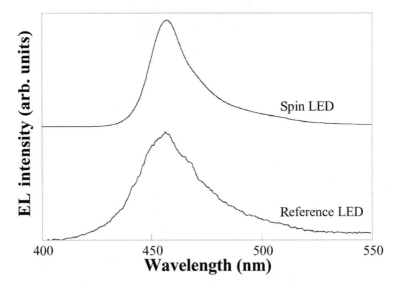

Figure 5.25. Typical EL spectra measured at room temperature from the Mn-doped LED annealed at 750°C and reference diode

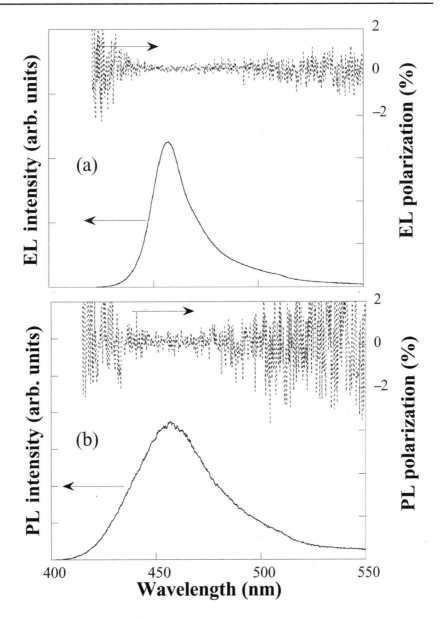

Figure 5.26. Comparison of the (a) EL and (b) PL spectra measured at room temperature from the same spin-LED structure, as well as their polarization properties at 5 T (the upper part of each figure)

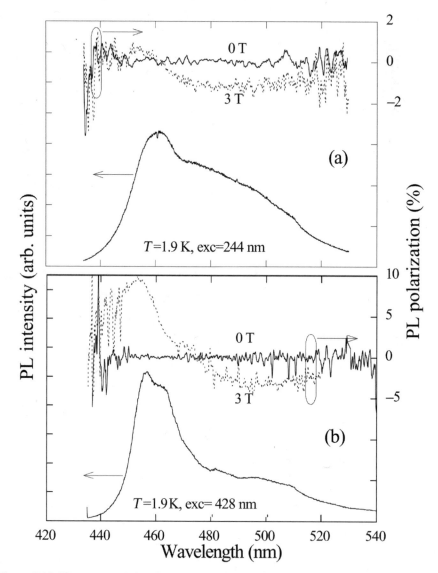

Figure 5.27. PL spectra at 1.9 K from the (Ga,Mn)N–InGaN diode, using optical injection from the (Ga,Mn)N layer and (a) direct optical excitation of the multi-quantum well. The upper part of each figure shows PL polarization at 0 T (solid lines) and 3 T (dashed lines)

As is obvious from Figure 5.28, polarization properties of the InGaN quantum well with and without carrier supply from the magnetic layer seem to be very similar, except for a consistent reduction of optical polarization of the InGaN quantum well upon optical excitation of GaMnN, detected for all structures investigated. One could speculate that this might be caused by injection of the electrons with the spin orientation opposite to that of the lowest spin state in the InGaN quantum well. As the degree of polarization is very low, a definite conclusion on this issue cannot be obtained and requires further investigations.

In order to shed light on the possible origin for the weak spin polarization of the InGaN quantum well, optical orientation experiments were performed where a chosen spin orientation of excitons/carriers in the InGaN quantum well was generated by the relevant circularly polarized excitation light resonantly pumping the InGaN quantum well. No spin polarization was observed at 0 T (Figure 5.28), showing that the spin orientation generated was completely lost during the energy relaxation process to the ground state of the excitons giving rise to the PL. This is in sharp contrast to the cases in II–VI materials and GaAs, where spin polarization can usually be observed in such optical orientation experiments. With an applied magnetic field, the spin polarization of the InGaN quantum well is independent of the polarization of the excitation light, which again proves that the spin relaxation in the InGaN quantum well is extremely efficient. The exact physical mechanism responsible for this fast spin relaxation is currently unknown and requires in-depth experimental and theoretical studies.

These findings imply the fast spin relaxation within the InGaN quantum well is enough to destroy any injected spin polarization, leading to the observed low spin injection efficiency in the (Ga,Mn)N–GaInN spin LEDs studied. This, therefore, calls for further in-depth studies of spin relaxation processes in GaN-based materials and quantum structures aiming at optimization of spin detectors as a necessary step in understanding spin injection processes and also in improving performance of possible spintronic devices based on Group III nitrides.

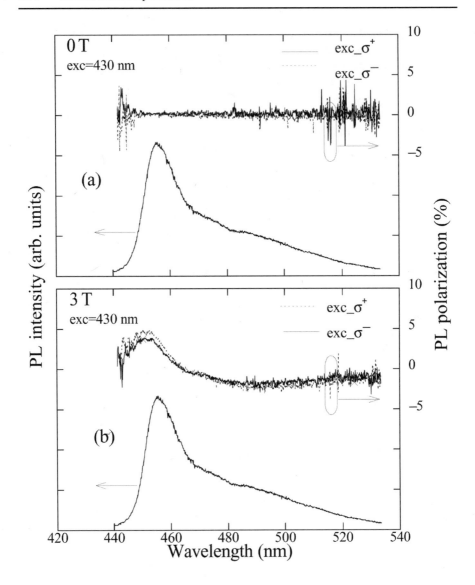

Figure 5.28. PL intensity and polarization at 1.9 K from the (Ga,Mn)N–InGaN diode obtained under optical excitation with the circularly polarized light specified in the figure, obtained at (a) 0 T and (b) 3 T

5.14 Issues to Be Resolved

The mean-field models consider the ferromagnetism to be mediated by delocalized or weakly localized holes in the p-type materials. The magnetic Mn ion provides a localized spin and acts as an acceptor in most III–V semiconductors, so that it can also provide holes. In these models, the T_C is proportional to the density of Mn ions and the hole density. Many aspects of the experimental data can be explained by the basic mean-field model. However, ferromagnetism has been observed in samples that have very low hole concentrations, in insulating material and more recently in n-type material.

More work is also needed to establish the energy levels of the Mn, whether there are more effective magnetic dopant atoms and how the magnetic properties are influenced by carrier density and type.[83] Even basic measurements, such as how the magnetism changes with carrier density or type and how the bandgap changes with Mn concentration in GaN, have not been performed. There is a strong need for a practical device demonstration showing spin functionality in a nitride-based structure, such as a spin-LED or a tunneling magneto-resistance device. The control of spin injection and manipulation of spin transport by external means, such as voltage from a gate contact or magnetic fields from adjacent current lines or ferromagnetic contacts, is at the heart of whether spintronics can be exploited in device structures, and these areas are still in their infancy.

References

1. Von Molnar S, Read D (2003) Proc IEEE 91: 715; Dietl T (2002) Semicond Sci Technol 17: 377; Ohno H (2000) J Vac Sci Technol B 18: 2039
2. Wolf SA, Awschalom DD, Buhrman RA, Daughton JM, von Molnar S, Roukes ML, Chtchelkanova AY, Treger DM (2001) Science 294: 1488
3. Ohno H, Matsukura F, Ohno Y (2002) JSAP Int 5: 4
4. Awschalom DD, Kikkawa JM (2000) Science 287: 473
5. Cho S, Choi S, Cha GB, Hong SC, Kim Y, Zhao Y-J, Freeman AJ, Ketterson JB, Kim BJ, Kim YC, Choi BC (2002) Phys Rev Lett 88: 257203-1
6. Medvedkin GA, Ishibashi T, Nishi T, Hiyata K (2000) Jp, J Appl Phys 39: L949

7. Medvedkin GA, Hirose K, Ishibashi T, Nishi T, Voevodin VG, Sato K (2002) J Cryst Growth 236: 609
8. Choi S, Cha GB, Hong SC, Cho S, Kim Y, Ketterson JB, Jeong S-Y, Yi GC (2002) Solid-State Commun 122: 165
9. Ueda K, Tahata H, Kawai T (2001) Appl Phys Lett 79: 988
10. Chambers SA (2002) Mater Today (April.) 34-39
11. Dietl T, Ohno H, Matsukura F, Cibert J, Ferrand D (2000) Science 287: 1019
12. Van Schilfgaarde M, Myrasov ON (2001) Phys Rev B 63: 233205
13. Dietl T, Ohno H, Matsukura F (2001) Phys Rev B 63: 195205
14. Dietl T (2001) J Appl Phys 89: 7437
15. Jungwirth T, Atkinson WA, Lee B, MacDonald AH (1999) Phys Rev B 59: 9818
16. Berciu M, Bhatt RN (2001) Phys Rev Lett 87: 108203
17. Bhatt RN, Berciu M, Kennett MD, Wan X (2002) J Supercond Incorp Novel Magn 15: 71
18. Litvinov VI, Dugaev VA (2001) Phys Rev Lett 86: 5593
19. Konig J, Lin HH, MacDonald AH (2001) Phys Rev Lett 84: 5628
20. Schliemann J, Konig J, MacDonald AH (2001) Phys Rev B 64: 165201
21. Reed ML, Ritums MK, Stadelmaier HH, Reed MJ, Parker CA, Bedair SM, El-Masry NA (2001) Mater Lett 51: 500
22. Reed ML, Ritums MK, Stadelmaier HH, Reed MJ, Parker CA, Bedair SM, El-Masry NA (2001) Appl Phys Lett 79: 3473
23. Theodoropoulou N, Hebard AF, Overberg ME, Abernathy CR, Pearton SJ, Chu S.G, Wilson RG (2001) Appl Phys Lett 78: 3475
24. Overberg ME, Abernathy CR, Pearton SJ, Theodoropoulou NA, McCarthy KT, Hebard AF (2001) Appl Phys Lett 79: 1312
25. Sonoda S, Shimizu S, Sasaki T, Yamamoto Y, Hori H (2002) J Cryst Growth 237–239: 1358
26. Kim KH, Lee KJ, Kim DJ, Kim HJ, Ihm YE, Djayaprawira D, Takahashi M, Kim CS, Kim CG, You SH (2003) Appl Phys Lett 82: 1775
27. Soo YL, Kioseoglou G, Kim S, Huang S, Koo YH, Kuwarbara S, Owa S, Kondo T, Munekata H (2001) Appl Phys Lett 79: 3926
28. Kuwarbara S, Kondo T, Chikyou T, Ahmet P, Munekata H (2001) Jpn J Appl Phys 40: L724
29. Thaler GT, Overberg ME, Gila B, Frazier R, Abernathy CR, Pearton SJ, Lee JS, Lee SY, Park YD, Khim Z,G, Kim J, Ren F (2002) Appl Phys Lett 80: 3964
30. Hori HS, Sasaki T, Yamamoto Y, Shimizu S, Suga K, Kindo K (2002) Physica B 324: 142

31. Pearton SJ, Abernathy CR, Norton DP, Hebard AF, Park YD, Boatner LA, Budai JD (2003) Mater Sci Eng R 40: 137
32. Pearton SJ, Abernathy CR, Overberg ME, Theodoropoulou N, Hebard A.F, Park YD, Ren F, Kim J, Boatner LA (2003) J Appl Phys 93: 1
33. Dhar S, Brandt O, Trampert, Daweritz AL, Friedland KJ, Ploog KH, Keller J, Beschoten B, Guntherhold G (2003) Appl Phys Lett 82: 2077
34. Park MC, Huh KS, Hyong JM, Lee JM, Chung JY, Lee KI, Han SH, Lee WY (2002) Solid-State Commun 124: 11
35. Lee JS, Lim JD, Khim ZG, Park YD, Pearton SJ, Chu SNG (2003) J Appl Phys 93: 4512
36. Ando K (2003) Appl Phys Lett 82: 100
37. Baik JM, Kim JK, Yang HW, Shon Y, Kang TW, Lee JL (2002) Phys Status Solidi B 234: 943
38. Sardar K, Raju AR, Basal B, Venkataraman, Rao CNR (2003) Solid-State Commun 125: 55
39. Baik JM, Yang HW, Kim JK, Lee J-L (2003) Appl Phys Lett 82: 583
40. Shon Y, Kwon YH, Yuldashev SU, Park YS, Fu DJ, Kim DY, Kim HS, Kang TW (2003) J Appl Phys 93: 1546
41. Dietl T, Ohno H, Matsukura F (2001) Phys Rev B 63: 195205
42. Dugaev VK, Litvinov VI, Barnes J, Viera M (2003) Phys Rev B 67: 033201
43. Schliemann J (2003) Phys Rev B 67: 045202
44. Rao BK, Jena P (2002) Phys Rev Lett 89: 185504
45. Korotkov RY, Gregie JM, Wessels BW (2002) Appl Phys Lett 80: 1731
46. Graf T, Gjukic M, Brandt MS, Stutzmann M, Ambacher O (2002) Appl Phys Lett 81: 5159
47. Polyakov AY, Govorkov AV, Smirnov NB, Pashkova NY, Thaler GT, Overberg ME, Frazier R, Abernathy CR, Pearton SJ, Kim J, Ren F (2002) J Appl Phys 92: 4989
48. Theodoropoulou NA, Hebard AF, Chu SNG, Overberg ME, Abernathy CR, Pearton SJ, Wilson RG, Zavada JM (2001) Appl Phys Lett 79: 3452
49. Pearton SJ, Overberg ME, Thaler G, Abernathy CR, Theodoropoulou N, Hebard AF, Chu SNG, Wilson RG, Zavada JM, Polyakov AY, Osinsky A, Park YD (2002) J Vac Sci Technol A 20: 583
50. Akinaga H, Nemeth S, De Boeck J, Nistor L, Bender H, Borghs G, Ofuchi H, Oshima M (2000) Appl Phys Lett 77: 4377
51. Hashimoto M, Zhou YZ, Kanamura M, Asahi H (2002) Solid-State Commun 122: 37

52. Yang SG, Pakhomov AB, Hung ST, Wong CY (2002) Appl Phys Lett 81: 2418
53. Wu SY, Liu HX, Gu L, Singh RK, Budd L, Schilfgaaarde M, McCartney MR, Smith DJ, Newman N (2003) Appl Phys Lett 83: 3255.
54. Liu C, Alves E, Ramos AR, da Silva MF, Soares JC, Matsutani T, Kiuchi M (2002) Nucl Instrum. Methods Phys B 191: 544
55. Ohno H, Shen S, Matsukura F, Oiwa A, Endo A, Katsumoto S, Iye Y (1996) Appl Phys Lett 69: 363
56. Janotti L, Wei S, Bellaiche L (2003) Appl Phys Lett 82: 766
57. Martin AL, Spalding CM, Dimitrova EI, Van Patten PG, Caldwell MC, Kordesch ME, Richardson HH (2001) J Vac Sci Technol A 19: 1894
58. Lu F, Carius R, Alam A, Heuken M, Buchal Ch (2002) J Appl Phys 92: 2457
59. Cho D-H, Shimizu M, Ide T, Ookita H, Koumwa H (2002) Jpn J Appl Phys 41: 4481
60. Hu X, Deng J, Pala N, Gaska R, Shur MS, Chen CQ, Yang J, Simin G, Khan MA, Rojo JC, Schwalker ZJ (2003) Appl Phys Lett 82: 1299
61. Serina F, Ng KYS, Huang C, Amer GW, Romni L, Naik R (2001) Appl Phys Lett 79: 3350
62. Lee SH, Lee JK, Yoon KH (2003) J Vac Sci Technol A 21: 1
63. Takagaki Y, Santos P, Wiebicke E, Brandt O, Schmerr JD, Ploog K (2002) Appl Phys Lett 81: 2538
64. Kipshidze G, Kuryatkov V, Zhu K, Vorizov B, Holtz M, Nikishin S, Temkin H (2003) J Appl Phys 93: 1363
65. Nishida T, Kobayashi N, Ban T (2003) Appl Phys Lett 82: 1
66. Gaska R, Chen C, Yang J, Kookstis E, Kahn MA, Tamulaitis G, Yilmog I, Shur MS, Rojo JC, Schowalter LJ (2002) Appl Phys Lett 81: 4658
67. Yang SG, Pakhomov AB, Hung CY, Wong CY (2002) Appl Phys Lett 81: 2418
68. Theodoropoulou N, Hebard AF, Overberg ME, Abernathy CR, Pearton SJ, Chu SNG, Wilson RG (2003) Phys Rev Lett 89: 107203
69. Kucheyev SO, Williams JS, Zou J, Jagadish C, Pophristic M, Guo S, Ferguson IT, Manasreh MO (2002) J Appl Phys 92: 3554
70. Li J, Nam KB, Nakarmi MC, Lin JY, Jiang HX (2002) Appl Phys Lett 81: 3365
71. Hashimoto M, Zhou Y-K, Kanamura M, Asahi H (2002) Solid-State Commun 122: 37
72. Nam KB, Li J, Kim KH, Lin KH, Jiang HX (2001) Appl Phys Lett 78: 3690

73. Ji Y, Stijkers GJ, Yang FY, Chien CC, Byers JM, Angvelovch A, Xiao G, Gupta A (2001) Phys Rev Lett 86: 5585
74. Inumarn K, Okamoto H, Yamanka SJ (2002) J Cryst Growth 237–39: 2050
75. Suzuki K, Kancho T, Yoshida H, Morita H, Fujimori H (1995) J Alloys Compds 224: 232
76. Sharma P, Gupta A, Rao KV, Owens FJ, Sharma R, Abuja R, Osorio Guillen JM, Johansson B, Gehring GA (2003) Nature Mater 2: 673
77. Frazier RM, Thaler G, Abernathy CR, Pearton SJ (2003) Appl Phys Lett 83: 1758
78. Pearton SJ, Overberg ME, Abernathy CR, Theodoropoulou N, Hebard AF, Chu SNG, Osinsky A, Zuflyigin V, Zhu LD, Polyakov AY, Wilson RG (2002) J Appl Phys 92: 2047
79. Ohno Y, Young DK, Beschoten B, Matsukura F, Ohno H, Awschalom DD (1999) Nature 402: 790
80. Jonker BT, Park YD, Bennet BR, Cheong HD, Kioseoglou G, Petrou A (2000) Phys Rev B 62: 8180
81. Park YD, Jonker BT, Bennet BR, Itzkos G, Furis M, Kioseoglou G, Petrou A (2000) Appl Phys Lett 77: 3989
82. Jonker BT, Hanbicki AT, Park YD, Itzkos G, Furis M, Kioseoglou G, Petrou A (2001) Appl Phys Lett 79: 3098
83. Graf T, Goennenwein STB, Brandt MS (2003) Phys Status Solidi B 239: 277

6 Novel Insulators for Gallium Nitride Metal-Oxide Semiconductor Field Effect Transistors and AlGaN–GaN Metal-Oxide Semiconductor High Electron Mobility Transistors

6.1 Abstract

The use of gate insulators for compound semiconductor electronics would alleviate many of the problems encountered in current Schottky-based devices. Further, circuit design can be simplified, since enhancement-mode MOSFETs can be used to form single supply voltage control circuits for power transistors. The use of MOSFETs also allows the use of complementary devices, thus producing less power consumption and simpler circuit design. A critical need is to develop reliable methods for deposition of these insulating films. This will enable the development of a new class of compound semiconductor electronics for high-speed communication and data processing applications. Both MgO and Sc_2O_3 are shown to provide low interface state densities (in the 10^{11} eV^{-1} cm^{-2} region) on n- and p-GaN, making them useful for gate dielectrics for MOS devices and also as surface passivation layers to mitigate current collapse in GaN–AlGaN HEMTs. Clear evidence of inversion has been demonstrated in gate-controlled MOS p-GaN diodes using both types of oxide. Charge pumping measurements on diodes undergoing a high-temperature implant activation anneal show a total surface state density of $\sim 3 \times 10^{12}$ cm^{-2}. On HEMT structures, both oxides provide effective passivation of surface states, and these devices show improved output power. The MgO–GaN structures are also found to be quite radiation resistant, making them attractive for satellite and terrestrial communication systems requiring a high tolerance to high-energy (40 MeV) protons.

6.2 Introduction

AlGaN–GaN HEMTs appear well suited to high-speed and high-temperature applications, including high-frequency wireless base stations and broad-band links, commercial and military radar and satellite communications.[1–34] These devices are capable of producing very high power densities (>12 W/mm), along with high breakdown voltage and low noise figures. A schematic of a typical HEMT structure is shown in Figure 6.1. The highest performance devices are typically grown on SiC substrates because of their high thermal conductivity. To achieve high switching speeds, the gates are patterned using e-beam lithography. A cross-section of a typical gate is shown in Figure 6.2, with a picture of the gate contact, source and drain contacts shown in Figure 6.3.

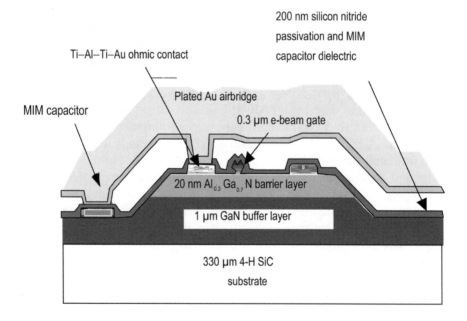

Figure 6.1. Schematic of typical microwave power HEMT MIM is metal-insulator-metal

HEMT processing: e-beem mushroom gates

Bilevel lithography:
  6% PMMA
  9% P(MMA-MAA)

Exposure:
  50 kV
  ~4 nA beam current
  225 mC/cm² (large area)
  375–800 mC/cm² (gate)

Develop:
  1:2::MIBK:IPA

Figure 6.2. SEM micrograph of mushroom-shaped gate fabricated on HEMT using e-beam lithography

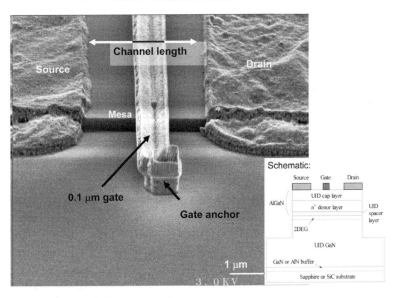

Figure 6.3. SEM micrograph of HEMT contacts. The inset shows a schematic cross-section view

These devices can achieve excellent power performance, well above that obtained with competing technologies such as GaAs and SiC MESFETs, as shown in Figure 6.4. The use of MOS or MIS gates for HEMTs produces a number of advantages over the more conventional Schottky metal gates, including lower leakage current and greater voltage swing.[2,22,26,28–43] The materials reported for gate oxide/insulators include SiO_2 [3,22–27,36], $Gd(Ga_2O_3)$ [22,23,27,34], AlN [35], SiN_x [14,32,33], MgO [24,37] and Sc_2O_3 [37]. These materials can also be employed as surface passivation layers on HEMTs. A commonly observed problem in these devices is the so-called "current collapse" in which the application of a high drain–source voltage leads to a decrease of the drain current and increase in the knee voltage.[4,6,12–17,20] This phenomenon can also be observed by a current dispersion between dc and pulsed test conditions or a degraded rf output power. The mechanisms include the presence of surface states on the cap layer or trapping centers in the resistive buffer underlying the active channel. The carriers in the 2-DEG can be lost either to the surface or buffer traps.[4 6,38] The former may be mitigated to a greater or lesser extent by use of appropriate surface passivation, most often SiN_x deposited by plasma-enhanced chemical vapor deposition (PECVD), while the latter is a function of the epitaxial growth conditions.

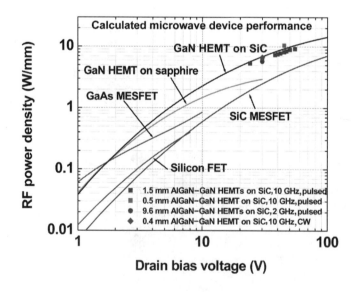

Figure 6.4. Calculated and experimental microwave power performance of various technologies (figure courtesy of Dr AP Zhang, GE Corporate R&D)

In this chapter we will describe progress on two alternative candidates for HEMT gate dielectrics and passivation, namely Sc_2O_3 and MgO.[39,41–43] Table 6.1 shows a comparison of the properties of these oxides with those of the other commonly used passivation materials on GaN. These novel oxides do not contain hydrogen and may have advantages over SiN_x in that respect, because atomic hydrogen diffuses rapidly and could enter the GaN or gate metal over extended periods of device operation.

Table 6.1. Material properties for SiC, GaN and various dielectrics. W = Wurtzite, A = amorphous, B = Bixbyite, N = NaC

	SiC	GaN	SiO_2	AlN	Gd_2O_3	Sc_2O_3	CaO	MgO	$Mg_{.5}Ca_{.5}O$	$Mg_{.75}Ca_{.25}O$
Structure	W (4H)	W	A	W or A	B	B	N	N	N	N
Lattice constant	3.07	3.186	–	3.113	10.813	9.845	4.799	4.2112	4.505	4.356
Atomic spacing in the (111) plane			–	–	3.828	3.4807	3.393	2.978	3.186	3.08
Mismatch to GaN (%)	–	–	–	2.3	20.1	9.2	6.6	–6.5	0	–
Mismatch to SiC (%)				1.1	24.3	13	10.2	–3.3		0.3
T_{MP} (K)		2800	1900	3500	2668	2678	2860	3073		
Bandgap (eV)	3.26	3.4	9	6.2	5.3	6.3	7	8		
Electron affinity (eV)		3.4	0.9	0–2.9	0.63			0.7		
Work function (eV)	4.5			0.9–1.2	2.1–3.3	4		3.1–4.4		
Dielectric constant	9.7–10.3	9.5	3.9	8.5	11.4	14	11.8	9.8		

6.3 Insulators for Gallium Nitride Metal-Oxide Semiconductor and Metal–Insulator–Semiconductor Field Effect Transistors

GaN MOSFETs are far less developed than for Si or even SiC. Most work in this area has focused on the amorphous dielectrics, and in particular on SiO_2. Silicon oxide deposited by PECVD has been reported to give interface state densities on the order of low 1×10^{11} $eV^{-1}cm^{-2}$. Silicon oxide deposited by electron beam evaporation has shown an interface state density of 5.3×10^{11} eV^{-1}cm^{-2}. This dielectric has also been employed as the insulator in an insulated gate heterostructure FET (IG-HFET) by Khan *et al.*[43] While these devices are promising, the low dielectric constant of SiO$_2$ and the absence of modulation at positive gate voltages (forward bias) above 3–4 V in SiO$_2$–GaN devices suggests that other alternatives should be considered.[44]

Silicon nitride is another amorphous dielectric that has been explored. Deposition by PECVD on GaN resulted in an interface state density of 6.5×10^{11} eV^{-1}cm^{-2}. Electrical measurements showed the MISFET structure had a large flat band voltage shift (3.07 V) and a low breakdown field strength (1.5 MV/cm) of the dielectric. A unique dielectric structure of SiO$_2$–Si$_3$N$_4$–SiO$_2$ (ONO) was reported to have a breakdown field strength of 12.5 MV/cm for temperatures as high as 300°C.[15] The ONO structure was deposited by jet vapor deposition to a thickness of 10 nm–20 nm–10 nm. This multilayer structure allows for unique engineering of a dielectric, and will most likely receive more attention in the future. However, the dielectric constant of the SiO$_2$ is still a disadvantage in this structure.

Gallium oxide is yet another material that has been investigated. Thermal oxidation of the GaN surface to form Ga$_2$O$_3$ has been performed in dry and wet atmospheres. Dry oxidation of GaN epilayers at temperatures below 900°C showed minimal oxidation. At temperatures above 900°C, a polycrystalline monoclinic Ga$_2$O$_3$ forms at a rate of 5.0 nm/h. This oxidation rate is too slow to be viable as a processing step. Wet oxidation of GaN also forms polycrystalline monoclinic Ga$_2$O$_3$, but at a rate of 50.0 nm/h at 900°C. From cross-sectional TEM (XTEM), the interface between the oxide and the GaN is found to be non-uniform. SEM shows that both films are rough and facetted. Electrical characterization of the oxide films shows the dry oxide dielectric has a breakdown field strength of 0.2 MV/cm and a wet oxide dielectric field strength of 0.05 to 0.1 MV/cm. As with GaAs, it does not appear that Ga$_2$O$_3$ will be a viable dielectric for GaN.

Owing to the recent success of gallium gadolinium garnet (GGG) as a dielectric in GaAs MOSFETs, attention has turned toward this as a dielectric material for GaN. This material is amorphous when deposited by e-beam evaporation. The large dielectric constant of this material is particularly attractive, and the bandgap, though significantly lower than that of SiO$_2$, is expected to be adequate for n-type material. Using e-beam evaporation in an MBE environment for deposition of the dielectric, a Ga$_2$O$_3$(Gd$_2$O$_3$)–GaN depletion mode MOSFET was fabricated.[23] While the device showed significantly less gate leakage at elevated temperature relative to a conventional MESFET fabricated on the same GaN layer, the device still could not be modulated at voltages above roughly 3 V. This again suggests problems with leakage at the dielectric–GaN interface.

In addition to the amorphous dielectrics, crystalline AlN has also been considered. AlN deposited by MBE and MOCVD has been used to create MISFET devices and IG-HFET devices. The AlN gate in the MISFET structure grown by MOMBE at 400°C was polycrystalline and resulted in a dielectric breakdown field of 1.2 MV/cm.[25] The AlN IG-HFET structure was grown at 990°C, forming a single-crystal film of 4.0 nm. This device operated in enhancement mode and had a pinch-off voltage of 0 V. Unlike the amorphous dielectrics, single-crystal AlN and polycrystalline AlN films are expected to suffer from defects and grain boundaries that reduce the breakdown field sustainable in the material.

An attractive oxide alternative to both GGG and SiO_2 is Gd_2O_3, which also has a high dielectric constant, 11.4, and a bandgap of 5.3 eV. This oxide exists in the Mn_2O_3 (Bixbyite) crystal structure, which exhibits similar atomic symmetry in the (111) plane as the GaN (0001) basal plane. This similarity in symmetry offers the possibility to grow an epitaxial dielectric. Single-crystal growth of Gd_2O_3 on GaAs has, for example, been demonstrated. However, in the case of GaAs the expected orientation between the Gd_2O_3 and the substrate is (100) rather than (111). Further, the bond length mismatch between the Gd_2O_3 and GaAs (100) planes is smaller than between the Gd_2O_3 (111) and GaN (0001), 4.2% vs. 20%. In spite of this mismatch, we have recently demonstrated that Gd_2O_3 can be deposited epitaxially on GaN using elemental Gd and an ECR oxygen plasma in gas-source MBE (GSMBE). The dislocations present in the film limit the breakdown field which can be sustained in the dielectric and necessitate the use of an SiO_2 overlayer in order to fabricate MOSFETs from this material. Transistors with 1 µm gate length fabricated using the SiO_2–Gd_2O_3 dielectric stack showed modulation up to a forward gate bias of 7 V.[28] This is the highest modulation yet reported for a GaN MOSFET. Thus, it appears that the amorphous dielectrics, in particular GGG, produce the best breakdown fields while the Gd_2O_3 appears to produce the best interface with GaN. This is not surprising, since in general one would expect the amorphous dielectrics to contain fewer defects and thus withstand higher fields before breakdown. By contrast, the crystalline dielectrics appear to be more amenable to the formation of a low defect interface and are better suited to passivation of the surface dangling bonds, but suffer from high concentrations of bulk defects which limit the breakdown field. Both GGG and Gd_2O_3 appear to be stable at temperatures up to 950–1000°C, suggesting that both the interface and the bulk microstructure are capable of withstanding high temperature processing. It should be noted that both GGG and Gd_2O_3 are not well suited to fabrication of devices on either p-SiC or p-GaN, since the valence band offset to these materials is expected to be either very small or in fact negative, as shown in Figure 6.5.

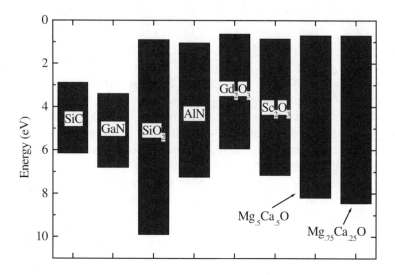

Figure 6.5. Band alignments for SiC, GaN and various dielectrics. The top of the bar corresponds to the conduction band edge, while the bottom corresponds to the valence band edge. The vacuum level is set at $E = 0$. Values for AlN are based on an average of the reported electron affinity values. The conduction band edge shown for Sc_2O_3 was calculated from the work function. Values for the MgCaO ternaries were calculated using Vegard's law assuming that the electron affinity of CaO is similar to that of MgO

6.4 Approach for Gallium Nitride

It is clear that substantial room for improvement remains in the choice of dielectrics for use with wide bandgap semiconductors. Based on recent work, it is believed there are some attractive materials which offer promise for this application. All of the proposed dielectrics are expected to be thermally stable based on their melting points and upon thermal stability studies we have performed on similar materials. Each of the candidate materials has a bandgap which is at least 2 eV larger than the semiconductor, and all except AlN have dielectric constants comparable to or larger than the underlying semiconductor.

6.4.1 Bixbyite Oxides

In the development of GaN MOSFET power transistors there have been investigations of the growth of Gd_2O_3 on GaN using elemental Gd and an ECR-generated oxygen plasma. It is known that growth initiation is a critical step in the formation of epitaxial films. Four different procedures for growth initiation were tested on the (1×3) GaN surface produced by heating in vacuum to 700°C. First, a low-temperature (300°C) exposure to the oxygen plasma was studied. As shown in Figure 6.6, the surface changes from a sharp (1×3) to a hazy (1×2). If the (1×3) GaN surface was exposed to the oxygen plasma at 650°C, then the surface also immediately changed to the (1×2) surface, then after an additional 5 min exposure to the oxygen plasma the surface became further reconstructed to a (4×4), as shown in Figure 6.7. This (4×4) surface remained after the oxygen plasma was turned off. When the (1×3) surface was exposed to a Gd atomic beam, the RHEED pattern, seen in Figure 6.8, is also a (1×2); however, the spacing of the main diffracted lines has changed, indicating that the spacing between the atoms has changed and that a different crystal structure has formed.

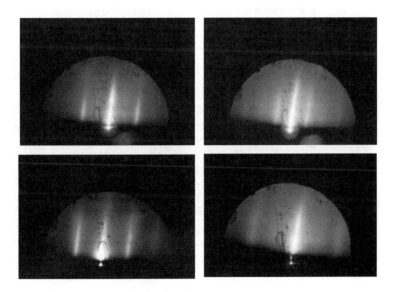

Figure. 6.6. RHEED images of (top) a (1×3) GaN surface at 700°C and (bottom) the same surface at 300°C under an O plasma showing a (1×2)

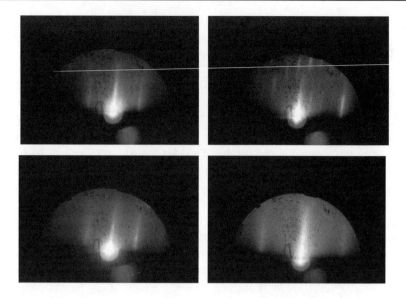

Figure 6.7. RHEED images of (top) a (1×3) GaN surface after receiving an O plasma exposure for 5 min at 700°C showing a faint (4×4), and (bottom) a (1×1) Gd$_2$O$_3$ surface grown at 610°C on the (4×4) oxygen-treated surface

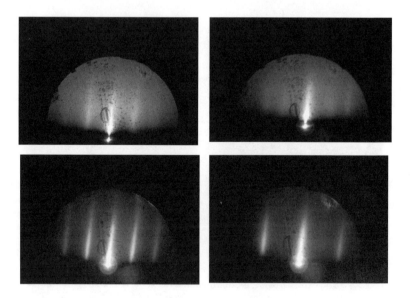

Figure 6.8. RHEED images of (top) a (1×3) GaN surface after receiving an O plasma exposure for 5 min at 700°C showing a faint (4×4), and (bottom) a (1×1) Gd$_2$O$_3$ surface grown at 610°C on the (4×4) oxygen-treated surface

Growth on the low-temperature oxygen plasma-exposed surface initially indicated a good Gd_2O_3 crystal structure, as indicated by RHEED, but after 10 min of growth a symmetric RHEED pattern emerged, as seen in Figure 6.9, indicating that the Gd_2O_3 crystal quality had been degraded. Reducing the growth rate produced the same symmetric pattern. Growth was also performed on the (1×2) GaN surface produced from the short oxygen exposure at 650°C. Here, the initial Gd_2O_3 RHEED pattern remained for the duration of the growth; however, extra spots were seen in the RHEED images (see Figure 6.10), indicating that the layer was partly polycrystalline.

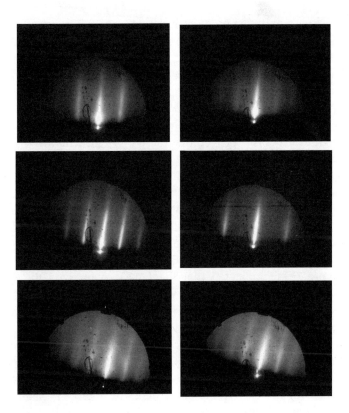

Figure 6.9. RHEED images of (top) a (1×2) 300°C oxygen-treated GaN surface, (middle) a (1×1) Gd_2O_3 surface after 10 min of growth at 650°C, and (bottom) a symmetric pattern after 90 min of growth. The <11-20> is at left and the <1-100> is at right

Figure 6.10. RHEED image of Gd_2O_3 (1×1) grown at 650°C at 0.9 nm/min indicating extra spots due to defects

Further growth experiments with reduced gadolinium effusion cell temperatures and a reduced substrate temperature of 605°C gave final RHEED patterns of the Gd_2O_3 crystal surface without the extra spots, indicating a uniform single-crystal film. Samples grown on the high-temperature oxygen plasma-treated surface (4×4) and the gadolinium-exposed surface showed sharp Gd_2O_3 crystal RHEED patterns and smooth surfaces on the order of 0.5 nm RMS roughness.

High-resolution XTEM proved that the Gd_2O_3 grew on the GaN in a planar fashion, as seen in Figure 6.11. There are dislocations visible in the image, indicating that the film is highly defective. However, large areas of dislocation-free material were seen in XTEM. From this image, it can be seen that the Gd_2O_3–GaN interface was clean. The dark layer between the Gd_2O_3 and the GaN is a thickness effect from the ion milling process used to fabricate the TEM sample. On the left side of the image, the Gd_2O_3 lattice can be seen in contact with the GaN lattice. A lower resolution TEM image (Figure 6.11) indicated that the Gd_2O_3 planarized the GaN quite well. In fact, the Gd_2O_3 filled in the void in the GaN surface with the same registry as the entire film. Thus, there was no polycrystalline morphology in the void. There were, however, some pockets of crystal lattice that were tilted and rotated, as seen in the image. This can be attributed to the dislocations arising from the single-crystal material in the void meeting single-crystal material on the surface. X-ray diffraction of this sample showed only one peak for the Gd_2O_3 that coincided with the (111) peak from published JCPDS cards. This peak, shown in Figure 6.12, had a shoulder to the left side and had a full width at half maximum (FWHM) of 883 arcseconds The intensity of the x-ray system used was too low to resolve this shoulder in triple axis mode, so the cause of this shoulder is unknown. X-ray diffraction of the sample that showed the extra RHEED spots (Figure 6.9) is shown in Figure 6.12. Here, a second peak was clearly seen, which corresponds to the (321) plane of Gd_2O_3. Also, the FWHM of this (111) peak was over 1600 arcseconds X-ray diffraction was also performed on the

6.4 Approach for Gallium Nitride   325

gadolinium-initiated sample and the 5 min high-temperature oxygen plasma-initiated sample, seen in Figure 6.13. The difference in the two samples was quite large. The gadolinium-initiated growth showed a FWHM of 1023 arcseconds for the (111) peak. The high-temperature oxygen plasma-initiated growth showed a very broad peak that was several thousand arcseconds wide. As a result of this study, it was determined that Gd initiation is the best method for growth initiation, and that single-crystal epitaxial Gd_2O_3 can be deposited on GaN. The optimized process was used to fabricate the devices mentioned in the previous section.

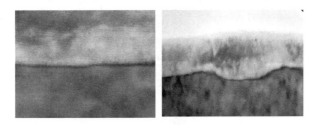

Figure 6.11. TEM images of Gd_2O_3 grown on GaN at 650C using elemental sources. The GaN is on the bottom and the Gd_2O_3 is on top. Notice the difference in lattice spacing.

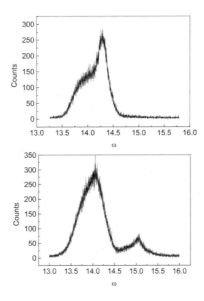

Figure. 6.12. X-ray diffraction plots of Gd_2O_3 grown on GaN at 650°C. The FWHM of the peak at left is 883 arcseconds and at right is 1623 arcseconds. The RHEED pattern from the film at right showed spots indicative of polycrystalline structure

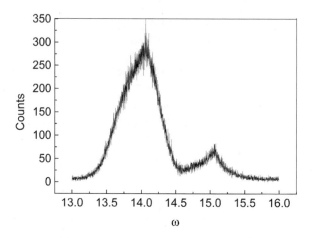

Figure 6.13. X-ray diffraction plot of samples grown on a Gd-exposed GaN surface or a 5 min high-temperature O plasma-exposed GaN surface. FWHM of the Gd-initiated sample is 1058 arcseconds

6.4.2 Scandium Oxide

Sc_2O_3 exists in the same Bixbyite structure as Gd_2O_3, but it has a much smaller lattice constant; see Table 6.1. This should make it less defective when deposited on GaN. Like Gd_2O_3 it has a high melting point, suggesting good thermal stability. The material is often used as an optical coating for high power photonic devices and has a bandgap of 6.3 eV. Based upon the reported work function of 4 eV, it is likely that the electron affinity is ~0.85 eV. If this estimate is correct, then, unlike Gd_2O_3, the band offsets for Sc_2O_3 should allow fabrication of devices on both n- and p-type material. The dielectric constant is also favorable, as it is slightly larger than that of Gd_2O_3, 14 vs. 11.

In light of the success we have had with Gd_2O_3, we believe that Sc_2O_3 has a high probability of showing improvement over existing dielectrics. By limiting the Sc_2O_3 thickness to ~10 nm, we should be able to minimize the defect concentration. However, a large mismatch still exists in this system, and we expect that an amorphous overlayer will still be required. We have found that reducing the Gd_2O_3 deposition temperature to 100°C is sufficient to produce amorphous material on GaN. However, when this material is deposited on crystalline Gd_2O_3 a polycrystalline layer is produced. It is possible that using an e-beam source of Gd_2O_3 will help to suppress the surface mobility and allow for the deposition of amorphous material, in spite of the template effect of the crystalline Gd_2O_3.

6.4.3 MgO–CaO

MgO is a rock salt dielectric which has been explored as an intermediate buffer layer for growth of ferroelectric materials on semiconductors or as a potential gate dielectric for GaAs or Si. While MgO deposition by MBE has been successfully demonstrated by a number of groups, the crystal quality of the films deposited on GaAs and Si has been poor due to the large lattice mismatch between the MgO and the semiconductor substrate. GaN has a smaller lattice constant than GaAs and is thus a much closer match to MgO. CaO is also a rock salt dielectric, but with a substantially larger lattice spacing than MgO. A ternary oxide consisting of MgO and CaO could thus, in principle, be closely matched to the GaN, resulting in a substantial reduction in defect density. Similarly, a lattice-matched alloy should also be possible for SiC. An additional advantage of this system is the large bandgap, and thus large band offsets are expected relative to either an n- or p-type semiconductor. Further, the dielectric constants for both MgO (9.8) and CaO (11.8) are substantially higher than for SiO_2.

One problem with this material system is the large difference in ionic radius between Mg and Ca, which causes severe immiscibility. At the temperature regime needed for this work the solubility of the compounds is minimal. However, the use of MBE as the deposition method often allows for the formation of metastable phases. In fact, it has been demonstrated that the MgCaO ternary can be grown with compositions covering the entire compositional range when deposited at a temperature of 300°C.

It is possible to use two approaches to form lattice, matched MgCaO on GaN and SiC. One employs low growth temperatures to overcome the immiscibility, while the other relies on alternating layers of MgO and CaO, often referred to as the digital alloy approach, to form a material with the properties of the average composition of the film. On GaN, the layers would be alternated sequentially, while on SiC the ratio of MgO layers to CaO layers would be 3:1. An additional consideration with the SiC substrates is the possibility of SiO_2 formation at the SiC–oxide interface. By initiating growth with a metal layer, the formation of SiO_2 can be suppressed.

6.4.4 Amorphous Aluminum Nitride

While crystalline AlN shows some promise as a gate dielectric, it still suffers from leakage and low breakdown due to dislocations and other defects. Presumably these arise from the small but finite lattice mismatch which exists between AlN and SiC or GaN. An alternative approach is the use of amorphous AlN. It is expected that reducing the growth temperature should suppress the surface mobility sufficiently to prevent nucleation of a crystalline phase. Rogers *et al.* (unpublished) have in fact shown that by using a growth temperature of 200°C in the CVD of AlN on Si it is possible to reduce the leakage and improve the breakdown relative to crystalline AlN grown on the same substrate material. The absence of X-ray diffraction peaks in their material suggests that the deposited material is amorphous.

Amorphous AlN has also been deposited using a gaseous Al precursor and an rf nitrogen plasma in an MOMBE system. As shown in Figure 6.14, it was possible to deposit material at 325°C in which approximately 30% of the volume is amorphous. On SiC, this has produced MIS capacitors with a gate breakdown of 6.8 MV/cm, and *C–V* analysis shows little or no evidence of hysteresis. Similar experiments on GaN were not as successful, with breakdowns of only 1 MV/cm attainable. This is most likely due to enhanced crystallinity from the template effect of the GaN substrate.

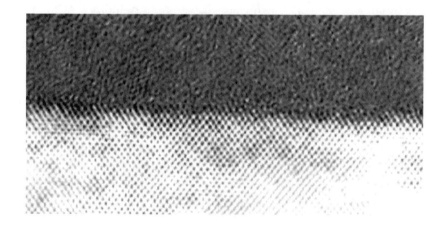

Figure 6.14. XTEM of AlN deposited on Si at 325°C showing crystalline pockets dispersed in an amorphous matrix

In order to improve the electrical behavior of the AlN–SiC and AlN–GaN structures it will be necessary to reduce the crystallinity of the material further. This can most likely be accomplished by further reducing the growth temperature. Hence, it will be necessary to switch from the gaseous Al precursor, which has a lower thermal decomposition temperature of ~325°C, to an elemental Al effusion oven which can, in principle, be used at much lower substrate temperatures. In addition to using the amorphous AlN as a sole dielectric, it can also be incorporated into stacked gate structures employing the crystalline oxides discussed previously.

6.5 Gate-Controlled Metal-Oxide Semiconductor Diodes

Fabrication of gate-controlled MOS diodes on GaN is an excellent method for establishing the quality of the dielectrics. The use of MOS gates should enhance the performance of GaN–AlGaN transistors, which already show excellent potential.[45–88] To fabricate gate-controlled diodes using the oxides as gate dielectrics, the starting sample was a 1 μm thickness p-GaN (hole concentration is $\sim 2 \times 10^{17}$ cm^{-3} at 25°C) layer grown on a 2 μm undoped GaN buffer grown on an Al_2O_3 substrate by MOCVD. Both MgO and Sc_2O_3 were used as the gate dielectrics. The Sc_2O_3 was deposited by rf plasma-activated MBE at 650°C using elemental Sc evaporated from a standard effusion cell at 1130°C and O_2 derived from an electron cyclotron plasma source operating at 200 W forward power (2.45 GHz) and 10^{-4} Torr. The Sc_2O_3 layers were ~400 Å thick. The smaller mismatch between the Sc_2O_3 and the GaN (0001) should lead to a lower defect density, and in fact the FWHM in the x-ray diffraction scan is substantially reduced at 514 arcseconds as compared with the 883 arcseconds found for comparable Gd_2O_3 layers, as shown in Figure 6.15.

The MgO was deposited by MBE at 100°C. The MgO precursors were elemental Mg and RF plasma-activated oxygen. The GaN samples were cleaned initially with a 3 min chemical etch in HCl/H_2O (1:1), H_2O rinse, UV–ozone exposure for 25 min, rinse in buffered oxide etch solution (6:1, NH_4F/HF) and rinsed in H_2O. The samples were then loaded into the MBE system and heated at 650°C to ensure oxide removal. A standard effusion cell operating at 380°C was used for evaporation of the Mg, while the O_2 source was operated at 300 W forward power (13.56 MHz) and 2.5×10^{-5} Torr. The MgO layers were ~800 Å thick. In separate measurements, we obtained interface state densities of $(2-3) \times 10^{11}$ cm^{2}eV^{-1} from the AC conductance and Terman methods.[39–43]

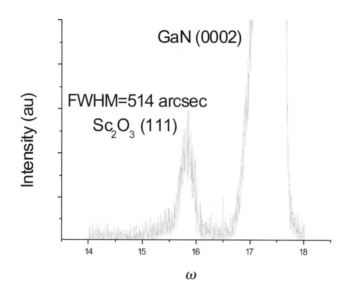

Figure 6.15. High-resolution x-ray diffraction scans of Sc_2O_3 on h-GaN

N^+ regions were created by selective implantation of Si^+ at multiple energies and doses (70 keV, 2×10^{13} cm^2, 195 keV, 6×10^{13} cm^{-2} and 380 keV, 1.8×10^{14} cm^{-2}). The junction depth was ~0.4 μm from ion range simulations. The samples were then annealed at 950°C under N_2 to activate the Si-implanted regions. Windows were etched into the oxide and e-beam deposited p-ohmic (Ni–Au), n-ohmic (Ti–Al–Pt–Au) and gate metal (Pt–Au) were patterned by lithography. The separation of the n^+ regions was ~60 μm. A schematic of the completed gate-controlled diode is shown in Figure 6.16, and Figure 6.17 shows a plan-view photograph of the completed device.

Figure 6.18 shows C–V characteristics of the MgO–GaN MOS-controlled diodes at 25°C in the dark as a function of the measurement frequency. In each case, –20 V was applied at the gated contact to provide a source of minority carriers. The frequency dispersion observed in inversion is due to the resistance of the inversion channel. At 5 KHz measurement

6.5 Gate-Controlled Metal-Oxide Semiconductor Diodes 331

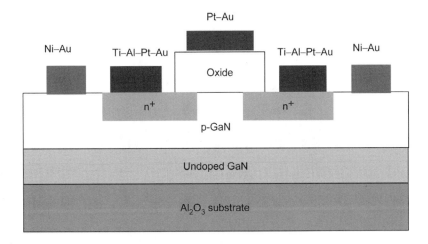

Figure 6.16. Schematic of Sc_2O_3–p-GaN gate-controlled diodes

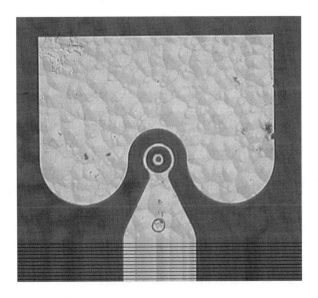

Figure 6.17. Photograph of completed gate-controlled diode

frequency we observe only deep depletion, since the characteristics are dominated by majority carriers. As the frequency is decreased, a clear inversion behavior is observed due to charge flow into and from the n^+ regions external to the gate. In diodes without the n^+ regions to act as an external supply of minority carriers, we could not observe inversion, even up to measurement temperatures of 300°C. Similar results were obtained for Sc_2O_3–GaN diodes, as shown in Figure 6.19. The shift in the "hooks" in the capacitance–voltage curves can be used to establish the total surface state density.[41,42]

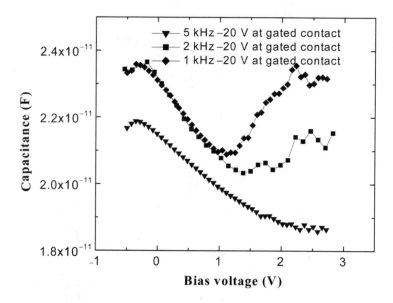

Figure 6.18. *C–V* characteristics of MgO/GaN MOS gate-controlled diodes at 25 °C as a function of measurement frequency (+15V bias in gated contact in each case)

Charge pumping experiments can also be used to measure the surface state density. This density, N_{SS}, can be related to the pulse frequency υ and charge-pumping current I_{CP} through the relation

6.5 Gate-Controlled Metal-Oxide Semiconductor Diodes

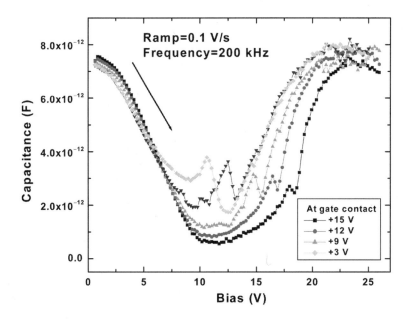

Figure 6.19. *C–V* characteristics of Sc_2O_3–GaN MOS gate-controlled diodes at 25 °C as a function of bias on the gate contact

$$N_{SS} = \frac{I_{CP}}{vA_ce} \qquad (6.1)$$

where A_C is the diode channel area and e is the electronic charge. Figure 6.20 shows typical data measured at $v = 50$ kHz. From this data, we obtain $N_{SS} = 3 \times 10^{12}$ cm^{-2} at the Sc_2O_3–p-GaN interface. Note that no effort has been made in our structures to passivate surface states through forming gas anneals or plasma hydrogenation. This is the first measurement of surface state density in GaN-based MOS structures that more fully approximate a transistor, *i.e.* the structures show inversion and have gone through an extensive device processing sequence.

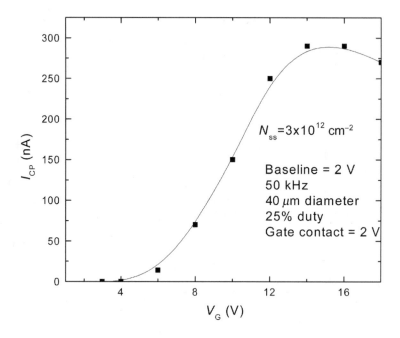

Figure 6.20. C–V characteristics of Sc_2O_3–GaN MOS gate-controlled diodes at 25 °C as a function of bias on the gate contact

6.5.1 Surface Passivation

We have employed gate lag measurements on the HEMTs as a metric for establishing the effectiveness of the oxide passivation.[38] In this method, the drain current I_{DS} response to a pulsed gate–source voltage V_G is measured. Figure 6.21 (top) shows the normalized I_{DS} as a function of drain–source voltage (V_{DS}) for both dc and pulsed measurements on unpassivated HEMTs. In the data, V_G was pulsed from –5 V to 0 V at 0.1 MHz and 10% duty cycle. The bottom of Figure 6.21 shows the normalized I_{DS} for the same device after MgO deposition. Note the much reduced current collapse. Figure 6.22 shows the normalized I_{DS} as a function of gate–source voltage V_G switched from –5 V to 0 V with V_{DS} held constant at a low value (3 V) to avoid the complications of device heating, for both unpassivated (top) and MgO-passivated HEMTs. The large differences between dc and pulsed drain currents in the case of the unpassivated devices are consistent with the presence of surface traps that deplete the channel in the access regions between the gate and drain contacts. However, after MgO deposition the HEMTs showed an increase in drain-source current of 20% in the dc mode, which is consistent with passivation of the surface state. This is clear evidence for the assumption that surface states are the cause of the gate-lag phenomenon and also that MgO passivation mitigates this problem. Once again, similar results were obtained with Sc_2O_3 passivation.

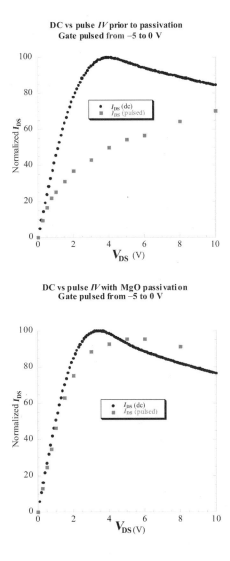

Figure 6.21. (Top) Gate lag measurements on unpassivated AlGaN–GaN HEMTs. V_G is switched from –5 V to 0 V; (Bottom) similar measurements after MgO passivation

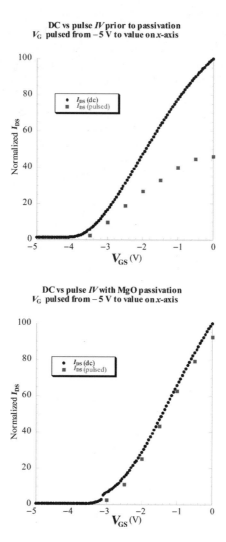

Figure 6.22. (Top) Gate lag measurements on unpassivated AlGaN–GaN HEMTs. V_G is switched from –5 V to the value shown on the x-axis; (Bottom) similar measurements after MgO passivation

Typical load-pull data for HEMTs before and after Sc_2O_3 passivation are shown in Figure 6.23 for a measurement frequency of 4 GHz. In all cases, the drain voltage V_D was held at 10 V, while the gate voltage V_G was –2 V. The wafer measurements employed mechanical tuners for matching and there was no harmonic termination under class A operation. The devices were matched for the power testing prior to passivation and were tested under these same conditions after oxide deposition. Note the increased power output. By contrast, the difference in output power before and after SiN_x passivation was <2 dB in all cases. This is consistent with the higher percentage recovery of I_{DS} during the gate-lag measurements with Sc_2O_3 passivation. The Sc_2O_3 passivation has shown excellent aging characteristics when measured under dc test conditions, with no change in HEMT performance over a period of >5 months.

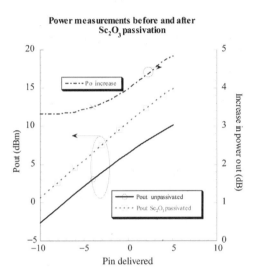

Figure 6.23. Output power characteristics before and after Sc_2O_3 passivation of 0.5×100 μm^2 devices measured at 4 GHz and a bias point of $V_D = 10$ V, $V_G = -2$ V.

Three different passivation layers (SiN_X, MgO and Sc_2O_3) were examined for their effectiveness in mitigating surface-state-induced current collapse in AlGaN–GaN HEMTs. The plasma-enhanced chemical vapor deposited SiN_x produced ~70–75% recovery of the drain-source current, independent of whether SIH_4–NH_3 or SiD_4–ND_3 plasma chemistries were employed. Both the Sc_2O_3 and MgO produced essentially complete recovery of the current in GaN-cap HEMT structures and ~80–90% recovery in AlGaN-cap structures. The Sc_2O_3 had superior long-term stability, with no change in HEMT behavior over 5 months aging.

Two different HEMT structures were used. The first employed a GaN undoped cap layer on top of an undoped $Al_{0.2}Ga_{0.8}N$ layer. Both Al_2O_3 and SiC substrates were used. The doping in this structure is basically due to piezo-induced carriers. The second type of HEMT employed an undoped $Al_{0.3}Ga_{0.7}N$ cap layer on top of a doped $Al_{0.3}Ga_{0.7}N$ donor layer. The doping in this structure is due to the intentional Si doping of the donor layer. The ohmic (Ti–Al–Pt–Au) and gate (Ni–Au) contacts were deposited by e-beam evaporation and patterned by lift-off. Schematics of the completed device structures are shown in Figure 6.24.

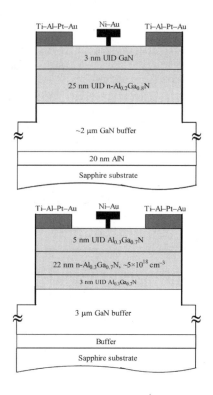

Figure 6.24. Schematic cross-sections of GaN-cap (top) and AlGaN-cap (bottom) HEMT structures

The 100 Å thickness MgO, Sc_2O_3 or SiN_x layers were deposited on completed devices (0.25–1.2 µm gate length, 100 µm gate width) using either MBE for the oxides or PECVD for the SiN_x, with deposition temperatures of 100°C in the oxide case and 250°C for the SiN_x case. The SiN_x films were deposited with either $SiH_4 + NH_3$ or $SiD_4 + ND_3$ to examine the effect of deuterated precursors.

The HEMT dc parameters were measured in dc and pulsed mode at 25°C, using a parameter analyzer for the dc measurements and pulse gen-

6.5 Gate-Controlled Metal-Oxide Semiconductor Diodes

erator, dc power supply and oscilloscope for the pulsed measurements. For the gate-lag measurements, the gate voltage V_G was pulsed from –5 V to 0 V at different frequencies with a 10% duty cycle.

Figure 6.25 (top) shows typical gate lag data for 0.5×100 μm^2 GaN-cap HEMT grown on sapphire substrates. The decrease in drain–source current becomes more pronounced at high measurement frequencies. The degradation in current was less significant when these same structures were grown on SiC substrates, as shown in Figure 6.25 (bottom). The defect density will be lower in the latter case due to the closer lattice match between GaN and SiC, and this appears to affect the resultant surface state density. This suggests that at least some of the surface traps are related to dislocations threading to the surface.

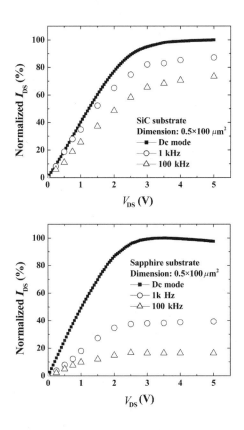

Figure 6.25. Gate-lag measurements on unpassivated 0.5 μm gate length, GaN-cap HEMTs grown on either sapphire (bottom) or SiC (top) substrates

The effects of both substrate type and HEMT gate length on the change in drain-source current are shown in Figure 6.26. The shorter the gate length, the more pronounced the degradation in current because of larger surface area and higher electric field in the channel between source S and drain D (the S–D distance was fixed in all cases).

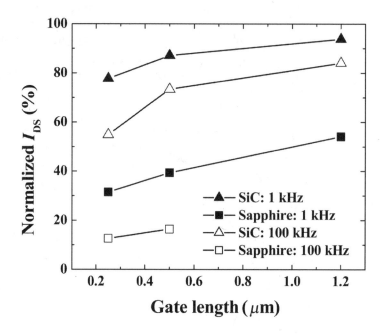

Figure 6.26. Normalized I_{DS} as a function of both gate length and substrate type for GaN-cap HEMTs at pulse frequencies of 1 kHz and 100 kHz

The gate-lag data is shown in Figure 6.27 (top) for the HEMTs before and after SiN_x passivation. Note that both the hydrogenated and deuterated precursor dielectric produces a recovery of 70–75% in the drain–source current. This range represents results obtained from five different devices with each type of dielectric. The unity current gain frequency f_T and maximum frequency of oscillation f_{MAX} data before and after passivation are shown at the bottom of Figure 6.27. Note that there is actually a slight increase in both parameters. There were no systematic differences between the results for hydrogenated and deuterated precursor SiN_x, even though the latter typically produces slightly denser films when deposited under the same PECVD conditions.

6.5 Gate-Controlled Metal-Oxide Semiconductor Diodes

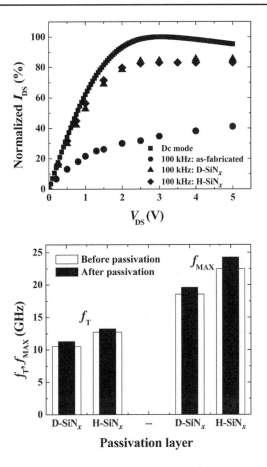

Figure 6.27. Normalized I_{DS} versus V_{DS} (top) for 1.2×100 µm², GaN-cap HEMTs before and after SiN$_x$ passivation using either hydrogenated or deuterated precursors. The panel at bottom shows the change in small signal rf performance

There was also an increase in forward and reverse gate currents after passivation with SiN$_x$. These results are also consistent with a decrease in surface depletion and decrease in surface trap density and indicate that there is no major degradation to the dc performance of the HEMTs upon addition of the SiN$_x$ layer. We believe the increase in gate leakage is not from the passivation itself, but may originate from degradation of the gate metallization during the oxide desorption step at 350°C.

Figure 6.28 shows the gate lag data for GaN-cap 0.5×100 µm² HEMTs. In the case at the top, V_G was pulsed from −5 to 0 V at 0.1 MHz with a 10% duty cycle and this pulsed data is compared with that from dc measurements. The bottom of Figure 6.28 shows the normalized I_{DS} for gate voltage switched from –5 V with V_{DS} held constant at 3 V to avoid any self-heating effects. This data establishes the baseline for measuring the effectiveness of the MgO and Sc_2O_3 passivation. These HEMTs showed an increase in drain–source current of ~20% after deposition of either type of oxide. In addition, the forward and reverse gate currents were slightly increased. These results are similar to those for SiN_x deposition. Once again, the increased reverse current may be due to gate metal degradation.

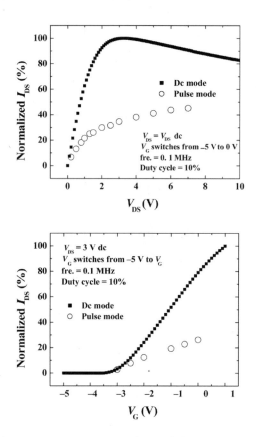

Figure 6.28. Gate-lag measurements on unpassivated GaN-cap HEMTs (1.2×100 μm² gate). At the top V_G was switched from –5 to 0 V, while at bottom it was switched from –5 V to the value shown on the x-axis

Figure 6.29 shows gate-lag measurement for MgO-passivated HEMTs immediately after MgO deposition and after 5 months aging without bias on the devices under room conditions (temperature and humidity). The I_{DS} increases ~20% upon passivation, as mentioned earlier, and there is almost complete mitigation of the degradation in I_{DS} immediately after MgO deposition. However, after 5 months aging there is a clear difference between the dc and pulsed data, indicating that the MgO passivation has lost some of its effectiveness. We are currently investigating possible mechanisms such as oxidation of the GaN surface through pinholes in the MgO, or reaction between the MgO and GaN.

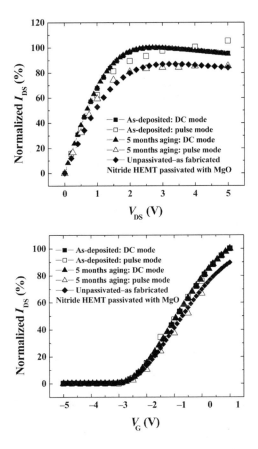

Figure 6.29. Gate lag measurements before and after MgO passivation and following 5 months aging of 1.2×100 μm², GaN-cap HEMTs. At the top V_G was switched from –5 to 0 V, while at the bottom it was switched from –5 V to the value shown on the x-axis

Similar data is shown in Figure 6.30 for Sc_2O_3-passivated HEMTs. In this data the I_{DS} also increases upon deposition of the oxide and there is also essentially complete mitigation of the degradation in drain–source current. However, the major difference is that there is no significant change in these characteristics after 5 months aging. This indicates that Sc_2O_3 provides more stable passivation than MgO. We have noticed in separate experiments that the MgO–GaN interface deteriorates over time if left uncapped, *i.e.* we see increases in interface state trap density in MOS diodes on which the metal is deposited a long period after the MgO was deposited on the GaN. However, if the MgO is immediately covered by the gate metal, these changes are not observed. In a real manufacturing process, the MgO-passivated HEMTs would be covered with a dielectric such as SiN_x to provide mechanical protection and, therefore, MgO should not be disqualified as a candidate to provide effective passivation of GaN–AlGaN HEMTs. Obviously, further work is needed to establish the reliability of Sc_2O_3 and MgO passivation on these devices, but the preliminary data with Sc_2O_3 looks very promising.

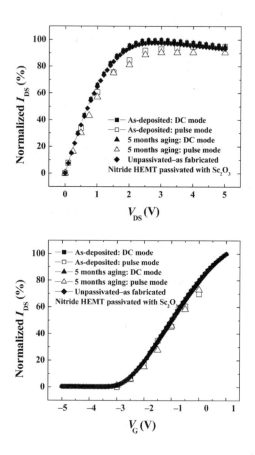

Figure 6.30. Gate-lag measurements before and after Sc_2O_3 passivation and following 5 months aging of 1.2×100 μm², GaN-cap HEMTs. At the top V_G was switched from –5 to 0 V, while at the bottom it was switched from –5 V to the value shown on the *x*-axis

Figure 6.31 shows the gate-lag data for the HEMT structures of Figure 6.24 (bottom), *i.e.* the structures with $Al_{0.3}Ga_{0.7}N$ as the top layer. As with the GaN-cap devices, there is a very significant degradation of drain–source current under pulsed conditions where the surface traps cannot completely empty.

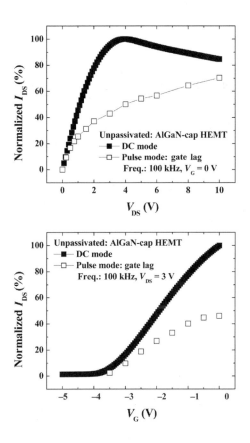

Figure 6.31. Gate-lag measurements on unpassivated 0.5 μm gate length, AlGaN-cap HEMTs. At the top V_G was switched from –5 to 0 V, while at the bottom it was switched from –5 V to the value shown on the *x*-axis

The effectiveness of the MgO and Sc_2O_3 passivation was more variable on the AlGaN-cap devices than for the GaN-capped HEMTs described in the previous section. Figure 6.32 shows typical gate-lag data from an Sc_2O_3 passivated AlGaN-cap HEMT. We found that the Sc_2O_3 deposition was typically able to restore 80–90% of the drain current loss relative to dc measurement conditions. Some devices showed essentially complete restoration of the current, but the 80–90% range was more typical. Similar results were obtained for MgO passivation, as shown in Figure 6.33. Once again, some individual HEMTs showed essentially full restoration of I_{DS} upon MgO deposition, but a more typical result was 80–90% effectiveness. We believe that our *in situ* cleaning procedure prior to deposition of these oxides in the MBE chamber is not able to completely remove the native oxide from the AlGaN surface. This limits the effectiveness of the resulting passivation and accounts for the more variable results we observe for the AlGaN-cap devices.

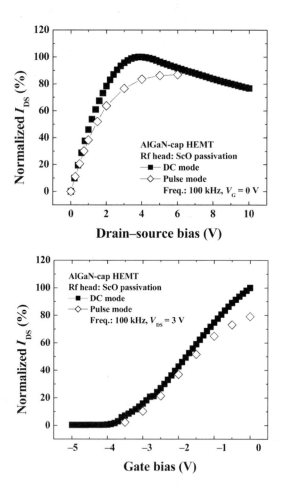

Figure 6.32. Gate-lag measurements before and after Sc_2O_3 passivation of 0.5×100 μm², AlGaN-cap HEMTs. At the top V_G was switched from −5 to 0 V, while at the bottom it was switched from −5 V to the value shown on the *x*-axis

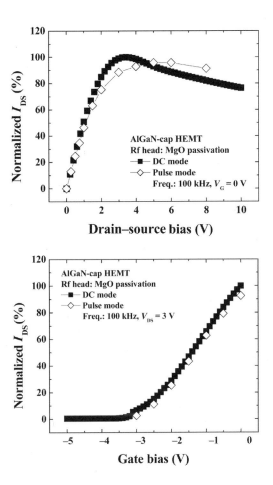

Figure 6.33. Gate-lag measurements before and after Sc_2O_3 passivation of 0.5×100 μm^2, AlGaN-cap HEMTs. At the top V_G was switched from −5 to 0 V, while at the bottom it was switched from −5 V to the value shown on the x-axis

MgO and Sc_2O_3 thin films deposited by MBE appear very promising as surface passivation layers on AlGaN–GaN HEMTs. In structures with a GaN-cap layer they provide more effective mitigation of drain current collapse than the conventional PECVD SiN_x films. The Sc_2O_3 provides stable passivation characteristics over a period of at least 5 months, while the MgO was found to lose some of its effectiveness under the same conditions. The passivation with both oxides is more variable on AlGaN-cap structures, which we believe is due to the greater difficulties in removing the native oxide from the AlGaN surface prior to deposition of the passivation films. With respect to SiN_x passivation, we found no obvious advantage to the use of deuterated precursors for the deposition. Finally, it showed be noted that both MgO and Sc_2O_3 can also be used as gate dielectrics on GaN to provide MOS diodes and transistors, and thus an MBE system equipped for their deposition can provide both the gate oxide and the device passivation in a manufacturing environment.

6.5.2 Radiation-Damage Experiments

MgO–GaN MOS diodes were irradiated with 40 MeV protons at a fluence of 5×10^9 cm^{-2}, simulating long-term (10 years) exposure in space-borne applications. The result of the proton irradiation was a decrease in device capacitance, consistent with the creation of deep electron traps that reduce the effective channel doping and also a decrease in breakdown field from ~10^6 V·cm^{-1} in control devices to 0.76×10^6 V·cm^{-1} in devices irradiated with the gate metal in place. The capacitance of the device irradiated with the contacts in place recovers to the same value as the contact diodes. The D_{it} values are decreased by the H_2 anneal in both the unirradiated and irradiated devices.

Figure 6.34 shows the current density J–voltage (top) and capacitance–voltage (C–V) characteristics (bottom) from the MgO–GaN devices with 500 Å oxide thickness before and after proton irradiation. We also irradiated devices both before and after deposition of their gate contacts in order to investigate the effect of the presence of the metal over the MgO. While we could not measure any significant change in reverse breakdown voltage as a result of the proton irradiation, the V_F values were decreased in both types of device (*i.e.* those irradiated before or after deposition of the gate metal). The D_{it} values extracted from the C–V characteristics using the Terman method (which is generally found to underestimate the trap density) were also essentially unchanged in the devices from which we could extract accurate date. Note that the forward breakdown field decreases from ~10^6 V·cm^{-1} in the control samples to ~0.76×10^6 V·cm^{-1} for the devices irradiated with contacts in place and for 0.46×10^6 V·cm^{-1} for the structures irradiated prior to deposition of the contacts. This result indicates that the protons make displacement damage in the MgO, which degrades its breakdown capabilities.

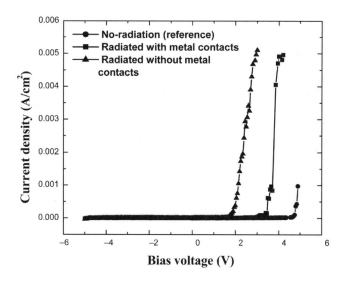

Figure 6.34. Current density–voltage (top) and capacitance–voltage characteristics (bottom) from MgO–GaN MOS devices before and after irradiation

6.6 Metal-Oxide Semiconductor Field Effect Transistors

6.6.1 n-Channel Depletion Mode Gallium Nitride Metal-Oxide Semiconductor Field Effect Transistor Using Stacked Gate Dielectric of SiO_2–Gd_2O_3

The nitride layer structure consisted of 2 μm unintentionally doped GaN and a 700 Å Si-doped GaN channel grown by MOCVD on a c-Al_2O_3 substrate. Hall measurement indicated a channel doping level in the low 10^{18} cm^{-3}. The wafer was prepared for oxide growth by a 3 min dilute HCl dip, 25 min UV–ozone exposure, and 5 min dilute HF dip. The Gd_2O_3 was grown in a Riber 2300 GSMBE system equipped with a Wavemat MPDR 610 ECR plasma source. After loading, the wafer was heated to 700°C to desorb the native oxides, then cooled to the growth temperature of 650°C. Growth proceeded at 1×10^{-4} Torr oxygen pressure in the ECR plasma head at 200 W forward power, with a Gd cell temperature of 1230°C. X-ray diffraction showed a single-crystalline Gd_2O_3 structure for the 700 Å film.

A schematic of the processing sequence and final device structure of the planar n-channel depletion mode GaN MOSFET is given in Figure 6.35.[78] Device fabrication began with a wet etch of Gd_2O_3 to open source and drain ohmic contact windows using HCl as the etchant and photoresist AZ-1818 as the mask. Ti(200 Å)–Al(600 Å)–Au(1000 Å) ohmic contacts were deposited by e-beam evaporation. Oxygen implantation was used for device isolation. A similar O^+ implantation scheme has previously been shown to yield sheet resistance in GaN on the order of 10^{12} $\Omega/\square$. To enhance the gate breakdown voltage, a 300 Å layer of SiO_2 was e-beam evaporated onto the Gd_2O_3 surface in the photolithographically-defined gate contact region. This was followed under vacuum by evaporation of Ti(100 Å)–Au(1000 Å) Schottky contacts to the SiO_2 surface.

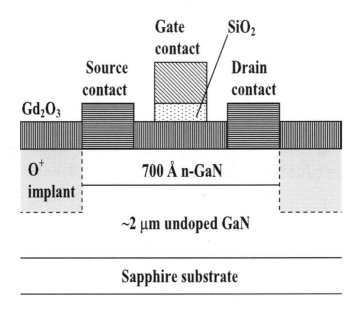

Figure 6.35. Schematic of stacked-gate GaN MOSFET

When deposited under optimum conditions, Gd_2O_3 was grown epitaxially on GaN, as verified by RHEED and XTEM. The interface was quite smooth despite a somewhat rough nitride surface. However, there was still a relatively large difference in lattice spacing between Gd_2O_3 and GaN, which contributed to a high number of dislocations in the oxide film. These dislocations act as current leakage paths and caused the breakdown voltage of the as-grown Gd_2O_3 to be rather low. In order to improve the breakdown for device fabrication, amorphous SiO_2 was deposited on the

Gd_2O_3 surface. The formation of a stacked-gate dielectric of SiO_2–Gd_2O_3 allowed the interfacial properties of Gd_2O_3–GaN to be maintained, while reducing leakage by terminating the dislocations in the crystalline oxide. Figure 6.36 (top) illustrates the significant reduction in gate leakage after SiO_2 deposition. The breakdown field of the insulator improved from 0.3 to 0.8 MV/cm. Gate reverse leakage current density was 5×10^{-6} A/cm^2 at a gate–source bias V_{GS} of –10V and remained below 10 nA at V_{GS} = –70 V for the 1×200 μm^2 device, clearly demonstrating the benefit of the insulated gate MOS structure.

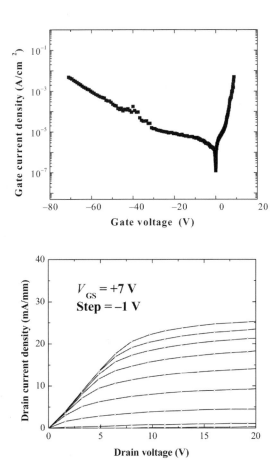

Figure 6.36. DC output characteristics of 1×200 μm^2 gate dimension GaN MOSFET

Charge modulation from accumulation to depletion was observed. The absence of surface inversion is due to the wide band gap and correspondingly low room temperature minority carrier generation rate of GaN, consistent with previous reports on wide bandgap MIS structures. Drain I–V characteristics of a 1×200 µm^2 gate dimension SiO$_2$–Gd$_2$O$_3$–GaN MOSFET were measured with an HP4145A and are shown in Figure 6.36 (bottom). The drain–source breakdown exceeds 80 V. For these high breakdown devices, charge modulation could be demonstrated for gate voltages from +2 to –5 V. Other depletion mode devices of identical geometry could be modulated at an accumulation bias of +7 V, verifying the high quality of the Gd$_2$O$_3$–GaN interface. A maximum intrinsic transconductance of 61 mS/mm was measured at V_{GS}=–0.5 V and V_{DS}=20 V.

6.6.2 MgO–p-GaN Enhancement Mode Metal-Oxide Semiconductor Field Effect Transistors

The sample structure consisted of ~2 µm thickness Mg-doped p-GaN ($p \approx 10^{16}$ cm^{-3}) grown on c-plane sapphire substrate. In order to create the n$^+$ source and drain-contact regions for MOSFET fabrication, these samples were implanted with ^{27}Si$^+$ ions at multiple energy conditions (10 keV, 1.6×10^{13} cm^{-2}; 30 keV, 3.6×10^{13} cm^{-2}; 70 keV, 7.2×10^{13} cm^{-2}; 170 keV, 1.3×10^{14} cm^{-2}; 340 keV, 2.2×10^{14} cm^{-2}) at room temperature. This implant scheme was designed to produce a uniform Si profile to depth of ~0.3 µm.[89] The samples were encapsulated with 500 nm of SiO$_2$ deposited by plasma-enhanced CVD and annealed at 1200°C under a flowing N$_2$ ambient. After HF removal of the SiO$_2$, the MgO was deposited by MBE at 100°C on the p-GaN layer as the gate dielectric. The MgO precursors were elemental Mg and RF plasma-activated oxygen. The GaN samples were cleaned initially with a 3 min chemical etch in HCl/H$_2$O (1:1), H$_2$O rinse, UV–ozone exposure for 25 min rinse in buffered oxide etch solution (6:1, NH$_4$F/HF) and rinsed in H$_2$O. The samples were then loaded into the MBE system and heated at 650°C to ensure oxide removal. A standard effusion cell operating at 380°C was used for evaporation of the Mg, while the O$_2$ source was operated at 300 W forward power (13.56 MHz) and 2.5×10^{-5} Torr. The deposited MgO layers were ~80 nm thickness. Following this step, the MgO was removed in the source, the drain and the p-epi contact areas, using H$_3$PO$_4$ etchant and an S-1818 resist mask. The e-beam deposited n-ohmic (Ti–Al–Pt–Au), p-ohmic (Ni–Au) and gate (Pt–Au) metals were followed by lift-off. A schematic of the final structure is shown in Figure 6.37.

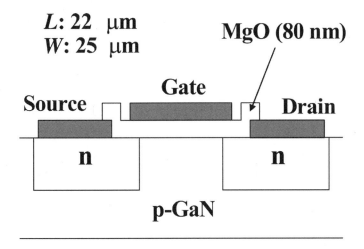

Figure 6.37. Schematic cross-section of the MgO–p-GaN MOSFET

Figure 6.38 shows the typical gate I–V characteristics of the MgO–GaN MOSFET. For the oxide thickness of 80 nm, the corresponding oxide breakdown field was 1.75 MV/cm at a current density of 5 mA/cm^2. The I–V characteristic is shifted due to the presence of oxide charge. The gate contact exhibited good rectification, demonstrating the MgO produces excellent insulator characteristics and a much larger breakdown than if the Pt–Au is placed directly on the p-GaN without any MgO. In that case, the forward turn-on voltage was <0.7 V and the reverse breakdown voltage was <1 V. In addition, the p–n junction between the p-GaN and the n-type source showed excellent rectifying characteristics (also shown in Figure 6.38), demonstrating that the Si activation efficiency was sufficient to produce an abrupt junction. Room-temperature capacitance–voltage characteristics measured at 1 MHz and with a sweep rate of 30 mV/s performed on companion diodes showed clear modulation from accumulation to depletion. Using the relation $C_{ox} = \varepsilon_o \varepsilon_{ox} A / T_{ox}$ (where ε_o is the permittivity in a vacuum, C_{ox} is the oxide capacitance, A is the cross-sectional area of oxide, and T_{ox} is the oxide thickness), the dielectric constant of the oxide ε_{ox} was calculated to be 9.9, a value which is in agreement with the tabulated value for MgO (9.8). A significant flat band voltage shift was observed, which indicated the existence of fixed charges in the oxide layer.

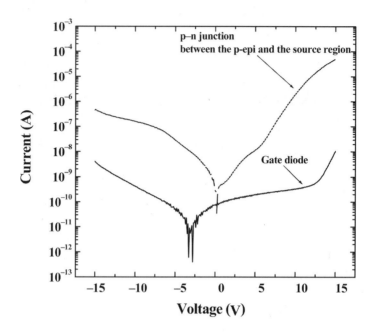

Figure 6.38. Transfer characteristics of MgO–p-GaN MOSFET at a drain–source voltage of 3V

The drain–source I–V characteristics from the MOSFET showed the channel was formed at a gate voltage of >+6 V and the device could be modulated to +12 V of gate voltage. There is no indication of negative resistance effects, and thus self-heating does not appear to be a problem under the biasing conditions employed here. The drain current I_{DS} can be expressed as [90]

$$I_{DS} = \mu_n C_i \frac{W}{L} (V_G - V_T) V_D \qquad (6.2)$$

where C_i is the oxide capacitance per unit area, L and W are the channel length and channel width respectively, μ_n is the low field mobility, V_G is the gate voltage, V_D is the drain voltage, and V_T is the threshold voltage. The transconductance of the MOSFET (*i.e.* the change in drain current for a change in gate voltage) is maximized for high electron mobilities, thinner MgO dielectric layers and larger values of W/L. MEDICI^{TM} simulation reported previously indicate that the drain current should be in the milliamp range for the structure used here, and the lower currents obtained show that we need to improve carrier mobility significantly [91] and probably also the N-doping level obtained in the source–drain regions, which affects the series resistance. The maximum drain current was ~0.2 µA at 12 V gate voltage and the transconductance was 5.4 µS.mm^{-1}. This is comparable to the first results for e-mode MOSFETs in GaAs, using $GdGa_2O_3$ gate oxides.[92] Clearly, much more work has to be done to increase the channel mobility. It is easier to activate n-type implants in GaN than p-type implanted dopants, but in principle it should also be possible to use n-type epi and create p-type source and drain regions

6.7 Conclusions

Rapid recent progress in the development of gate dielectrics and surface passivation films for GaN electronic devices has been reported. This research shows the promise of MOS-HEMTs with appropriate passivation for applications in low-noise amplifiers with high dynamic range and also in high-efficiency power amplifiers.

References

1. Nguyen NX, Micovic M, Wong W-S, Hashimoto P, McCray L-M, Janke P, Nguyen C (2000) Electron Lett 36: 468
2. Wu Y-F, Keller BP, Keller S, Kapolnek D, Kozodoy P, DenBaars SP, Mishra UK (1997) Solid-State Electron 41: 1569
3. Khan MA, Hu X, Simin G, Lunev A, Yang J, Gaska R, Shur MS (2000) IEEE Electron Dev Lett 21: 63
4. Green BM, Chu KK, Chumbes EM, Smart JA, Shealy JR, Eastman LF (2000) IEEE Electron Dev Lett 21: 268
5. Binari SC, Kruppa W, Dietrich HB, Kelner G, Wickenden AE, Freitas JA Jr, (1997) Solid-State Electron 41: 1549
6. Chu KK, Smart JA, Shealy JR, Eastman LF (1998) Proc State-of-the-Art Program on Compound Semiconductors (SOTAPOCS XX/X, Electrochem Soc, Pennington, NJ); M.3-12
7. Eastman LF, Tilak V, Smart J, Green BM, Chumbes EM, Dimitrov R, Kim H, Ambacher OS, Weimann N, Prunty T, Murphy M, Schaff WJ, Shealy JR (2001) IEEE Trans Electron Dev 48: 479
8. Wu Y-F, Kapolnek D, Ibbetson JP, Parikh P, Keller BP, Mishra UK (2001) IEEE Trans Electron Dev ED48: 586

9. Johnson JW, Baca AG, Briggs RD, Shul RJ, Monier C, Ren F, Pearton SJ, Dabiran AM, Wowchack AM, Polley CJ, Chow PP (2001) Solid-State Electron 45: 1979
10. Lu W, Yang J, Khan MA, Adesida I (2001) IEEE Trans Electron Dev ED48: 581
11. Pearton SJ, Zolper JC, Shul RJ, Ren F (1999) J Appl Phys 86
12. Simin G, Hu X, Ilinskaya N, Kumar A, Koudymov A, Zhang J, Asif Khan M, Gaska R, Shur MS (2000) Electron Lett 36: 2043
13. Kohn E, Daumiller I, Schmid P, Nguyen NX, Nguyen CN (1999) Electron Lett 35: 1022
14. Lee JS, Vescan A, Wieszt A, Dietrich R, Leier H, Kwon Y-S (2001) Electron Lett 37: 130
15. Hu X, Koudymov A, Simin G, Yang J, Asif Khan M, Tarakji A, Shur MS, Gaska R (2001) Appl Phys Lett 79: 2832
16. Daumiller I, Kirchner C, Kamp M, Ebeling KJ, Kohn E (1999) IEEE Electron Dev Lett 20: 448
17. Tarakji A, Simin G, Ilinskaya N, Hu X, Kumar A, Koudymov A, Yang J, Asif Khan M, Shur MS, Gaska R (2001) Appl Phys Lett 78: 2169
18. Chumbes EM, Smart JA, Prunty T, Shealy KM (2001) IEEE Trans Electron Dev 48: 416
19. Luo B, Johnson JW, Ren F, Allums KK, Abernathy CR, Pearton SJ, Dwivedi R, Fogarty TN, Wilkins R, Dabiran AM, Wowchack AM, Polley CJ, Chow PP, Baca AG (2001) Appl Phys Lett 79: 2196
20. Pearton SJ, Ren F, Zhang AP, Lee KP (2000) Mater Sci Eng Rep R 30: 55
21. Binari SC, Ikossi K, Roussos JA, Kruppa W, Park D, Dietrich HB, Koleske DD, Wickenden AE, Henry RL (2001) IEEE Trans Electron Dev 48: 465
22. Simin G, Koudymov A, Tarakji A, Hu X, Yang J, Asif Khan M, Shur MS, Gaska R (2001) Appl Phys Lett 79: 2651
23. Ren F, Hong M, Chu SNG, Marcus MA, Schurman MJ, Baca A, Pearton SJ, Abernathy CR (1998) Appl Phys Lett 73: 3893
24. Johnson JW, Luo B, Ren F, Gila BP, Krishnamoorthy V, Abernathy CR, Pearton SJ, Chyi JI, Nee TE, Lee CM, Chuo CC (2000) Appl Phys Lett 77: 3230
25. Gila BP, Johnson JW, Lee KN, Krishnamoorthy V, Abernathy CR, Ren F, Pearton SJ (2001) ECS Proc Vol 2001-1: 71 (Electrochemical Society, Pennington, NJ 2001)
26. Asif Khan M, Hu X, Tarakji A, Simin G, Yang J, Gaska R, Shur MS (2001) Appl Phys Lett 77: 1339

27. Simin G, Hu X, Ilinskaya N, Zhang J, Tarakji A, Kumar A, Yang J, Asif Khan M, Gaska R, Shur MS (2001) IEEE Electron Dev Lett 22: 53
28. Morkoç H, Di Carlo A, Cingolani R (2002) Solid-State Electron 46: 157
29. Ohno Y, Kuzuhara M (2001) IEEE Trans Electron Dev ED48: 517
30. Wu Y-F, Kapolnek D, Ibbetson JP, Parikh P, Keller BP, Mishra UK (2001) IEEE Trans Electron Dev ED48: 586
31. Arulkumaran S, Egawa T, Ishikawa H, Jimbo T, Umeno M (1998) Appl Phys Lett 73: 809
32. Irokawa I, Nakano Y (2002) Solid-State Electron 46: 1467
33. Lay TS, Hong M, Kwo J, Mannaerts JP, Hung WH, Huang DJ (2001) Solid-State Electron 45: 1679
34. Hashizume T, Alekseev E, Pavlidis D, Boutros KS, Redwing J (2000) J Appl Phys 88: 1983
35. Zhang N-Q, Keller S, Parish G, Heikman S, DenBaars SP, Mishra UK (2000) IEEE Electron Dev Lett 21: 421
36. Gila BP, Johnson J, Mehandru R, Luo B, Onstine AH, Allums KK, Krishnamoorthy V, Bates S, Abernathy CR, Ren F, Pearton SJ (2001) Phys Status Solidi A 188: 239
37. Klein PB, Binari SC, Ikossi K, Wickenden AE, Koleske DD, Henry RL (2001) Appl Phys Lett 79: 3527
38. Luo B, Johnson JW, Kim J, Mehandru RM, Ren F, Gila BP, Onstine AH, Abernathy CR, Pearton SJ, Baca AG, Briggs RD, Shul RJ, Monier C, Han J (2002) Appl Phys Lett 80: 1661
39. Mehandru R, Gila BP, Kim J, Johnson JW, Lee KP, Luo B, Onstine AH, Abernathy CR, Pearton SJ, Ren F (2002) Electrochem Solid-State Lett 5: G51
40. Kim J, Mehandru R, Luo B, Ren F, Gila BP, Onstine AH, Abernathy CR, Pearton SJ, Irokawa Y (2002) Appl Phys Lett 81: 373
41. Kim J, Mehandru R, Luo B, Ren F, Gila BP, Onstine AH, Abernathy CR, Pearton SJ, Irokawa Y (2002) Appl Phys Lett 80: 4555
42. Kim J, Gila B, Mehandru R, Johnson JW, Shin JH, Lee KP, Luo B, Onstine AH, Abernathy CR, Pearton SJ, Ren F (2002) J Electrochem Soc 149: G482
43. Asif Khan M, Chen Q, Shur MS, Dermott BT, Higgins JA, Burm J, Schaff WJ, Eastman LF (1997) Solid-State Electron 41: 1555
44. Nguyen NX, Nguyen C, Grider DE (1999) Electron Lett 35: 1356
45. Tarakji A, Simin G, Ilinskaya N, Hu X, Kumar A, Koudymov A, Yang J, Asif Khan M, Shur MS, Gaska R (2001) Appl Phys Lett 78: 2169

46. Luo B, Johnson JW, Ren F, Allums KK, Abernathy CR, Pearton, Dwivedi R, Fogarty TN, Wilkins R, Dabiran AM, Wowchack AM, Polley CJ, Chow PP, Baca AG (2001) Appl Phys Lett 79: 2196
47. Pearton SJ, Ren F, Zhang AP, Lee KP (2000) Mater Sci Eng Rep R 30: 55
48. Binari SC, Ikossi K, Roussos JA, Kruppa W, Park D, Dietrich HB, Koleske DD, Wickenden AE, Henry RL (2001) IEEE Trans Electron Dev 48: 465
49. Shur MS (1998) Solid-State Electron 42: 2131
50. Morkoç H, Di Carlo A, Cingolani R (2003) Solid-State Electron 47: 358
51. Ohno Y, Kuzuhara M (2001) IEEE Trans Electron Dev ED48: 517
52. Morkoç H (1999) *Nitride Semiconductors and Devices*, Springer, Berlin
53. Gaffey B, Guido LJ, Wang XW, Ma TP (2001) IEEE Trans Electron Dev ED48: 458
54. Arulkumaran S, Egawa T, Ishikawa H, Jimbo T, Umeno M (1998) Appl Phys Lett 73: 809
55. Lay TS, Hong M, Kwo J, Mannaerts JP, Hung WH, Huang DJ (2001) Solid-State Electron 45: 1679
56. Hashizume T, Alekseev E, Pavlidis D, Boutros KS, Redwing J (2000) J Appl Phys 88: 1983
57. Zhang N-Q, Keller S, Parish G, Heikman S, DenBaars SP, Mishra UK (2000) IEEE Electron Dev Lett 21: 421
58. Therrien R, Lucovsky G, Davis RF (1999) Phys Status Solidi A 176: 793
59. Gila BP, Johnson J, Mehandru R, Luo B, Onstine AH, Allums KK, Krishnamoorthy V, Bates S, Abernathy CR, Ren F, Pearton SJ (2001) Phys Status Solidi A 188: 239
60. Klein PB, Binari SC, Ikossi K, Wickenden E, Koleske DD, Henry RL (2001) Appl Phys Lett 79: 3527
61. Klein PB, Freitas JA Jr, Binari SC, Wickenden AE (1999) Appl Phys Lett 75: 4016
62. Klein PB, Binari SC, Freitas JA Jr, Wickenden AE (2000) J Appl Phys 88: 2843
63. Klein PB, Binari SC, Ikossi-Anastasiou K, Wickenden AE, Koleske DD, Henry PL, Katzer DS (2001) IEEE Electron Lett 37: 661
64. Luo B, Johnson JW, Gila BP, Onstine A, Abernathy CR, Ren F, Pearton SJ, Baca AG, Dabiran AM, Wowchack AM, Chow PP (2002) Solid-State Electron 46: 467
65. Simin G, Tarakji A, Hu X, Koudymov A, Yang J, Khan MA, Shur MS, Gaska R (2001) Phys Status Solidi A 188: 219

66. Tarakji A, Hu X, Koudymov A, Simin G, Yang J, Khan MA, Shur MS, Gaska R (2002) Solid-State Electron 46: 1211
67. Koudymov A, Hu X, Simin K, Simin G, Ali M, Yang J, Khan MA (2002) IEEE Electron Dev Lett 23 : 449
68. Simin G, Koudymov A, Fatima H, Zhang JP, Yang J, Khan MA, Hu X, Tarakji A, Gaska R, Shur MS (2002) IEEE Electron Dev Lett 23 : 458
69. Asif Khan M, Hu X, Tarakji A, Simin G, Yang J, Gaska R, Shur MS (2000) Appl Phys Lett 77: 1339
70. Hu X, Koudymov A, Simin G, Yang J , Khan MA, Tarakji A, Shur MS, Gaska R (2001) Appl Phys Lett 79: 2832
71. Ren F, Hong M, Chu SNG, Marcus MA, Schurman MJ, Baca A, Pearton SJ, Abernathy CR (1998) Appl Phys Lett 73: 3839
72. Johnson JW, Luo B, Ren F, Gila BP, Krishnamoorthy V, Abernathy CR, Pearton SJ, Chyi JI, Nee TE, Lee CM, Chuo CC (2000) Appl Phys Lett 77: 3230
73. Zhang N-Q, Keller S, Parish G, Heikman S, DenBaars SP, Mishra UK (2000) IEEE Electron Dev Lett 21: 421
74. Casey HC Jr, Fountain GG, Alley RG, Keller BP, DenBaars SP (1996) Appl Phys Lett 68: 1850
75. Therrien R, Lucovsky G, Davis F (1999) Phys Status Solidi A 176: 793; Appl Surf Sci 166: 513
76. Hashizume T, Alekseev E, Pavlidis D, Boutros KS, Redwing J (2000) J Appl Phys 88: 1983
77. Khan MA, Hu X, Simin G, Lunev G, Yang J, Gaska R, Shur MS (2000) IEEE Electron Dev Lett 21: 63
78. Johnson JW, Gila BP, Luo B, Lee KP, Abernathy CR, Pearton SJ, Chyi JI, Nee TE, Lee CM, Chou CC, Ren F (2001) J Electrochem Soc 146: 4303
79. Arulkumaran S, Egawa T, Ishikawa H, Jimbo T, Umeno M (1998) Appl Phys Lett 73: 809
80. Ivanov PA, Levinshtein ME, Simin G, Hu X, Yang J, Khan MA, Rumyantsev SC, Shur MS, Gaska R (2001) Electron Lett 37: 1479
81. Chen P, Wang W, Chua SJ, Zheng YD (2001) Appl Phys Lett 79: 3530
82. Mistele D, Rotter T, Ferretti R, Fedler F, Klausiag H, Semchinova OK, Stemmer J, Aderhold J, Graul J (2001) Mater Res Soc Symp Proc 639: G11.42.1 (MRS, Pittsburgh, PA 2001).
83. Gila BP, Johnson JW, Mehandru R, Luo B, Onstine AH, Allums KK, Krishnamoorthy V, Bates S, Abernathy CR, Ren F, Pearton SJ (2002) Phys Status Solidi B 201: 437

84. Hong M, Anselm KA, Kwo J, Ng HM, Baillargeon JN, Kortan AR, Mannaerts JP, Cho AY, Lee CM, Chyi JI, Lay TS (2000) J Vac Sci Technol B 18: 1453
85. Irokawa Y, Nakano Y (2002) Solid-State Electron 46: 1467
86. Nakano Y, Kachi T, Jimbo T (2003) J Vac Sci Technol B 21: 2220
87. Fu DJ, Kwon YH, Kang TW, Park CJ, Baek KH, Cho HY, Shin DH, Lee CH, Chung KS (2002) Appl Phys Lett 80: 446
88. Mehandru R, Gila BP, Kim J, Johnson JW, Lee KP, Luo B, Onstine AH, Abernathy CR, Pearton SJ, Ren F (2002) Electrochem Solid-State Lett 5: G51
89. Irokawa Y, Kim J, Ren F, Baik KH, Gila BP, Abernathy CR, Pearton SJ, Pan C, Chen GT, Chyi JI (2003) Appl Phys Lett 83: 4987
90. Shur M (1990) *Physics of Semiconductor Devices*, Prentice Hall, Eaglewood Cliffs, NJ
91. Cho H , Lee KP, Gila BP, Abernathy CR, Pearton SJ, Ren F (2003) Solid-State Electron 47: 1597
92. Ren F, Hong M, Hobson WS, Kuo JM, Lothian JR, Mannaerts JP, Kuo J, Chu SNG, Chen YK, Cho AY (1996) Tech Dig Intl Electron Devices Meeting; p 943 (IEEE, New York, NY 1996).

Index

AES. *see* Auger electron spectroscopy
AFM. *see* Atomic force microscopy
AlGaN
   Co+-implantation, 295
   Cr+-implantation, 295, 298
   Mn-implantation, 296, 297
AlGaN–GaN
   HEMTs, 214, 218, 314
      MIS gate, 316
      passivation, 347
   MOS diodes, 316
      hydrogen gas sensors, 222–226
Aluminum nitride (AlN), 108, 122, 287, 291, 293, 294, 319, 327, 328
   amorphous, 328, 329
   cap layers, 17
   Curie temperature, 287
   encapsulant, 17
   etch rate, 108
   fabrication of, 287
      AlMnN, 287
   implanted, 291
      band-edge emission, 291
      band-edge luminescence, 294
      Co-implanted, 291
      Cr-implanted, 291, 293
      Mn-implanted, 291, 294
   wide band-gap, 2
Amorphous dielectric. *see also* Dielectric
   gallium gadolinium garnet (GGG), 318, 319
   gallium oxide, 318
   silicon nitride, 318
   SiO_2, 317, 318
   SiO_2–Si_3N_4–SiO_2 (ONO), 318

Annealing, 17–22, 80, 125, 131, 139, 140, 148
Anomalous Hall effect, 266, 267
As-implanted resistance, 36, 39
Atomic force microscopy, 18
Auger electron spectroscopy, 66
Auger recombination, 186, 195
Avalanche breakdown, 179, 187
Avalanche multiplication, 186. *see also* Reverse breakdown voltage

Bandgap
   Curie temperature, 261
   gallium nitride, 181
   6H-SiC, 181
Barrier height, 15, 282, 286
Biosensors, 245–247. *see also* Sensors
BJT (bipolar junction transistors), 61
Bixbyite oxides. *see also* GaN MOSFET
   Gd_2O_3 growth initiations, 321
      full width at half maximum (FWHM), 324, 325
      RHEED pattern, 322–324
BMP (bound magnetic polaron), 264
   effect of temperature, 265
Breakdown voltage, 4
   for AlGaN–GaN HEMTs, 314
   punch-through junction diode, 190
   reverse breakdown voltage, 179
      6H-SiC, 190
      avalanche multiplication, 186
      electron–hole pair (EHP), 187
      GaN, 190
      impact ionization coefficients, 187, 188

Index

CAIBE (chemically assisted ion beam etching), 101–102. *see also* Reactive ion beam etching
Chlorine-based plasmas, 104–116. *see also* Dry etching
   AlN, 122
   etch rates, 104
   etch selectivity, 107
   GaN, 122
   InN, 122
   secondary gas additions, 111
      H_2, 111
      N_2, 113
Co+-implantation. *see also* Cr+-implantation; Mn-implantation
   (Ga,Mn)N, 278
   AlGaN, 295
   AlN, 291, 294
Cr+-implantation. *see also* Co+-implantation; Mn-implantation
   AlGaN, 295, 298
   AlN, 291, 293
Crystalline AlN, 319
Curie temperature, 262, 263, 287
   bandgap, 265
   BMP, 265
   for AlN, 287
   Mn concentration, 264, 294, 298
Current collapse, 316
Cyclotron resonance, electron. *see* Electron cyclotron resonance

DLTS (deep level transient spectroscopy), 278
Dielectric
   amorphous
      AlN, 328
      gallium gadolinium garnet (GGG), 318, 319
      gallium oxide, 318
      silicon nitride, 318
      SiO_2, 317, 318
      SiO_2–Si_3N_4–SiO_2 (ONO), 318
   crystalline
      AlN, 319, 327
   gate dielectric, 329–349
      MgO, 329
      Sc_2O_3, 329
      Gd_2O_3
      Mn_2O_3 (bixbyite) crystal structure, 319
Diodes
   Schottky, 141–148
      annealing, 148
      Cl_2–Ar discharges, 141, 142
      dry etching, 141
      etch damage thermal stability, 144
      n-type GaN, 144
      p-GaN diodes, 145, 147
      vs. p–n junction diodes, 201–203
   high breakdown lateral, 204–206
      edge termination techniques, 204
      junction-barrier-controlled Schottky (JBS) rectifiers, 204
      p-floating field rings, 204
      p-guard ring, 204
DMSs (dilute magnetic semiconductors), 263
   BMP, 265
   Curie temperatures, 266
   ferromagnetic second phase, 263
   ferromagnetism mechanisms, 263, 265
   mean-field theory, 263, 265
   spin FETs, 285
   superconducting quantum interference device, 264
   transport properties, 281
Doping. *see* implantation
Dry etching, 15, 97–169. *see also* Gallium nitride
   damage, 1, 60–86, 89
      determination of damage profile, 82–86
      effect of etching chemistries on damage, 66–71
      plasma damage, 61–66, 75–80
      thermal stability of damage, 71–75, 80–82
   device processing, 152–166
      field effect transistors, 161–165
      heterojunction bipolar transistors, 157–160
      Microdisk lasers, 152
      ridge waveguide lasers, 153–156
      ultraviolet detectors, 166

Index 363

etch profile and etched surface
morphology, 122
plasma chemistries, 104–121
chlorine-based plasmas, 104–115
iodine- and bromine-based plasmas,
116–120
methane–hydrogen–argon plasmas,
121
plasma reactors, 97–103
chemically assisted ion beam etching,
101
high-density plasmas, 100
low-energy electron enhanced etching,
103
reactive ion etching, 98–99, 102
plasma-induced damage, 124–148
n-gallium nitride, 126–133
p-gallium nitride, 133–140
p-n junctions, 148–151
Schottky diodes, 141–147
ultraviolet detectors, 166–169. *see also*
dry etching
AlGaN MSM, 169
AlGaN, 166, 168
GaN p-i-n, 167, 168
metal semiconductor metal (MSMs),
167
nitride-based, 166
e-beam evaporation
gate contacts, 338
ohmic, 338

ECR. *see* Electron cyclotron resonance
Edge termination design, 195–200. *see also*
GaN Schottky rectifiers
field plate termination, 195–197
Auger recombination, 195
breakdown voltages, 196, 197
MEDICITM code, 195
Shockley–Read–Hall, 195
SiO_2 thickness, 196
junction termination, 198
Edge termination techniques, 204. *see also*
GaN Schottky rectifiers
junction-barrier-controlled Schottky (JBS)
rectifiers, 204
p-floating field rings, 204
p-guard ring, 204

EHP (electron–hole pair), 187
Electron cyclotron resonance, 16, 100
etch rates, 100, 103
ELOG. *see* Epitaxial lateral overgrowth
Energy band gap, 41
Enhancement-mode MOSFETs, 313
Epitaxial lateral overgrowth, 155
Etch profile and etched surface morphology,
122–124
Etch rate, 104
dc bias dependence, 105, 106, 117
for AlN, 108
for GaN, 108, 109
for InN, 108, 109, 117, 118
for the binary nitrides, 117
ICP source power dependence, 107
nitride, 120
plasma flux dependence, 107
pressure dependence, 104, 105
substrate temperature dependence, 109
volatility of the etch products dependence,
109
Etch selectivities, 107
for InN
EXAFS (Extended x-ray absorption fine
structure), 264, 267, 281

FC. *see* Field cooled
FC magnetization, 295
Ferromagnetism, 263–265, 291. *see also*
Semiconductor spintronics
aluminum nitride-based semiconductors,
87–90, 287–290
electrical and optical properties, 273–281
FC magnetization, 295
ferromagnetic second phase, 269–273
magnetic properties, 261
carrier concentration, 261
mean-field approach, 264
bound magnetic polaron (BMP), 264
FET. *see* field effect transistor
Field cooled, 270
Field effect transistor, 157, 161–166.
see also Dry etching
GaAs, 316
group III-nitride, 166
heterostructure FETs (HFETs), 161
IG-HFET, 319

Field effect transistor (*cont.*)
  InAlN channel structures, 161, 162
  ISFET, 242
  MESFETs, 316
  MOSFET, 242, 349–355
  Schottky contact, 164
  SiC, 316
Field plate termination, 195–197. *see also* Edge termination design
  Auger recombination, 195
  breakdown voltages, 196
  MEDICI™ code, 195
  reverse breakdown voltage, 197
  Shockley–Read–Hall, 195
  SiO_2 thickness, 196
Field gating of transistor structures, 261
  Cr-implanted n-AlGaN, 298
  hole-induced interactions, 273
  transition-metal-implanted, 291
    Co-implanted AlN, 291
Forward bias, 201. *see also* Reverse bias

GaAs MOSFETs
  dielectric
    gallium gadolinium garnet, 318
Gallium gadolinium garnet, 318
Gallium nitride (GaN), 1–89
  amorphous aluminum nitride, 327
    bixbyite oxides, 321
    MgO–CaO, 327
    scandium oxide, 326
    SiO_2–Gd_2O_3 dielectric stack, 319
  bandgap temperature dependence, 179
  based MODFETs, 12
  based solid-state sensors, 213–260
    chemical, gas, biological, and pressure, 213–260
  breakdown voltage
  punch through diode, 190, 191
  bulk, 179
  charged dislocations, 282
  critical electric field, 189
  design and fabrication of, 178–212
  dry etch damage, 60–82
    determination of damage profile, 82–85
    effect on damage, 66–70
    plasma damage in n-gallium nitride, 61–65, 71–74

  plasma damage in p-gallium nitride, 75–81
  dry etching, 97–177
    device processing, 152–169
    etch profile and etched surface morphology, 122–124
    plasma chemistries, 104–122
    plasma reactors, 97–98
    plasma-induced damage, 124–152
  ECR etch, 101
  edge termination methods, 179
  effective density of states, 181
  electric field breakdown strength, 178
  electrical contacts, 41–51
    effects of interfacial oxides on Schottky contact, 44–48
    interfacial insulator model, 49–50
    tungsten and tungsten silicide contacts on p-gallium nitride, 54–59
    tungsten-based ohmic contact, 51–53
  epi-layer GaN, 179
  ferromagnetism, 261
  gas sensors, 215
  high-power rectifiers, 79–210
  implant doping, 16–31
    dopant redistribution, 26–30
    high-temperature annealing and aluminum nitride encapsulation, 17–22
    n-type implant doping, 22–24
    p-type implant doping, 25
    residual damage, 31
  implant isolation, 32–38
    oxygen implantation for selective area isolation, 34–37
    Ti, iron, and chromium implantation, 38–40
  insulators, 313–352
  intrinsic carrier concentration, 181
  ion implantation doping, 88
  MOSFET, 317–320, 349
    amorphous aluminum nitride, 327
    bixbyite oxides, 321
    MgO–CaO, 327
    scandium oxide, 326
    SiO_2–Gd_2O_3 dielectric stack, 319
    stacked-gate, 349–351

Index 365

on-state resistance, 179
reverse blocking voltages, 178
reverse breakdown voltage, 179
  6H-SiC, 190
  avalanche multiplication, 186
  electron–hole pair (EHP), 187
  GaN, 190
  impact ionization coefficients, 187, 188
room-temperature ferromagnetism, 273
Schottky rectifiers, 192
  bulk diode arrays, 207–210
  edge termination design, 195–200
  generation and recombination, 186
  high breakdown lateral diodes, 204–206
  incomplete ionization of impurities, 183
  mobility models, 184–185
  on-state resistance, 191–194
  reverse breakdown voltage, 186
  Schottky and p-n junction diodes comparison, 201
(Ga,Mn)N, 266–269, 285–286. *see also* Semiconductor spintronics
  anomalous Hall effect, 266, 267
  bandgap, 286
  barrier height, 286, 282
  coercivities, 266
  Co-implanted sample, 278
  deep level transient spectroscopy (DLTS), 278
  defects, 283
  dislocation scattering, 285
  extended x-ray absorption fine structure (EXAFS), 264, 267
  FC and ZFC magnetization, 272
  ferromagnetic second phase, 269–273
  hysteresis, 266, 267
  I–V characteristics, 282
  magnetic properties, 269
    Mn concentration, 269, 270, 273, 275
    spin-glass-type transition, 273
  mean-field theories, 266
  Micro Callodo luminescence, 278
  optical absorption spectra, 276
  optical transmission spectrum, 276
  resistivity, 273, 275
  room-temperature ferromagnetism, 266, 267

  semi-insulating behavior, 276
  transition-metal-doped AlGaN, 294
    Co+-implanted, 278, 295
    Cr+-implanted, 295
    polarized light emitters or spin transistors, 295
  vertical electron mobilities, 284, 285
(Ga,Mn)N–GaInN LED
  InGaN quantum well, 305
  energy relaxation process, 305
  magnetic field, 305
  spin relaxation, 305
  weak spin polarization, 305
Gallium nitride high-power rectifiers, 79–210
  design and fabrication of, 178–210
    Auger recombination, 186
    bulk diode arrays, 207
    comparison of Schottky and p-n junction diodes, 201
    edge termination design, 195
    effective density of states, 182
    field plate termination, 195–197
    forward bias, 201
    generation and recombination, 186
    high breakdown lateral diodes, 204–206
    incomplete ionization of impurity atoms, 183
    intrinsic carrier concentration, 182
    junction termination, 198
    mobility models, 184
    on-state resistance, 179, 191–194
    reverse bias, 201
    Shockley–Read–Hall lifetime, 186
    temperature dependence of bandgap, 180
  edge termination methods, 179
  electric field breakdown strength, 178
  reverse blocking voltages, 178
  reverse breakdown voltage, 179, 186
    6H-SiC, 190
    avalanche multiplication, 186
    electron–hole pair (EHP), 187
    impact ionization coefficients, 187, 188
GaN device processing, 1–96
  AlN cap layers, 17
  AlN encapsulant, 17

GaN device processing (cont.)
  aluminum nitride encapsulation, 17–22
  and AlGaN, 12
  based laser diode, 10
  based metal semiconductor field reflect transistor (MESFETs), 13
  based MODFETs, 12
  contact resistances, 15
  dislocation-free, 8
  dopant redistribution, 26
    secondary ion mass spectroscopy (SIMS), 26
  dry etch damage, 1, 60–86
    determination of damage profile, 82–86
    effect of etching chemistries on damage, 66–71
    plasma damage, 61–66, 75–80
    thermal stability of damage, 71–75, 80–82
  electrical contacts, 41
    effects of interfacial oxides on Schottky contact, 44–48
    interfacial insulator model, 49–50
    thermally stable tungsten-based ohmic contact, 51–54
    tungsten and tungsten silicide, 54
  electron transport, 4
  Hall mobilities, 9, 10
  heteroepitaxial growth, 6
  high breakdown fields, 3
  high-resistivity gallium nitride
    by Ti, iron, and chromium implantation, 38–41
  high-temperature annealing, 17–22
  ideal contact layer, 10
  implant doping
    junction FET, 17
    planar homojunction LED, 17
    ultra-high-temperature activation, 16–32
  implant isolation mechanism, 14, 15, 32–33, 38
    chemically induced isolation, 33
    damage-induced isolation, 33
    transmission line method (TLM), 33
  ion implantation, 1, 14, 16
  lateral epitaxial overgrowth (LEO), 7
  LED, 3, 13
  microwave rectifiers, 4
  MOCVD, 7, 8
  modulation-doped FETs (MODFETs), 11
  n-type doping, 9
  n-type implant doping, 22–25
    S+ activation, 23
    sheet carrier concentration, 23
    Te+ implantation, 25
  oxygen implantation for selective area isolation, 34–37
  p-type conductivity, 9
  p-type doping, 9, 14
  rapid thermal processing (RTP), 17
  residual damage, 31–32
  Schottky barrier height (SBH), 15
  Schottky contacts, 1, 15, 41
    effects of interfacial oxides on, 44–48
    on p-GaN, 43
    surface cleanliness, 43
    to n-GaN, 42
  sheet carrier concentrations, 26
  specific contact resistance, 11
  surface cleanliness, 1
  wide bandgap, 2, 3, 14
  with low defect densities, 8
GaN spin relaxation, 305
GaN–AlGaN transistors
  MOS gates, 329
GaN–InGaN LED, 300
Gas sensing, 250. see also Sensors
Gas source MBE, 319
Gate contacts, 338
Gate insulators, 313
Gate-controlled metal-oxide semiconductor diodes, 329
  radiation-damage experiments, 348
    capacitance–voltage characteristics, 349
    current density–voltage, 349
    device capacitance, 348
  surface passivation, 334
Gateless AlGaN–GaN HEMT
  block copolymers, 219
  drain current, 219
  polystyrene oxide (PEO), 219
  response to block co-polymers, 219–222
  source–drain current, 222

Generation and recombination, 186
    Auger recombination, 186
    Shockley–read–hall lifetime, 186
GGG. *see* Gallium gadolinium garnet
Giant magnetic resistant, 240
GMR. *see* Giant magnetic resistant
Group III nitrides
    dry plasma etching, 97
    RIE, 99
GSMBE. *see* Gas source MBE
6H-SiC. *see also* Reverse breakdown voltage
    bandgap, 180
    critical electric field, 189
    effective density of states, 181
    incomplete ionization of impurities, 183
    intrinsic carrier concentration, 181

Hall effect, anomalous, 266, 267
HBT. *see* Hetrojunction bipolar transistors
HEMT. *see* High electron-mobility transistors
Heteroepitaxial growth, 6, 87
    high-density dislocations, 11
    hydride vapor phase epitaxy (HVPE), 7
    lateral epitaxial overgrowth (LEO), 7
    metalorganic chemical vapor deposition (MOCVD), 7
    molecular beam epitaxy (MBE), 7
    substrate selection, 6
Heterojunction bipolar transistors, 16, 157–160. *see also* Dry etching
    and FET, 157
    device fabrication, 158
    GaN–AlGaN system, 157, 158
        fabrication process, 158
        Mg doping, 160
    GaN–SiC system, 157
    MBE, 160
    MOCVD-grown material, 160
HFET (Heterostructure FETs), 161
High electron-mobility transistors, 107, 213, 314
    AlGaAs–GaAs, 214
    AlGaN–GaN, 214, 218, 314
        MIS gate, 316
        MOS gate, 316
        passivation, 347
    drain–source current voltage, 226

gas sensor, 223
MOS-gate, 223
pressure sensor, 214
sheet carrier concentration, 214
High-density plasmas etch systems, 100–101. *see also* Dry etching
    ECR plasmas, 100
    ICP, 101
    magnetron RIE (MRIE), 100
    MRIE, 101
HVPE (hydride vapor phase epitaxy), 7
Hydrogen gas sensors, 222–226. *see also* MOS
    based on metal-oxide semiconductor, 222
    MOSFET, 224
    time response, 224
Hysteresis loops, 288, 291, 296
    (Ga,Mn)N, 266, 271
    Co-implanted AlN, 294
    Cr-implanted AlN, 291, 293, 294
    disorder effects, 291
    Mn-implanted AlN, 291, 294

ICP. *see* Inductively coupled plasma
IG-FET. *see* Insulated gate heterostructure FET
IG-HFET
    crystalline AlN, 319
III–V nitrides, 86
    AlN, GaN, InN, 2
    breakdown fields, 3
    metalorganic chemical vapor deposition (MOCVD), 7
Immunosensors
    ISFET, 242
    MOSFET, 242
Implant doping. *see also* Galim nitride
    junction FET, 17
    planar homojunction LED, 17
    ultra-high-temperature activation, 16–32
Implant isolation, 14, 15, 32–33
    chemically induced isolation, 33
    damage-induced isolation, 33
    transmission line method (TLM), 33
Implantation
    (Ga,Mn)N
        Co-implanted sample, 278

368 Index

Implantation (*cont.*)
  AlGaN, 294
    Co+-implantation, 295
    Cr+-implantation, 295, 298
    Mn-implantation, 296, 297
  AlN, 291
    Co-implanted AlN, 291
    Cr-implanted AlN, 291, 293
    Mn-implanted AlN, 291, 294
Implanted AlN, 291
  band-edge emission, 291
  band-edge luminescence, 294
  Co-implanted, 291
Inductively coupled plasma, 16, 101
  $Cl_2–H_2–CH_4–Ar$ plasmas etch rate, 103
InGaN quantum well
  spin relaxation, 305
  weak spin polarization, 305
    energy relaxation process, 305
    magnetic field, 305
InN, 115
  high electron-mobility transistors (HEMTs), 107
  wide band-gap, 2
Insulated gate heterostructure FET, 318
Insulators, 313–352
  for AlGaN–GaN HEMT, 313–355
  for MISFET, 317–320
  approach for gallium nitride, 320
    amorphous aluminum nitride, 327
    bixbyite oxides, 321–325
    MgO–CaO, 327
    MOS, 317–320
    scandium oxide, 326
  gate-controlled MOS diodes, 329–333
    radiation-damage experiments, 348
    surface passivation, 334–347
  MOSFET, 349–355
    for gallium nitride, 313–355
    MgO–p-GaN enhancement mode, 352
    stacked-gate GaN MOSFET, 349–351
Inter-device isolation. *see* Implant isolation
Iodine- and bromine-based plasmas, 116–121
  etch rates, 117
  rf chuck power, 120

Ion implantation, 16, 87, 287. *see also* Doping
  inter-device isolation, 14
Ion-assisted plasma etching, 98
ISFET. *see* Silicon-based ion-sensitive FET

JBS (Junction-barrier-controlled Schottky), 204
JTE (junction termination extension), 200. *see also* Junction termination
  equipotential p+ guard rings, 199
  SiO_2 thickness, 198

Lateral diodes, high breakdown, 204–206. *see also* GaN Schottky rectifiers
  edge termination techniques, 204
  junction-barrier-controlled Schottky (JBS) rectifiers, 204
  p-floating field rings, 204
  p-guard ring, 204
Lateral epitaxial overgrowth, 7
LE4. *see* Low-energy electron-enhanced etching
Leakage current mechanism, 286
LED. *see also* GaN device processing
  (Ga,Mn)N, 301
  (Ga,Mn)N–GaInN LED
    InGaN quantum well, 305
  GaN–InGaN LED, 300
  Mn implantation, 300, 302
    GaMnN layers, 300
    PL polarization, 302
  spin injection efficiency, 301
  spin-LED, 300
  ternary and quaternary materials, 3
  unintentionally doped layer, 301
LEO. *see* Lateral epitaxial overgrowth
Liquid sensing. *see also* Sensors
  flexural plate wave device, 251
Magnetization, 288, 290
  FC, 295
  for AlMnN and AlN, 288
  ZFC, 295
    Cr+-implanted AlGaN, 295
    polarized light emitters or spin transistors, 295

Magnetron RIE, 100, 101
MBE. *see* Molecular beam epitaxy
MCL. *see* Micro Callodo luminescence
Mean-field approach, 268
    BMP, 264
    Mn concentration, 298
    spin–spin coupling, 264
MEM. *see* Micro-electro-mechanical
MESFETs (metal semiconductor field reflect transistor), 13
    GaAS, 316
    SiC, 316
Metal organic MBE, 18
Metal semiconductor metal, 167
Metal–insulator–semiconductor, 285
Metalorganic chemical vapor deposition, 7, 18, 22
    grown GaN–AlGaN HBT, 159
    p-type conductivity, 9
    TEM of, 8
Metal-oxide semiconductor
    gate-controlled, 329–349
        radiation-damage experiments, 348
        surface passivation, 334
    hydrogen gas sensors, 222–226
Methane–hydrogen–argon plasmas, 121–122. *see also* Plasma chemistries
    dc bias, 121
MgO passivation, 346, 347. *see also* Passivation
    gate lag measurements, 343
    interface state trap density, 344
    SiN_x covering, 344
MgO–p-GaN enhancement mode MOSFET, 352
    channel mobility, 355
    drain–source voltage, 354
    gate contact, 353
    gate-lag measurement, 342
    I–V characteristic, 353
    rectifying characteristics, 353
    self-heating, 354
    transfer characteristics, 354
Micro Callodo luminescence, 278
Microdisk lasers, 152. *see also* Dry etching
Micro-electro-mechanical, 236
MIS. *see* metal–insulator–semiconductor

MISFET, 317–320. *see also* Insulators
    crystalline AlN, 319
Mn-implantation. *see also* Co+-implantation; Cr-implantation
    AlGaN, 296, 297
    AlN, 291, 294
Mobilities, 284, 285
    vertical electron, 284, 285
    models, 184
        analytical mobility model, 184
        field-dependent mobility model, 185
MOCVD. *see* Metalorganic chemical vapor deposition
MODFETs (modulation-doped FETs), 11, 12
Molecular beam epitaxy, 7
    e-beam evaporation, 318
MOMBE. *see* Metal organic MBE
MOS. *see* Metal-oxide semiconductor
MOSFET, 349–355
    MgO–p-GaN enhancement mode, 352
        channel mobility, 355
        drain–source voltage, 354
        gate contact, 353
        gate-lag measurement, 342
        I–V characteristic, 353
        rectifying characteristics, 353
        self-heating, 354
        transfer characteristics, 354
    stacked-gate GaN MOSFET, 349–351
        charge modulation, 352
        gate leakage reduction, 351
        Gd_2O_3 wet etch, 349, 350
        SiO_2 deposition, 350
MRIE. *see* Magnetron RIE
MSM. *see* Metal semiconductor metal

n-gallium nitride (n-GaN), 126–133, 144. *see also* Plasma-induced damage
    annealing, 131
    damage depth, 131
    electrical quality degradation, 127, 130, 131
    electron-mobility in, 9
    I–V characteristics, 68
    plasma-induced damage, 126–133
    rectifying contact diameters, 45
    rf chuck power, 129

n-gallium nitride (n-GaN) (cont.)
  Schottky diodes, 141–147
  Cl_2–Ar and Ar ICP exposure, 67
  vertical electron mobilities, 282
Nitride-based spintronics. see
      Semiconductor spintronics
n-type implant doping, 22–25

OES. see Optical emission spectroscopy
Ohmic contact, 41, 338
  surface cleanliness, 43
  to p-GaN, 42, 43
On-state resistance, 179, 191–194
Optical emission spectroscopy, 113
Oxygen implantation, 34. see also
      Selective area isolation
  sheet resistance, 36

Passivation
  for AlGaN-cap devices, 345
  GaN-capped HEMTs, 345
  MgO passivation, 343
    gate lag measurements, 343
    interface state trap density, 344
    SiN_x covering, 344
  MgO vs. Sc_2O_3, 343
  Sc_2O_3 passivation
    aging characteristics, 337
    drain–source current, 339, 340, 342
    gate metal degradation, 342
    gate-lag measurements, 339, 340
    HEMT structures, 337
    output power characteristics, 337
Pauli ferromagnetic, 295
PCB. see Printed circuit board
Peak photoluminescence, 125
PECVD (plasma-enhanced chemical vapor deposition)
  silicon nitride deposition, 318
PEO. see Polystyrene oxide
p-gallium nitride (p-GaN), 46, 54–59, 75–81, 133–140, 145, 147. see also
      Plasma-induced damage
  annealing, 139, 140
  behavior of tungsten and tungsten silicide contacts on, 54
  breakdown voltage, 134, 135
  damage depth, 136, 138
  etch rate, 136
  Fermi level position, 57
  plasma-induced damage, 133–140
p-guard ring, 204
PL. see Peak photoluminescence
Plasma chemistries, 104–122. see also Dry etching
  chlorine-based plasmas, 104–116
    etch rates, 104
    etch selectivity, 107
    GaN, AlN and InN, 122
    secondary gas additions, 111, 113
  iodine- and bromine-based plasmas, 116–121
  methane–hydrogen–argon plasmas, 121–122
Plasma damage
  in n-GaN, 61–65, 89
    effect on damage, 66–70
  in p-GaN, 75–80, 89
    thermal stability of damage, 80–81
Plasma etching, ion-assisted, 98
Plasma reactors, 97–98. see also Dry etching
  chemically assisted ion beam etching, 101–102
  high-density plasmas, 100–101
  low-energy electron-enhanced etching, 103
  reactive ion etching, 98–99, 102–103
Plasma-enhanced chemical vapor deposition, 316
Plasma-induced damage, 124–152. see also Dry etching
  Auger spectra, 125
  n-gallium nitride, 126–133
    annealing, 131
    damage depth, 131
    electrical quality degradation, 127, 130, 131
    rf chuck power, 129
  peak photoluminescence (PL), 125
  p-gallium nitride, 133–140
    annealing, 139, 140
    breakdown voltage, 134, 135
    damage depth, 136, 138
    etch rate, 136
  p-n junctions, 148–151

post-etch annealing, 125
Schottky diodes, 141–148
  annealing, 148
  Cl_2–Ar discharges, 141, 142
  dry etching, 141
  etch damage thermal stability, 144
  n-GaN, 144
  p-GaN diodes, 145, 147
  vs. p-n junction diodes, 201–203
  surface stoichiometry, 125
Plasmas, chlorine-based, 104–116. *see also* Dry etching
  AlN, 122
  etch rates, 104
  etch selectivity, 107
  GaN, 122
  InN, 122
  secondary gas additions, 111
    H_2, 111
    N_2, 113
p-n junctions, 148–152, 201–203. *see also* Dry etching
  dc bias, 149, 150
  vs. Schottky diodes, 201–203
  reverse leakage current, 149, 150, 151, 152
Polystyrene oxide, 219
Pressure sensor, 231
  micro-electro-mechanical (MEM), 236
Printed circuit board, 230
p-type AlGaN
  Mn implantation, 296, 297
p-type implant doping, 25–26

QMS (Quadrupole mass spectrometry), 111

Rapid thermal processing, 17
Reactive ion beam etching, 98–99, 101–103. *see also* Chemically assisted ion beam etching
  Cl_2–Ar plasmas etch rate, 103
  removal rates, 102
Rectifiers, gallium high-power
  bulk diode arrays, 207–210
  design and fabrication of, 178–212
  edge termination design, 195–200
  electric field breakdown strength, 178
  generation and recombination, 186
  high breakdown lateral diodes, 204–206
  incomplete ionization of impurities, 183
  mobility models, 184–185
  on-state resistance, 179, 191–194
  reverse blocking voltages, 178, 179, 186
  Schottky vs. p-n junction diodes, 201
Reverse bias, 201
Reverse breakdown voltage, 179, 186–191
  6H-SiC, 190
  bandgap, 180
  critical electric field, 189
  effective density of states, 181
  incomplete ionization of impurities, 183
  intrinsic carrier concentration, 181
  avalanche multiplication, 186
  electron–hole pair (EHP), 187
  GaN, 190
  impact ionization coefficients, 187, 188
rf chuck power, 119, 129
RHEED. *see* Reflection high-energy electron diffraction
RHEED pattern, 322, 323, 324
RIBE. *see* Reactive ion beam etching
Ridge waveguide lasers, 153–157. *see also* Dry etching
  AlGaN cladding layers, 154
  epitaxial lateral overgrowth (ELOG), 155
  green InGaN laser diodes, 154
  sidewall smoothness, 155
  vapor phase epitaxy, 155
RIE. *see* Reactive ion etch
RMS (root-mean-square), 18
RTP. *see* Rapid thermal processing

SAW. *see* Surface acoustic wave
SBH (Schottky barrier height), 15
Sc_2O_3 passivation, 347
  aging characteristics, 337
  AlGaN-cap devices, 345
  drain–source current, 339, 340, 342
  gate metal degradation, 342
  gate-lag measurements, 339, 340
  HEMT structures, 337
  output power characteristics, 337

Scanning electron microscopy, 18
    GaN surfaces, 18
Schottky diodes, 141–148, 201–203.
        see also Dry etching
    annealing, 148
    barrier height, 15
    Cl_2–Ar discharges, 141, 142
    contacts, 1, 15, 41
        effects of interfacial oxides on, 44–48
        on p-GaN, 43
        surface cleanliness, 43
        to n-GaN, 42
    dry etching, 141
    etch damage thermal stability, 144
    forward bias, 201
    n-GaN diodes, 144
    p-GaN diodes, 145, 147
    vs. p-n junction diodes, 201–203
    reverse bias, 201
    Schottky and p-n junction diodes comparison, 201
        forward bias, 201
        reverse bias, 201
Schottky rectifiers, GaN, 178, 192
    bulk diode arrays, 207–210
    edge termination design, 195–200
        junction termination, 198–200
    generation and recombination, 186
        Auger recombination, 186
        Shockley–read–hall lifetime, 186
    high breakdown lateral diodes, 204–206
        edge termination techniques, 204
        junction-barrier-controlled Schottky (JBS) rectifiers, 204
        p-floating field rings, 204
        p-guard ring, 204
    incomplete ionization of impurities, 183
    mobility models, 184–185
        analytical, 184
        field-dependent, 185
    on-state resistance, 191–194
    reverse breakdown voltage, 186
Secondary ion mass spectroscopy, 26–29. see also GaN device processing

Secondary ion mass spectroscopy (*cont.*)
    implanted Be+, 30
    implanted C+, 29
    implanted Mg, 29
    implanted S+, 28
    implanted Si, 27
    implanted Te, 28
Selective area isolation
    oxygen implantation, 34–37
SEM. *see* Scanning electron microscopy
Semiconductor spintronics, 261–307
    (Ga,Mn)N, 266–269, 285–286
        anomalous Hall effect, 266, 267
        bandgap, 286
        barrier height, 286, 282
        coercivities, 266
        dislocation scattering, 285
        EXAFS, 267
        hysteresis, 266, 267
        magnetic properties, 269
        mean-field theories, 266
        Mn concentration, 269, 270, 273, 275
        room-temperature ferromagnetism, 266, 267
        vertical electron mobilities, 284, 285
    AlN-based ferromagnetic semiconductors, 287–290
    Co+-implanted AlGaN, 278, 295
    Cr+-implanted AlGaN, 295
    defects, 283
    DMS, 263–265
    electrical and optical properties, 273–280
        Co-implanted sample, 278
        deep level transient spectroscopy, 278
        Micro Callodo luminescence, 278
        optical absorption spectra, 276
        optical transmission spectrum, 276
        resistivity, 273, 275
        semi-insulating behavior, 276
    FC and ZFC, 272
    ferromagnetic second phase, 269–273
    ferromagnetism, 263–265, 295
        BMP, 264
        implanted AlGaN films, 294–298

implanted aluminum nitride films, 291–293
I–V characteristics, 282
magnetic ordering temperatures, 262
  wide bandgap semiconductors, 263
material selection, 262–263
polarized light emitters or spin transistors, 295
potential device applications, 299–306
spin-glass-type transition, 273
spin-polarized transport, 262
transition-metal-doped AlGaN, 294
  Co+-implanted AlGaN, 295
transport properties, 281–284
Sensors, 213
  based on AlGaN–GaN heterostructures, 219–247
    biosensors, 240–244
    effect of external strain on conductivity, 230–236
    gateless AlGaN–GaN HEMT response to block co-polymers, 219–222
    hydrogen gas sensors, 222–226
    hydrogen-induced reversible changes, 226–228
    pressure sensor fabrication, 236–238
    selective-area substrate removal, 239–240
    surface acoustic wave-based biosensors, 245
  liquid sensing
    flexural plate wave device, 251
  surface acoustic wave device
    array, 251
    biosensors, 245–247
    fabrication, 247–249
    gas sensing, 250
  wireless sensor network, 252
Sheet resistance
  annealing temperature dependence, 39
Shockley–Read–Hall, 186
Sidewall striations, 123. *see also* Dry etching
Silicon-based ion-sensitive FET, 242
SIMS. *see* Secondary ion mass spectroscopy
Spin FETs
  rectifying contacts on (Ga,Mn)N, 285

Spin LEDs, 300
  EL and PL structure, 303
  GaMnN layers, 300
  spin-polarized carriers, 281
Spin–spin coupling, 264
Spintronics, nitride-based. *see* Semiconductor spintronics
SQUID. *see* Superconducting quantum interference device
SRH. *see* Shockley–Read–Hall
Stacked-gate GaN MOSFET. *see also* MOSFET
  charge modulation, 352
  gate leakage reduction, 351
  Gd_2O_3 wet etch, 349, 350
  SiO_2 deposition, 350
Superconducting quantum interference device, 264
Surface acoustic wave device, 213. *see also* Sensors
  array, 251
  biosensors, 245–247
  fabrication, 247–249
  gas sensing, 250
Surface passivation, 334. *see also* Gate-controlled metal-oxide semiconductor diodes

TEM (transmission electron microscopy), 8
TLM (transmission line method), 33
Transition metal doping
  Co+-implantation
    (Ga,Mn)N, 278
    AlGaN, 295
  Cr+-implantation
    AlGaN, 295, 298
    AlN, 291, 293
  Mn-implantation
    AlGaN, 296, 297
    AlN, 291, 294
TRIM (transport-of-ions-in-matter), 35

UID (unintentionally doped layer), 301
Ultraviolet detectors, 166–169. *see also* Dry etching
  AlGaN MSM, 169
  AlGaN, 166, 168

Ultraviolet detectors (*cont.*)
  GaN p-i-n, 167, 168
  metal semiconductor metal (MSMs), 167
  nitride-based, 166

VPE (vapor phase epitaxy), 155

Wet chemical etching, 15
Wet etch depth, 86

Wide bandgap materials, 263
  GaN, 3, 17, 214
  ohmic contact, 42
  SiC, 3, 17, 214

Zero field-cooled, 270
ZFC magnetization, 291, 295
  Cr+-implanted AlGaN, 295
  polarized light emitters or spin transistors, 295